Methods in Molecular Biology

Series Editor
John M. Walker
School of Life and Medical Sciences
University of Hertfordshire
Hatfield, Hertfordshire, AL10 9AB, UK

For further volumes:
http://www.springer.com/series/7651

Systems Chemical Biology

Methods and Protocols

Edited by

Slava Ziegler

Department of Chemical Biology, Max Planck Institute of Molecular Physiology, Dortmund, Germany

Herbert Waldmann

Department of Chemical Biology, Max Planck Institute of Molecular Physiology, Dortmund, Germany

Humana Press

Editors
Slava Ziegler
Department of Chemical Biology
Max Planck Institute of Molecular Physiology
Dortmund, Germany

Herbert Waldmann
Department of Chemical Biology
Max Planck Institute of Molecular Physiology
Dortmund, Germany

ISSN 1064-3745 ISSN 1940-6029 (electronic)
Methods in Molecular Biology
ISBN 978-1-4939-8890-7 ISBN 978-1-4939-8891-4 (eBook)
https://doi.org/10.1007/978-1-4939-8891-4

Library of Congress Control Number: 2018959856

This Humana Press imprint is published by the registered company Springer Science+Business Media, LLC, part of Springer Nature.
The registered company address is: 233 Spring Street, New York, NY 10013, U.S.A.

Preface

Advances in organic synthesis and screening approaches have spurred interdisciplinary research in Chemical Biology allowing the identification of small-molecule modulators of biomolecules and thus biological phenomena. In parallel, Systems Biology aims to understand entire cellular networks, i.e., cell signaling, metabolism, and developmental processes. Progress in cell culture techniques like stem cell and 3D cell culture allows to exploit physiologically relevant models for the study of diseases, thus increasing the significance and validity of the obtained findings. Moreover, several methodologies were established to detect the localization of small molecules or the interaction of small molecules with their targets in cells. High-content imaging and analysis can capture hundreds of cellular parameters to generate morphology-based fingerprints of bioactive compounds. Developments in *–omics* technologies like genomics, transcriptomics, proteomics, etc. have yielded valuable data collections on small molecules-perturbed living systems. Today scientists have access to gene and protein expression and gene mutation data for a multitude of cell lines and can readily explore gene expression and proteome profiles after compound perturbation. Collecting these data in public databases allows to compare compound-induced profiles to deduce similarities in the mode-of-action or to identify unique signatures. Phosphoproteomics contributes to the more global view on cell signaling networks and their perturbation. Efficient sequencing techniques allow to identify drug resistance-conferring mutations among thousands of gene transcripts to elucidate the targets of bioactive compounds. Advances in gene editing approaches provide tools to perturb genes at the DNA level in a genome-wide fashion, which can be employed in chemogenomic profiling of small molecules to understand their mode of action.

Apart from achievements in wet-protocol approaches, researchers have access to a variety of cheminformatic tools to characterize and explore small molecules with regard to biology. Furthermore, computational strategies were developed to explore and analyze the cumulative data on compounds' activity in cells. Implementation of bioinformatic and cheminformatic approaches in the study of bioactive small molecules allows to integrate the available data on compounds, targets, genes, metabolic pathways, diseases, side effects, etc. to analyze, comprehend, or predict the cellular targets and networks of these compounds. These integrated networks provide valuable information for assessing compound selectivity and safety and may help uncovering off-targets and repurposing drugs.

This book compiles an array of available wet-lab techniques and computational methods to study small-molecule behavior and interactions with their targets in cells and is meant to assist scientists in their attempts to initiate or continue their chemical biology studies at a systems level. Since this is a relatively young research field, methodologies are still being developed or optimized. This book is not aimed to be (and cannot be) comprehensive, but rather contains a selection of modern and, in our opinion, relevant approaches to the field. In addition, the chapter collection also reflects the group of scientists who agreed to make their experience available to the wider scientific community and contribute to this project by submission of a protocol. Unfortunately, not all invitations finally were matched by a contribution, such that the reader should be aware that additional methodology may be available. We would like to express our gratitude to all authors for the time and effort in writing and editing their chapters. We also thank John M. Walker for his guidance and assistance throughout this endeavor, which was both educational and illuminating to us.

Dortmund, Germany

Slava Ziegler
Herbert Waldmann

Contents

Contributors

MASIN ABO-RADY • *Center for Regenerative Therapies Dresden (CRTD), Technische Universität Dresden, Dresden, Germany*
TERO AITTOKALLIO • *Institute for Molecular Medicine Finland (FIMM), University of Helsinki, Helsinki, Finland; Department of Mathematics and Statistics, University of Turku, Turku, Finland*
JUN ANDO • *Department of Applied Physics, Osaka University, Suita, Osaka, Japan*
MAHENDRA AWALE • *Department of Chemistry and Biochemistry, National Center of Competence in Research NCCR TransCure, University of Berne, Berne, Switzerland*
MICHAEL BARBIER • *Laboratory of Cell Biology and Histology, Department of Veterinary Sciences, University of Antwerp, Wilrijk, Belgium*
JESSICA BELLMANN • *Center for Regenerative Therapies Dresden (CRTD), Technische Universität Dresden, Dresden, Germany*
CATARINA BRITO • *iBET, Instituto de Biologia Experimental e Tecnológica, Oeiras, Portugal; Instituto de Tecnologia Química e Biológica António Xavier, Universidade Nova de Lisboa, Oeiras, Portugal*
ANDREAS BROCKMEYER • *Department of Chemical Biology, Max Planck Institute of Molecular Physiology, Dortmund, Germany*
DANIEL BUCIO-NOBLE • *Department of Molecular Sciences, ARC Training Centre for Molecular Technology in the Food Industry, Macquarie University, Sydney, Australia*
RYAN BYRNE • *Department of Chemistry and Applied Biosciences, Swiss Federal Institute of Technology (ETH), Zurich, Switzerland*
SRIRAM CHANDRASEKARAN • *Department of Biomedical Engineering, University of Michigan, Ann Arbor, MI, USA*
YOLANDA T. CHONG • *Recursion Pharmaceuticals, Salt Lake City, UT, USA*
PAUL A. CLEMONS • *Chemical Biology and Therapeutics Science Program, Broad Institute of Harvard and MIT, Cambridge, MA, USA*
KENNETH M. COMESS • *AbbVie, Inc., North Chicago, IL, USA*
CESEAR R. CORONA • *Promega Biosciences Incorporated, San Luis Obispo, CA, USA*
RONALD DE HOOGT • *Janssen Pharmaceutica N.V., Beerse, Belgium*
KOSUKE DODO • *Synthetic Organic Chemistry Laboratory, RIKEN Cluster for Pioneering Research, Wako, Saitama, Japan*
MARTA F. ESTRADA • *iBET, Instituto de Biologia Experimental e Tecnológica, Oeiras, Portugal; Instituto de Tecnologia Química e Biológica António Xavier, Universidade Nova de Lisboa, Oeiras, Portugal*
KATSUMASA FUJITA • *Department of Applied Physics, Osaka University, Suita, Osaka, Japan*
KJERSTI GJERDE • *Janssen Pharmaceutica N.V., Beerse, Belgium*
MICHAEL GLATZA • *Department of Cell and Developmental Biology, Max Planck Institute for Molecular Biomedicine, Münster, Germany*
RALPH GRAESER • *Boehringer Ingelheim Pharma GmbH & Co. KG, Biberach an der Riß, Germany*
TING HAN • *Department of Biochemistry, UT Southwestern Medical Center, Dallas, TX, USA; National Institute of Biological Sciences, Beijing, China*

DOMINIC HOEPFNER • *Novartis Institutes for BioMedical Research, Novartis Pharma AG, Basel, Switzerland*
GREGORY R. HOFFMAN • *Novartis Institutes for BioMedical Research, Cambridge, USA*
ROBIN HURST • *Promega Corporation, Fitchburg, WI, USA*
MICHIO IWATA • *Department of Bioscience and Bioinformatics, Faculty of Computer Science and Systems Engineering, Kyushu Institute of Technology, Iizuka, Fukuoka, Japan*
ALOK JAISWAL • *Institute for Molecular Medicine Finland (FIMM), University of Helsinki, Helsinki, Finland*
PETRA JANNING • *Department of Chemical Biology, Max Planck Institute of Molecular Physiology, Dortmund, Germany*
MANISHA A. JHALA • *AbbVie, Inc., North Chicago, IL, USA*
LARA MARRONE • *Center for Regenerative Therapies Dresden (CRTD), Technische Universität Dresden, Dresden, Germany*
GREGORY MCALLISTER • *Novartis Institutes for BioMedical Research, Cambridge, MA, USA*
MALTE METZ • *Department of Chemical Biology, Max Planck Institute of Molecular Physiology, Dortmund, Germany*
MARK P. MOLLOY • *Department of Molecular Sciences, ARC Training Centre for Molecular Technology in the Food Industry, Macquarie University, Sydney, Australia; Australian Proteome Analysis Facility, Macquarie University, Sydney, Australia; Bowel Cancer and Biomarker Research, Kolling Institute, The University of Sydney, Sydney, Australia*
MAKOTO MUROI • *Chemical Biology Research Group, RIKEN CSRS, Wako, Saitama, Japan*
DEEPAK NIJHAWAN • *Department of Biochemistry, UT Southwestern Medical Center, Dallas, TX, USA; Department of Internal Medicine, UT Southwestern Medical Center, Dallas, TX, USA; Harold C. Simmons Comprehensive Cancer Center, UT Southwestern Medical Center, Dallas, TX, USA*
RACHEL FRIEDMAN OHANA • *Promega Corporation, Fitchburg, WI, USA*
HIROYUKI OSADA • *Chemical Biology Research Group, RIKEN CSRS, Wako, Saitama, Japan*
ANNIKA OSSWALD • *Boehringer Ingelheim RCV, GmbH & Co. KG, Vienna, Austria*
AXEL PAHL • *Max Planck Institute of Molecular Physiology, Dortmund, Germany*
ELENA S. RECKZEH • *Department of Chemical Biology, Max Planck Institute of Molecular Physiology, Dortmund, Germany*
MATTHEW G. REES • *Cancer Biology Program, Broad Institute of Harvard and MIT, Cambridge, MA, USA*
LYDIA REINHARDT • *Center for Regenerative Therapies Dresden (CRTD), Technische Universität Dresden, Dresden, Germany; Department of Cell and Developmental Biology, Max Planck Institute for Molecular Biomedicine, Münster, Germany*
JEAN-LOUIS REYMOND • *Department of Chemistry and Biochemistry, National Center of Competence in Research NCCR TransCure, University of Berne, Berne, Switzerland*
MATTHEW B. ROBERS • *Promega Corporation, Fitchburg, WI, USA*
VÍTOR E. SANTO • *iBET, Instituto de Biologia Experimental e Tecnológica, Oeiras, Portugal; Instituto de Tecnologia Química e Biológica António Xavier, Universidade Nova de Lisboa, Oeiras, Portugal*
GISBERT SCHNEIDER • *Department of Chemistry and Applied Biosciences, Swiss Federal Institute of Technology (ETH), Zurich, Switzerland*
BRINTON SEASHORE-LUDLOW • *Department of Oncology-Pathology, Karolinska Institutet, Stockholm, Sweden*

CRYSTAL SEMAAN • *Department of Molecular Sciences, ARC Training Centre for Molecular Technology in the Food Industry, Macquarie University, Sydney, Australia*
SONJA SIEVERS • *Max Planck Institute of Molecular Physiology, Dortmund, Germany*
MIKIKO SODEOKA • *Synthetic Organic Chemistry Laboratory, RIKEN Cluster for Pioneering Research, Wako, Saitama, Japan*
WOLFGANG SOMMERGRUBER • *Boehringer Ingelheim RCV, GmbH & Co. KG, Vienna, Austria*
JARED STERNECKERT • *Center for Regenerative Therapies Dresden (CRTD), Technische Universität Dresden, Dresden, Germany*
SANTIAGO TENA • *Center for Regenerative Therapies Dresden (CRTD), Technische Universität Dresden, Dresden, Germany*
JAMES D. VASTA • *Promega Corporation, Fitchburg, WI, USA*
SUZANA VIDIC • *Faculty of Medicine, Institute of Biochemistry, University of Ljubljana, Ljubljana, Slovenia*
HERBERT WALDMANN • *Department of Chemical Biology, Max Planck Institute of Molecular Physiology, Dortmund, Germany*
KRISTER WENNERBERG • *Institute for Molecular Medicine Finland (FIMM), University of Helsinki, Helsinki, Finland*
KEITH V. WOOD • *Promega Corporation, Fitchburg, WI, USA*
BHAGWAN YADAV • *Hematology Research Unit Helsinki (HRUH), University of Helsinki, Helsinki, Finland*
YOSHIHIRO YAMANISHI • *Department of Bioscience and Bioinformatics, Faculty of Computer Science and Systems Engineering, Kyushu Institute of Technology, Iizuka, Fukuoka, Japan; PRESTO, Japan Science and Technology Agency, Kawaguchi, Saitama, Japan*

Chapter 1

PREDECT Protocols for Complex 2D/3D Cultures

Suzana Vidic, Marta F. Estrada, Kjersti Gjerde, Vítor E. Santo, Annika Osswald, Michaël Barbier, Yolanda T. Chong, Wolfgang Sommergruber, Ronald de Hoogt, Catarina Brito, and Ralph Graeser

Abstract

PREDECT, a European IMI consortium, has assumed the task to generate robust 2D and 3D culture platforms. Protocols established for 2D and 3D monoculture and stromal coculture models of increasing complexity (spheroid, stirred-tank bioreactor, Matrigel- and collagen-embedded cultures) have been established between six laboratories within academia, biotech, and pharma. These models were tested using three tumor cell lines (MCF7, LNCaP, and NCI-H1437), covering three pathologies (breast, prostate, and lung), but should be readily transferable to other model systems. Fluorescent protein tagged cell lines were used for all platforms, allowing for online measurement of growth curves and drug responses to treatments. All methods, from culture setup to phenotypic characterization and gene expression profiling are described in this chapter.

The adaptable methodologies and detailed protocols described here should help to include these models more readily to the drug discovery pipeline.

Key words 3D cell culture, Bioreactor 3D cell culture, Matrix-embedded 3D cell culture, Fluorescence-labeled cell culture, 3D image analysis, Tumor models, Tumor microenvironment, Heterotypic interactions, Drug screening

1 Introduction

In vitro tumor models are essential for drug discovery. However, the ability, particularly of two-dimensional (2D) tumor cell monocultures on plastic to model complex human tumors for target identification and validation or the prediction of response and resistance to drugs has been increasingly questioned over the past decades [1, 2].

Suzana Vidic, Marta F. Estrada, Kjersti Gjerde, Vítor E. Santo, Annika Osswald, and Michael Barbier share first authorship.

Slava Ziegler and Herbert Waldmann (eds.), *Systems Chemical Biology: Methods and Protocols*, Methods in Molecular Biology, vol. 1888, https://doi.org/10.1007/978-1-4939-8891-4_1,

Tumors are complex structures that contain stromal cells, vasculature, extracellular matrix (ECM)—they have rightly been described as pathological organs [3]. More complex models have been developed over the years in order to represent this complexity and address its impact on drug discovery. Admixing of stromal cells, or stromal-cell derived growth factors, to tumor cells leads to resistance toward certain targeted drugs of genotypically sensitive tumor cells. For example, hepatocyte growth factor (HGF) produced by some stromal cells can lead to vemurafenib resistance of BRAF V600E mutant tumor cells [4] or gefitinib resistance of triple-negative breast cancer cell lines [5]. When moving to 3D, there is ample literature suggesting that tumor cell spheroids show altered treatment responses compared to 2D cultures. Cytotoxic compounds like adriamycin or gemcitabine, on the one hand, appear to penetrate less effectively into spheroids, which most likely contributes to the increased resistance observed in these cultures compared to 2D [6–8]. Epidermal growth factor receptor (EGFR) or mesenchymal epithelial transition (MET) receptor tyrosine kinase targeted compounds, on the other hand, showed greater efficacy in 3D grown lung tumor cells, which correlated well with increased pathway activities in spheroid cultures [9] (for a review, see [10]).

Last but not least, most cells in the body are firmly embedded in a 3D structure, the ECM. Whereas epithelial cells normally reside on a laminin-rich basement membrane (BM), fibroblasts are embedded in a collagen-rich matrix. In multiple experimental model systems, the BM has been shown not only to have an architectural but also a functional role in regulating epithelial tissue differentiation, and the loss of this interaction may be involved in tumorigenesis [11]. Matrix interactions of 3D cultures embedded in either a mouse-derived basal membrane extract (BME), commonly referred to as Matrigel, or collagen have also been shown to affect drug sensitivities, for example in some models of breast cancer [12].

The 2D/3D platform of the PREDECT consortium has assumed the task to compare and characterize in vitro models for oncology research [13]. The focus was set on models that should incorporate the complexity and heterogeneity of human cancers, but still remain robust and reproducible. Results from these studies have recently been published [14]. However, this article was focussed on the results from the comparison between the various model platforms, rather than on providing detailed, comprehensive protocols. The aim of this book chapter is, therefore, to provide access to these technologies for nonspecialists, and thus hopefully allow for these technologies to find a wider user community.

Within the PREDECT consortium, we started with simple 2D monocultures as the current gold-standard, then increased the complexity stepwise to include stromal cells in 2D cocultures, and finally to monocultures and cocultures in 3D. 3D cultures included free-floating spheroids ("floaters"), cultures microencapsulated into alginate, an inert hydrogel, and grown in bioreactors ("alginate-BR"), or embedded in ECM, all in the presence or absence of stromal cells. The alginate hydrogel capsules provide an inert structure to colocalize tumor spheroids and stromal cells and let them deposit their own, coculture-specific ECM [15]. In contrast, ECM embedded cultures are set up in prespecified matrices: Matrigel, a basal membrane extract that reflects a localized tumor environment, collagen I as an interstitial stroma matrix component, providing an invasive growth environment, and a 1:1 mix of both. Besides protocols for basic culture setup, methods to label tumor and stromal cells fluorescently and then monitor their growth online are presented. Methods to analyse spheroid sizes and proliferation patterns within spheroids more in-depth, as well as to molecularly characterize 3D cultures via immunocytochemistry and gene expression analysis will also be shown.

The detailed, step-by-step protocols established by this collaborative effort should help to pave the way for a better incorporation of complex models into the drug discovery process. An overview over the methods covered in this chapter is shown in Fig. 1.

2 Materials

2.1 Cell Culture

1. 1× phosphate buffered saline (PBS) without calcium and magnesium: 2.7 mM KCl, 1.5 mM KH_2PO_4, 138 mM NaCl, 8 mM Na_2HPO_4-$7H_2O$. To 800 mL water add 0.2 g KCl, 0.2 g KH_2PO_4, 8 g NaCl and 2.16 g Na_2HPO_4-$7H_2O$ and adjust the pH to 7.4 with 1 N HCl. Add water to a total volume of 1 L and sterilize using a 0.22 μm filter.
2. Regular growth medium used for cell culture maintenance (e.g., 1× Roswell Park Memorial Institute 1640 (RPMI 1640, RPMI), Dulbecco's Modified Eagle's Medium (DMEM), or Alpha modification of Minimum Essential Medium (α-MEM), phenol red-free if required (*see* **Note 1**), supplemented with 10% (v/v) fetal calf serum (FCS) (*see* **Note 2**), 2 mM GlutaMAX, 1× penicillin/streptomycin (optional).
3. 0.25% (v/v) trypsin–EDTA
4. Freezing medium: 95% (v/v) fetal calf serum (FCS) and 5% (v/v) dimethyl sulfoxide (DMSO 99.9% pure): add 5 mL DMSO to 95 mL FCS.

2.2 Cell Transduction

1. Either generate virus supernatant, or buy from a commercial source, e.g., IncuCyte® NucLight Red Lentivirus Reagent (EF1a, Puro), IncuCyte® NucLight Green Lentivirus Reagent (EF1a, Bleo).

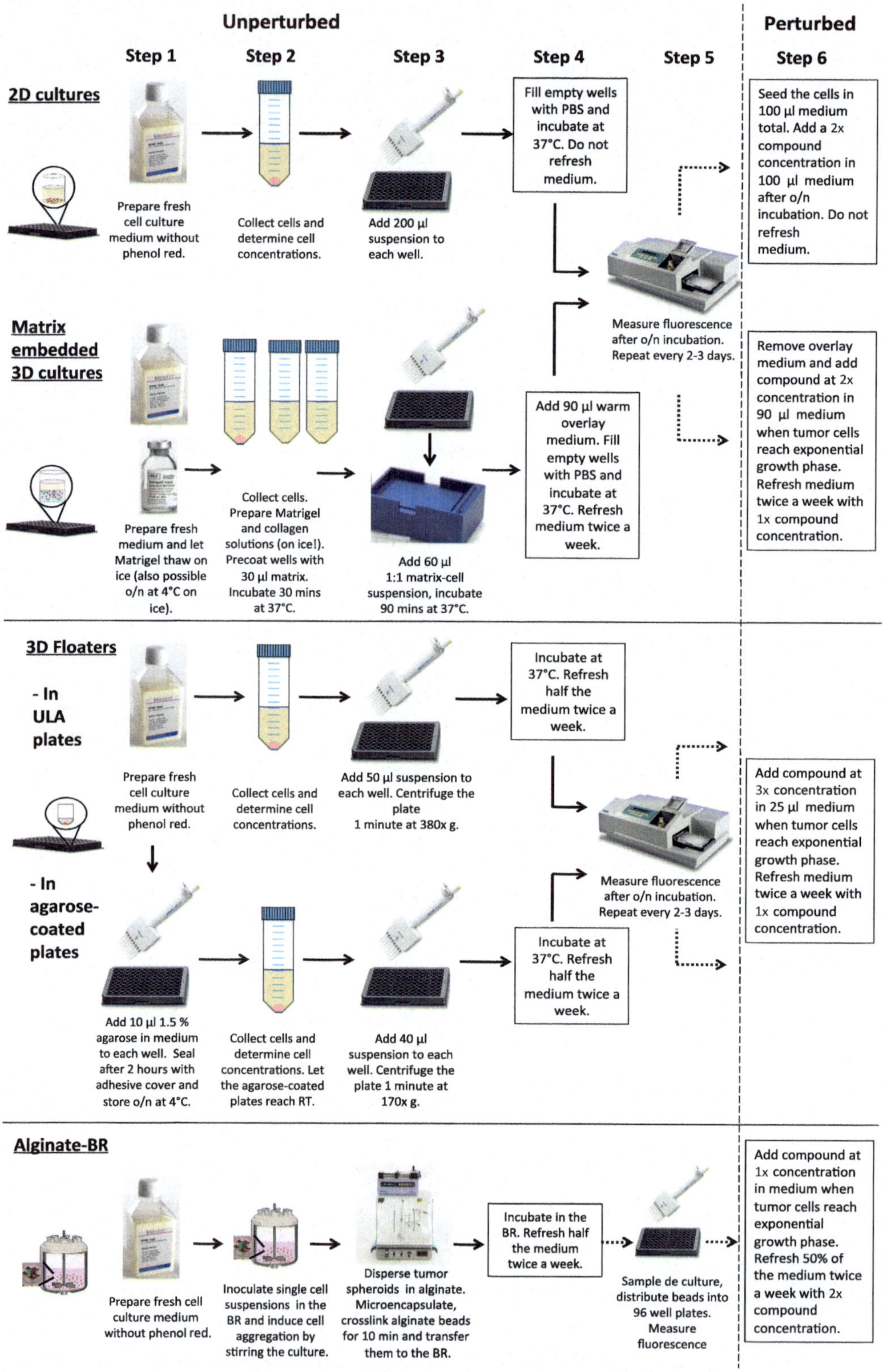

Fig. 1 Flow diagram illustrating the setup of the 2D and 3D cultures covered in this chapter

2. 10 mg/mL polybrene: Add 100 mg polybrene to 10 mL water and sterilize using a 0.22 μm filter.
3. Tissue culture T75 flasks.
4. 6-well plates with flat clear bottom.
5. Fluorescence microscope equipped with camera.
6. Fluorescence-activated cell sorter.
7. FACS buffer: 2% fetal calf serum (FCS) in PBS: Add 20 mL FCS to 980 mL PBS.

2.3 2D Cell Culture

1. Black 96-well clear-bottom microplates.
2. 96-well plate fluorescence reader (plate reader).

2.4 3D Floater Culture

1. Black 384-well ultralow attachment plates with clear round bottom.
2. Plate reader.

2.5 3D Alginate-Bioreactor (BR) Cultures

1. BR Device: Wall-baffled spinner-vessels (125 or 500 mL) with straight blade paddle impeller (Corning®Life Sciences).
2. Electrostatic bead generator (Nisco VarV1).
3. Alginate solving solution: 0.9% (w/v) NaCl. Dissolve 9 g NaCl in 1 L water and autoclave the solution.
4. Solution A: 1.1% (w/v) of Ultrapure Ca^{2+} MVG alginate (UP MVG NovaMatrix, Pronova Biomedical, Oslo, Norway) in alginate bead dissolving solution. Dissolve 0.44 g alginate (maintain the sterility) in 40 mL sterile alginate bead dissolving solution. Dissolve at 4 °C for 1–2 days in a shaker.
5. Cross-linking solution 1: 100 mM $CaCl_2$ in 10 mM HEPES, pH 7.4 (*see* **Note 3**). Dissolve 14.7 g $CaCl_2.2H_2O$ in 900 mL water; dissolve 2.38 g HEPES in the previous solution. Adjust the pH to 7.4. Adjust the volume to 1 L and autoclave the solution.
6. Cross-linking solution 2: 20 mM $BaCl_2$, adjusted to 290–300 mOsm, using NaCl, buffered at pH 7.4 with 5 mM histidine (*see* **Note 3**). Dissolve 4.8856 g $BaCl_2.2H_2O$, 6.7206 g NaCl, and 0.7758 g L-Histidine in 900 mL water. Adjust the pH to 7.4 and the osmolarity to 300 mOsm using a 10% (w/v) NaCl solution. Adjust the volume to 1 L and filter the solution with a 0.22 μm filter.
7. Alginate bead dissolving solution 1: 50 mM sodium citrate, 100 mM NaCl. Dissolve 0.303 g NaCl in 50 mL water. Add 0.735 g sodium citrate dihydrate. Sterilize using a 0.22 μm filter.
8. Alginate bead dissolving solution 2: 50 mM EDTA, 10 mM HEPES. Dissolve 9.306 g EDTA, 1.1915 g HEPES in 400 mL water. Adjust the pH to 7.4 with HCl, and adjust the volume to 500 mL. Sterilize using a 0.22 μm filter.

2.6 3D Matrix-Embedded Cultures

1. Black 96 wells μClear microtiter plate.
2. Analog heat block (plate heater).
3. Plate reader.
4. 2.5 M D-glucose: Dissolve 45 g D-glucose in 50 mL water. Fill up to 100 mL, and sterilize using a 0.22 μm filter.
5. 2× Matrigel/BME (*see* **Note 4**) working solution (keep on ice!): 8 mg/mL Matrigel (growth factor reduced, phenol red free in low glucose DMEM (1 g/L) containing 50 μg/mL gentamycin), 2 mM L-glutamine, 12 mM D-glucose, 2% (v/v) fetal calf serum (FCS) (*see* **Note 5**), RPMI: Add 135 μL 200 mM L-glutamine, 35 μL 2.5 M D-glucose, 270 μL FCS, and 3060 μL 1× RPMI to a 10 mL vial of Matrigel stock (10.8 mg/mL) to generate 8 mg/mL Matrigel. May be refrozen if not fully used.
6. 1× Matrigel/BME: Dilute 2× Matrigel/BME 1:1 with 1× RPMI.
7. 1 M HEPES: Dissolve 23.8 g HEPES in 80 mL water. Fill up to 100 mL and sterilize using a 0.22 μm filter.
8. 7.5% (w/v) $NaHCO_3$: Dissolve 7.5 g $NaHCO_3$ in 90 mL water. Fill up to 100 mL and sterilize using a 0.22 μm filter.
9. 2× collagen solution (keep on ice!): 3 mg/mL Collagen Type I from rat tail (colorless aqueous solution in 20 mM acetic acid, sterilized by filtration through 0.22 μm filter), 1× RPMI with 2 mM L-glutamine but lacking HEPES and phenol red, 30 mM HEPES, 0.2% $NaHCO_3$, 2% (v/v) fetal calf serum (FCS) (*see* **Note 5**), pH 7.2–7.4: Add 1 mL 10× RPMI containing L-glutamine, 300 μL 1 M HEPES, 300 μL 7.5% (w/v) $NaHCO_3$, 200 μL FCS, 40 μL water, and 160 μL 1 N NaOH to 8 mL 3.8 mg/mL collagen stock to generate 3 mg/mL collagen solution [16]. The solution may be refrozen if not fully used.
10. 1× collagen solution: Dilute 2× collagen solution 1:1 with 1× RPMI.
11. 1:1 collagen–Matrigel mix: Mix equal volumes of 2× Matrigel and 2× collagen solutions.

2.7 Imaging

1. Fluorescence microscope.
2. Plate type for imaging: It is critical to use plates that are suitable for imaging. These are either glass bottom with a coverslip thickness of 0.17 μm or plastic film bottoms that are 0.19 μm. For matrix-embedded 3D cultures, plastic bottom plates may work as well (*see* **Note 6**). For the washing of floaters or alginate BR cultures, V-shaped 96 or 384-well plates should be used.

2.7.1 Software that May Be Used for Image Analysis

1. The dedicated ellipsoid approximation approach used for the quantification of 3D fluorescent image stacks by the PREDECT consortium [17] is available as MATLAB code from https://github.com/mbarbie1/ellipsoids-analysis-paper (*see* **Note 7**).
2. For 2D size and shape analysis of 2D brightfield images there exists AnaSp [18, 19] and ReViSP [19], available as MATLAB executable and code.
3. CellProfiler [20] is a general image analysis tool that allows nonexperts to generate functional image analysis pipelines for batch analysis in a trivial fashion (although restricted to mostly 2D analysis). CellProfiler pipelines were used in the PREDECT consortium to extract spheroid size and shape features from widefield images.
4. FIJI [21] is a general and very versatile image analysis tool capable of both analysis in 2D and 3D (although batch analysis for a dedicated pipeline such as this might require an image analysis expert).

3. EdU Alexa Fluor 647 HCS Assay Kit: 10 μM EdU, Click-iT® reaction cocktail; 20 μM EdU solution (2× working solution) in RPMI containing 2% FCS; Click-iT® reaction rinse buffer (*see* **Note 8**).
4. Hoechst 33342 counterstain: 10 mg/mL Hoechst 33342 in PBS: Add 10 mg Hoechst 33342 to 1 mL PBS.
5. Fixing solution A: 4% paraformaldehyde: Dissolve 4 g paraformaldehyde in 90 mL PBS by heating to 60 °C until solution clears. Fill up to 100 mL with PBS.
6. 25% glutaraldehyde solution: Dissolve 25 g glutaraldehyde in 100 mL water.
7. Fixing solution B: 4% paraformaldehyde/0.25% glutaraldehyde in PBS (*see* **Note 9**): Dissolve 4 g paraformaldehyde in 90 mL PBS by heating to 60 °C until solution clears. Add 1 mL 25% glutaraldehyde solution. Fill up to 100 mL with PBS.
8. Permeabilization solution: 0.5% Triton X-100 in PBS: Add 5 mL Triton X-100 to 995 mL PBS.

2.8 RNA Expression Analysis

1. Chloroform.
2. QIAzol lysis reagent (Qiagen).
3. RNeasy 96 kit (Qiagen).
4. 96 well collection block (Qiagen).
5. Photospectrometer.

3 Methods

3.1 Cell Culture

Routine cell culture has been described elsewhere in detail and will not be covered in this method section.

3.2 Generation of Fluorescence-Labeled Cell Lines via Lentiviral Transduction

An example of lentiviral constructs used is summarized schematically in Fig. 2.

3.2.1 Viral Transduction

1. Transfer 50,000 cells per transduction in a 15 mL Falcon tube and spin down for 3 min at 530 × *g*.
2. Aspirate the medium gently with a pipette.
3. Resuspend the cells in 1 mL growth medium containing 16 μg/mL polybrene, add 1 mL lentiviral supernatant (5 × 10^5 TU/mL), and transfer the mix into a 6-well plate.
4. Spin for 90 min at 400 × *g* at room temperature.
5. Transfer the plate to the incubator at 37 °C and 5% CO_2 and leave the virus on the cells for 2–3 days.
6. Replace the medium with fresh growth medium containing the selection antibiotic (*see* **Note 10**).
7. Once selection is completed, check the fluorescence signal under the microscope—if not clearly above background, retransduce the cells.

3.2.2 Sort the Cells with a Fluorescence-Activated Cell Sorter (FACS) for Homogenous Fluorescence Intensity

1. Transfer at least 500,000 cells in 10 mL FACS buffer.
2. Dissociate cell aggregates by careful pipetting before FACS sorting them according to high, medium and low fluorescence intensity (for an example, see Fig. 3).

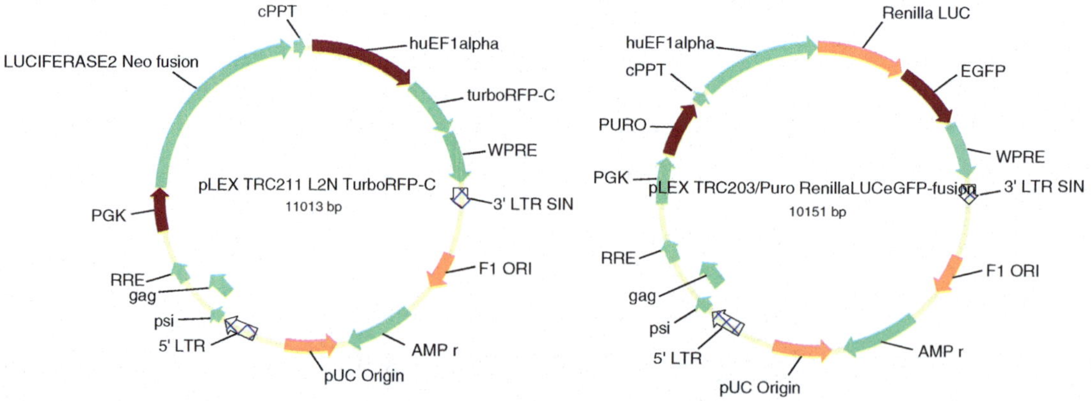

Fig. 2 Dual function lentiviral constructs to label tumor and stromal cells. (**a**) firefly luciferase (Fluc) with RFP, and (**b**) renilla luciferase (RLuc) with eGFP. Either fluorescent or luminescent tags allowed to differentiate between tumor and stromal cells

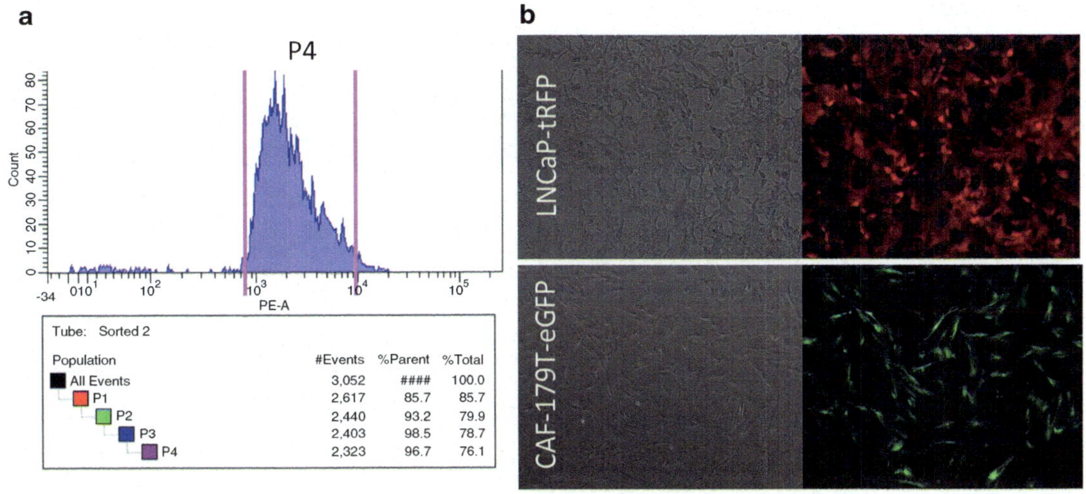

Fig. 3 Tumor and stromal cells expressing tRFP or eGFP were transduced with respective constructs and FACS sorted for even fluorescence. (**a**) A pool of cells with middle range fluorescence intensity was collected and (**b**) expanded for further experiments

3. Only cells with high or medium fluorescence intensity should be used for further experiments.
4. Create a master bank consisting of 5–10 vials of the labeled cell line (100,000 cells/mL freezing medium).

3.3 2D Cell Culture

Assays with 2D cell cultures with fluorescence-labeled cells should be performed in black 96-well clear-bottom microplates.

1. Collect cells, determine the concentration for each cell line, and prepare adequate dilutions for monocultures and cocultures in medium, calculating 200 μL per well (*see* **Note 11**).
2. Seed the desired number of cells per well in 200 μL medium with all supplements required, adding 200 μL medium only to control wells (*see* **Note 12**). If perturbation experiments are planned, seed cells in 100 μL medium per well and, after overnight incubation, add 100 μL 2× compound concentration to each well (*see* **Note 13**).
3. Place the plates in a humidified incubator at 37 °C and 5% CO_2. Do not refresh the medium during cultivation as this may cause cells to detach and disturb the fluorescence growth measurements.
4. Measure the baseline fluorescence values after overnight incubation with a plate reader (*see* **Note 14**).
5. Repeat measurements every 2–3 days for growth curve construction until the cultures reach stationary phase.

3.4 3D Floater Cultures

1. Collect cells, and resuspend cells in 5 mL fresh medium containing the appropriate FCS concentration (*see* **Note 5**).
2. Count the cells and prepare dilutions for each cell line at desired cell concentrations calculating 50 μL per well. For cocultures, mix the cell types in 50 mL tube at the appropriate ratio (*see* **Note 15**).
3. Seed the cells in the 384-well ultralow attachment clear round bottom plates.
4. Centrifuge the plate at 380 × *g* for 1 min at room temperature to aggregate the cells.
5. Place the cells at 37 °C, 5% CO_2 to grow spheroids, "3D floaters." These cultures may be maintained in culture for several weeks.
6. Carefully replace half of the medium with fresh medium twice a week using a multichannel pipet.
7. Measure the baseline fluorescence values after overnight incubation with a plate reader (*see* **Note 14**).
8. Repeat measurements every 2–3 days for growth curve construction until the cultures reach stationary phase.
9. For perturbation experiments, add 25 μL medium including the compound at 3x concentration. Refresh half of the medium with 1× compound concentration twice a week.

3.5 3D Alginate Bioreactor (BR) Cultures

1. In order to prevent cell attachment to the walls, the glass bioreactor (BR) should be silanized before the experiment. Precoat the vessels with 2–3 mL dimethyldichlorosilane, wash them with 2–3 mL toluene, and leave them to dry overnight. Before use the spinner vessels should be washed with abundant water and autoclaved.
2. Prepare the following material to autoclave prior to encapsulation: Exchangeable nozzle, electrode, magnetic stirrer with beaker, nozzle holder with autoclavable arm, and silicon tubes (*see* **Note 16**).
3. Collect tumor cells into 50 mL tubes and count viable cells.
4. Inoculate the single cell suspension (200,000 cells/mL) in the BR in fresh medium containing the appropriate FCS concentration and allow cell aggregates to form over 24–96 h (*see* **Note 17**).
5. For alginate microencapsulation, collect tumor cell aggregates from the BR and disperse them in 3 mL solution A either alone (monocultures) or together with fibroblasts, in a 1:1 proportion (*see* **Note 18**).
6. The alginate beads should be cross-linked either in cross-linking solution 1 or 2 (*see* **Note 3**). After the last droplet of

alginate falls, keep the beads under agitation in the cross-linking solution for 10 min in order to guarantee optimal cross-linking.

7. Wash the beads three times with alginate solving solution and finally equilibrate in culture medium before transferring to the BR.
8. For perturbation experiments, the microencapsulated cultures may be transferred to 96-well plates and treated as described for the 2D cultures (*see* **Note 13**), or the drug at the appropriate concentration may be added directly to the BR.
9. Keep the cultures on a magnetic stirrer at 80 rpm, in a humidified incubator, with 5% CO_2, for 20 days with 50% medium exchange every 3–4 days (*see* **Note 19**).
10. Sample the culture at day 1 to measure the reporter fluorescence or any other parameter at baseline.
11. Continue to sample every 5 days to monitor growth of the culture.
12. For some analyses (see also below Subheading 3.8.1.), the beads need to be dissolved. For this, alginate microcapsules should be treated with alginate bead dissolving solution 1 for calcium-cross-linked beads or solution 2 for barium-cross-linked beads, for 5 min at room temperature.

3.6 3D Matrix-Embedded Cultures

This protocol describes the setup of a homogeneous 3D assay for embedding cells in Matrigel, collagen and a 1:1 mix of Matrigel and collagen.

1. Preincubate the 96-well microtiter plates at 37 °C using a plate warmer at 37 °C to set up the assay. Warm RPMI medium to 37 °C.
2. Prepare sufficient 1× Matrigel, collagen, and 1:1 Matrigel–collagen solutions to precoat 96-well plates with 30 μL per well of the relevant matrix, and precoat the plates (*see* **Note 20**).
3. Centrifuge the plates at 400 × g for 5 min at room temperature.
4. Place the plates in a humidified incubator at 37 °C for at least 30 min to allow for the matrix to set.
5. Collect cells, resuspend cells in 5 mL assay medium (*see* **Notes 1** and **5**), and count them with a cell counter.
6. Dilute the cells to the desired concentration in assay medium and transfer sufficient cell suspension to a container on ice (30 μL required per well; *see* **Notes 11** and **21**).
7. Add the same volume of 2× Matrigel, 2× collagen, or 2× Matrigel–collagen (1:1) mixture to the above container

(30 μL required per well). For matrix control wells, use the relevant 1× matrix solution.

8. Take the precoated 96-well plates out of the incubator and place them on a plate heater in the laminar flow hood (*see* **Note 12**).
9. Pipette 60 μL of the matrix–cell suspension to the respective wells.
10. Incubate the plate at in a humidified incubator at 5% CO_2/ 37 °C for 2 h to allow the matrix to set.
11. Add 90 μL prewarmed RPMI medium supplemented with 2–10% FCS to the wells (*see* **Note 5**).
12. Fill any empty wells with 180 μL PBS.
13. Measure the baseline fluorescence values after overnight incubation with a plate reader (*see* **Note 14**).
14. Repeat measurements every 2–3 days for growth curve construction until the cultures reach the stationary phase.
15. For perturbation experiments, add 50 μL medium including the compound at 3× concentration (*see* **Note 13**). Refresh half of the medium with 1× compound concentration twice a week.

3.7 Imaging

3.7.1 General Considerations

Potential readouts from the imaging approaches described here include quantification of the size, shape or proliferation of 3D tumor spheroids. However, imaging options crucially depend on 3D culture specifics, like spheroid sizes, densities, and the expected number of proliferating cells (*see* **Note 22**). Once the ultimate readout has been decided upon, the instrument setup required to acquire adequate images will become clear: Size and shape analysis require only limited resolution, while the analysis of cell proliferation is based on the identification of single EdU-positive nuclei, and thus requires single-cellular resolution within the spheroids.

1. Fluorescent widefield images: Acquisition is relatively straightforward and the instrumentation set-up should not be complicated, although an automated 96-well plate set-up would reduce hands-on imaging acquisition time. Analysis from these kinds of images is mostly restricted to a single plane and resolution in axial direction is minimal, and therefore, proliferation analysis of single cells does not make sense and analysis should be limited to size and shape (*see* **Note 23**).
2. 3D confocal image stacks: This type of images allows for the quantification in full 3D, but the process is very computationally intense. Also, due to a lack of resolution in the axial dimension, image analysis may be limited to projections of the 3D image stacks.

3.7.2 EdU-Proliferation Staining and Fixation Procedure for 3D Cultures

1. Place fixing solution A or B (*see* **Note 8**) in a water bath and preheat to 37 °C.
2. Use EdU Kit in the following steps.
3. Add 100 μL 2× EdU working solution to 100 μL cell culture medium (*see* **Note 5**). Incubate cell cultures for at least 2 h at 37 °C.
4. After incubation with EdU, gently aspirate the cell culture medium tilting the plate to one side and keeping the pipette tip close to the wall in order to avoid damaging the cultures.
5. Add double the original medium volume of fixing solution A or B (*see* **Note 8**) to the side of the well and incubate 30 min at room temperature.
6. Collect floaters and alginate-BR cultures by spinning 1000 × *g* for 1 min. Transfer samples to a V-bottom plate in order to ease washing procedures. All steps are performed at room temperature, unless stated otherwise.
7. Gently aspirate the fixing solution and wash the sample three times with 100 μL PBS. Add the wash solutions slowly to the side of the well to prevent damaging the samples.
8. Aspirate wash solution, add 100 μL permeabilization solution to the cells and incubate for 20 min.
9. Gently aspirate the solution from the well and wash sample with 100 μL PBS in two short washing steps (2 × 1 min) and two long washing step (2 × 15 min). Remove the wash solution.
10. Add 100 μL Click-iT® reaction cocktail to each well. Make sure it is evenly distributed over the sample. Incubate the plate for 30 min at room temperature, protected from light. Remove the reaction cocktail and wash each well once with 100 μL Click-iT® reaction rinse buffer. Then carefully aspirate the Click-iT® reaction rinse buffer.
11. Wash the sample with 100 μL PBS.
12. Apply 100 μL Hoechst counterstain and incubate 30 min at room temperature.
13. Wash with PBS three times for 10 min each.
14. Fill up the well with PBS, seal the plate to prevent liquid evaporation and wrap with aluminium foil.
15. Analyse samples as described in Subheading 3.7.3. At this point the sample can be stored at 4 °C for a few weeks.

3.7.3 Image Acquisition of Floater and Alginate-BR Cultures

Widefield microscope images are recommended for size and shape analyses of large spheroids for which analysis of EdU-positive cells is not feasible due to signal attenuation. Using a 10× magnification objective, a representative number of spheroids (3 or more) per field may be imaged.

Confocal stacks using a 20× magnification objective may allow for a visualization of EdU-positive cells in the inner parts of smaller spheroids, while still covering whole spheroids in the *xy*-plane. The *z*-step should be set very small (e.g., 1 μm), in order to resolve EdU-positive cells. If correlation studies between size, shape and proliferation are required, the same spheroids should be imaged at lower (10×) and higher (20×) magnifications.

3.7.4 Image Acquisition of 3D Matrix-Embedded Cultures

A compromise needs to be found between resolution, coverage in *z*-axis, and scan speed/memory usage: A 10× magnification objective reaches a depth of ca. 1 mm along the *z*-axis, and with a z-step of 10 μm is sufficient to resolve the EdU signal while preserving scan-speed and memory. If spheroids contain a large percentage of EdU-positive cells, objectives or microscopy setups with a smaller axial resolution will have to be used.

3.7.5 Image Analysis of 3D Spheroid Cultures

There are a number of commercial solutions, linked to various imaging platforms, that can be used to perform the analysis of 3D confocal images in either full 3D or based on image maximum intensity projections. In the materials section there is a list of open source tools that are available to perform this type of analyses (Subheading 2.7).

3.8 RNA Expression Analysis

Besides imaging, transcriptomics may be applied to characterize cultures. Analysis of untreated cultures may help to understand pathways active in a given culture system, and when compared to treated cultures, used to investigate treatment-induced alterations and mode-of-action-related pharmacodynamic changes.

3.8.1 Floater and Alginate-BR Cultures

1. Collect 3D cultured cells (*see* **Note 24**). Alginate microcapsules should be dissolved using chelating solution 1 for calcium-cross-linked beads or 2 for barium-cross-linked beads, for 5 min at room temperature.
2. Rinse the spheroid pellet with PBS and centrifuge at 300 × *g* for 5 min.
3. Remove the supernatant and snap-freeze the cell pellet using liquid nitrogen. Make sure this process is quick. Store the cell pellets at −80 °C until RNA extraction.
4. For RNA extraction, thaw the cell pellets and add QIAzol lysis reagent followed by RNA extraction with Qiagen RNeasy mini kit columns, following the manufacturer's protocol.

3.8.2 3D Matrix-Embedded Cultures

1. Prepare collection microtubes (Qiagen) with 750 μL QIAzol lysis reagent (Qiagen) for each 96-plate well. Remove supernatant from 3D-embedded cultures by aspiration and immediately add 150 μL QIAzol.

2. Gently resuspend the 3D culture and pipet the solution back in its corresponding microtube containing the remaining QIAzol lysis reagent. Use another 150 μL of the solution to rinse the well.
3. Shake for 3 min at maximum frequency in the TissueLyser II (Qiagen). Quickly spin the tubes to clear the QIAzol from the inside of the caps. Leave the tubes containing the homogenates on the bench at room temperature for 5 min (*see* **Note 25**).
4. Add 150 μL chloroform and shake the tube vigorously for 15 s (*see* **Note 26**). Leave on the bench at room temperature for 2–3 min, and then centrifuge at 6000 × *g* for 15 min at 4 °C (*see* **Note 27**). Transfer the upper, aqueous phase to a Qiagen 96 wells S-block (Qiagen). Add 1 volume (usually 600 μL) of 70% ethanol, cover the S-block with an adhesive cover and mix thoroughly by shaking vigorously by hand for 15 s. Do not centrifuge, even if precipitates are visible. Transfer up to 700 μL of the sample to a Qiagen RNeasy 96 well plate placed on a 96 wells collection block, and follow the manufacturer's protocol.
5. Assess RNA purity via optical density at 280/260/230 nm and the RNA integrity by electrophoresis measuring the 28S–18S rRNA ratio. Or use equipment of choice capable of measuring the above.
6. Analyze the expression of RNA, either by quantitative real-time PCR, gene expression microarray, or RNA sequencing.

4 Notes

1. Phenol red activates some hormone receptors. If effects of androgens or estrogens are to be measured, phenol-free medium should be used.
2. FCS batch-to-batch variations may affect experiments. If possible, use the same batch of FCS for all cell culture manipulations during a given project.
3. Some culture media, such as RPMI, have components that will chelate the Ca^{2+} ions leading to capsule fragmentation. In these cases please use the $BaCl_2$ cross-linking solution.
4. Matrigel is a brand name; however, since it is very commonly used in the literature, we will call it Matrigel throughout the text. Similar preparations are sold as for example basal membrane extract (BME) from other vendors. Matrigel preparations vary in growth-factor content, glucose, etc., and thus, it is important to choose the preparation that best suits the needs of the planned assay. Also, Matrigel preparations may show significant lot-to-lot variations, so it is of utmost importance

to keep the same Matrigel lot during a given project. See also discussion in [14].

5. Lower FCS concentrations may reveal coculturing effects, but not all cell lines can be maintained at FCS concentration lower than 10% (e.g., LNCaP cells tolerate 2% FCS, while MCF7 do not—in our hands). If experiments are to be conducted at lower than 10% FCS concentrations (e.g., 2%), cells should be washed twice in serum-free medium prior to seeding.
6. Greiner plates are not suitable for imaging with short working distance objectives due to the low skirt and may result in damage to your objectives.
7. This ellipsoid approximation approach is dedicated to the analysis of spheroids that reached a size for which light attenuation is just becoming an issue (ca. 200 μm diameter). Another approach to tackle the problem of light attenuation is suggested by Strasters and colleagues [22], via an image restoration technique.
8. 10 μM EdU solution is our suggested starting concentration, but depending on the model system, it may be necessary to test a range of EdU concentrations in order to determine the optimal concentration. Also, the 2 h incubation time may be adjusted according to the proliferation rate of the cell line investigated.
9. Fixation of Matrigel cultures in fixing solution A results in matrix dissolution. This can be prevented by the addition of 0.25% glutaraldehyde (fixing solution B). Glutaraldehyde may mask antigenic sites and interfere with antibody labeling, but does not interfere with the EdU protocol.
10. The antibiotic concentration required for effective killing of nontransduced cells may have to be determined beforehand. In case of hypersensitivity of transduced cells to the selective antibiotic (e.g., significantly reduced growth rate), remove the antibiotic and sort fluorescent cells with FACS. Some cell lines tend to inactivate promoters driving transgene expression. If this is observed, expression of transgene and antibiotic resistance marker may have to be more tightly coupled, e.g., by generating a fusion gene (e.g., eGFP-Neomycin resistance gene), or by using an internal ribosomal entry site for the resistance marker rather than an own promoter.
11. The cell number and ratio for every new cell line/combination will have to be optimized. 1000–20,000 tumor cells per well is a good starting point. For tumor:stroma ratios, a useful range is between 10:1 and 1:10.
12. When performing experiments with several 96-well plates, drawing the layout of each plate on the lids simplifies and

speeds up the pipetting process. The outer wells should not be used for the experiment as these suffer more evaporation during long term culturing, and should be filled with 200 μL PBS. Pipetting with a multichannel pipet from a reservoir is an easy and fast approach to fill the plates.

13. The desired compound concentrations should be prepared in DMSO and the unperturbed controls should receive medium with the same DMSO concentration as the perturbed cells. Cell lines may react differently to DMSO, so maximally tolerated DMSO concentrations should be determined beforehand.

14. Excitation (ex) and emission (em) settings depend on the construct the cells have been transduced with for example ex/em 540/587 nm for RFP and ex/em 488/525 nm for GFP. The excitation and emission wavelengths of the reporters should not overlap. Even though, it may be worthwhile confirming experimentally that there is no interference between the two signals.

15. The optimal cell number and ratio need to be determined experimentally for every cell line and combinations. A good starting point is 10,000 cells each, tumor and stromal cells. The optimal ratio may vary from 10:1 to 1:10.

16. Since the microencapsulator is not autoclavable, it should be sterilized by UV light inside the laminar flow hood for at least 30 min.

17. The stirring speed and aggregation time will depend on the cell line. Stirring speeds may be varied from low (30–50 rpm) to medium (60–80 rpm) and high (≥100 rpm), for 1–4 days [23].

18. Tumor spheroids and fibroblasts should be very well mixed in the alginate solution in order to guarantee a homogeneous microencapsulation. During alginate handling, avoid bubbles as these will introduce artifacts in the bead formation.

19. Every time culture medium is exchanged, fresh drug should also be added to the BR.

20. Precoating the wells with matrix prevents cells from growing in 2D on the bottom of the wells.

21. Fibroblasts may cause collagen gel contraction. This depends on the fibroblasts, but numbers should not exceed 4000 cells/well at plating.

22. For the sample types under consideration, the main culture properties influencing or limiting the image acquisition and analysis procedures are:
    (a) Size of the spheroids: In 3D image stacks, the signal coming from the inner and deeper parts of larger spheroids (>100 μm) suffers from light attenuation, rendering

these spheroid parts invisible. Although 2D size and shape analysis is nearly unaffected, image analysis of single proliferating cells becomes impossible. Therefore, the analysis of spheroid cultures with large spheroids should be restricted to size and shape analysis.

(b) Spheroid shape: Spheroids can lose their natural spherical structure by aggregating or being very invasive. Once spheroids cannot be discerned from each other, any definition of size or shape becomes invalid.

(c) Density of spheroids: Discerning distinct spheroids from merged spheroids becomes more difficult with increasing spheroid density, and large enough numbers of spheroids are required for statistical analysis.

(d) Density of proliferating cells inside spheroids: When proliferating cells become omnipresent in spheroids, resolution restrictions become higher.

23. However, it is possible to recover the axial resolution from widefield z-stacks using an image deconvolution strategy, which removes out of focus light and re-enables the possibility of single-cell analysis [24]. Image deconvolution requires some conditions to apply though; for example, signal attenuation in the image should be negligible or slowly varying in spatial directions. Commercial software packages are available for this task such as Huygens Professional (Scientific Volume Imaging) and Autoquant X3 (Media Cybernetics), as well as some open source tools, such as the Parallel Iterative Deconvolution plugin available for FIJI.

24. For gene expression analysis, the chelating solutions should be used cold (4 °C). The pellet should be collected after centrifugation at 50 × g for 5 min. This step also enables to separate tumor aggregates (pelleted) from stromal cells that remain in suspension in the supernatant. Wash twice the cell population of interest for analysis with cold PBS.

25. This step promotes dissociation of nucleoprotein complexes.

26. Thorough mixing is important for subsequent phase separation.

27. After centrifugation, the sample separates into three phases: An upper, colorless, aqueous phase containing RNA; a white interphase; and a lower, red, organic phase.

References

1. Ravi M, Paramesh V, Kaviya SR, Anuradha E, Solomon FD (2015) 3D cell culture systems: advantages and applications. J Cell Physiol 230 (1):16–26. https://doi.org/10.1002/jcp.24683

2. Edmondson R, Broglie JJ, Adcock AF, Yang L (2014) Three-dimensional cell culture systems and their applications in drug discovery and cell-based biosensors. Assay Drug Dev Technol

12(4):207–218. https://doi.org/10.1089/adt.2014.573

3. Egeblad M, Nakasone ES, Werb Z (2010) Tumors as organs: complex tissues that interface with the entire organism. Dev Cell 18 (6):884–901. https://doi.org/10.1016/j.devcel.2010.05.012
4. Straussman R, Morikawa T, Shee K, Barzily-Rokni M, Qian ZR, Du J, Davis A, Mongare MM, Gould J, Frederick DT, Cooper ZA, Chapman PB, Solit DB, Ribas A, Lo RS, Flaherty KT, Ogino S, Wargo JA, Golub TR (2012) Tumour micro-environment elicits innate resistance to RAF inhibitors through HGF secretion. Nature 487(7408):500–504. https://doi.org/10.1038/nature11183
5. Mueller KL, Madden JM, Zoratti GL, Kuperwasser C, List K, Boerner JL (2012) Fibroblast-secreted hepatocyte growth factor mediates epidermal growth factor receptor tyrosine kinase inhibitor resistance in triple-negative breast cancers through paracrine activation of Met. Breast Cancer Res 14(4):R104. https://doi.org/10.1186/bcr3224
6. Kwok TT, Twentyman PR (1985) The response to cytotoxic drugs of EMT6 cells treated either as intact or disaggregated spheroids. Br J Cancer 51(2):211–218
7. Kerr DJ, Wheldon TE, Kerr AM, Freshney RI, Kaye SB (1986) The effect of adriamycin and 4′-deoxydoxorubicin on cell survival of human lung tumour cells grown in monolayer and as spheroids. Br J Cancer 54(3):423–429
8. Longati P, Jia X, Eimer J, Wagman A, Witt MR, Rehnmark S, Verbeke C, Toftgard R, Lohr M, Heuchel RL (2013) 3D pancreatic carcinoma spheroids induce a matrix-rich, chemoresistant phenotype offering a better model for drug testing. BMC Cancer 13:95. https://doi.org/10.1186/1471-2407-13-95
9. Ekert JE, Johnson K, Strake B, Pardinas J, Jarantow S, Perkinson R, Colter DC (2014) Three-dimensional lung tumor microenvironment modulates therapeutic compound responsiveness in vitro--implication for drug development. PLoS One 9(3):e92248. https://doi.org/10.1371/journal.pone.0092248
10. Hirschhaeuser F, Menne H, Dittfeld C, West J, Mueller-Klieser W, Kunz-Schughart LA (2010) Multicellular tumor spheroids: an underestimated tool is catching up again. J Biotechnol 148(1):3–15. https://doi.org/10.1016/j.jbiotec.2010.01.012
11. Howlett AR, Bailey N, Damsky C, Petersen OW, Bissell MJ (1995) Cellular growth and survival are mediated by beta 1 integrins in normal human breast epithelium but not in breast carcinoma. J Cell Sci 108 (Pt 5):1945–1957
12. Huang C, Park CC, Hilsenbeck SG, Ward R, Rimawi MF, Wang YC, Shou J, Bissell MJ, Osborne CK, Schiff R (2011) beta1 integrin mediates an alternative survival pathway in breast cancer cells resistant to lapatinib. Breast Cancer Res 13(4):R84. https://doi.org/10.1186/bcr2936
13. Hickman JA, Graeser R, de Hoogt R, Vidic S, Brito C, Gutekunst M, van der Kuip H, Consortium IP (2014) Three-dimensional models of cancer for pharmacology and cancer cell biology: capturing tumor complexity in vitro/ex vivo. Biotechnol J 9(9):1115–1128. https://doi.org/10.1002/biot.201300492
14. Stock K, Estrada MF, Vidic S, Gjerde K, Rudisch A, Santo VE, Barbier M, Blom S, Arundkar SC, Selvam I, Osswald A, Stein Y, Gruenewald S, Brito C, van Weerden W, Rotter V, Boghaert E, Oren M, Sommergruber W, Chong Y, de Hoogt R, Graeser R (2016) Capturing tumor complexity in vitro: Comparative analysis of 2D and 3D tumor models for drug discovery. Sci Rep 6:28951. https://doi.org/10.1038/srep28951
15. Estrada MF, Rebelo SP, Davies EJ, Pinto MT, Pereira H, Santo VE, Smalley MJ, Barry ST, Gualda EJ, Alves PM, Anderson E, Brito C Modelling the tumour microenvironment in long-term microencapsulated 3D co-cultures recapitulates phenotypic features of disease progression. Biomaterials 78:50–61 epub before print. https://doi.org/10.1016/j.biomaterials.2015.11.030
16. Artym VV, Matsumoto K (2010) Imaging cells in three-dimensional collagen matrix. Current Protoc Cell Biol. / editorial board, Juan S Bonifacino [et al] Chapter 10:Unit 10 18 11-20. https://doi.org/10.1002/0471143030.cb1018s48
17. Barbier M, Jaensch S, Cornelissen F, Vidic S, Gjerde K, de Hoogt R, Graeser R, Gustin E, Chong YT, Consortium IP (2016) Ellipsoid segmentation model for analyzing light-attenuated 3D confocal image stacks of fluorescent multi-cellular spheroids. PLoS One 11(6):e0156942. https://doi.org/10.1371/journal.pone.0156942
18. Zanoni M, Piccinini F, Arienti C, Zamagni A, Santi S, Polico R, Bevilacqua A, Tesei A (2016) 3D tumor spheroid models for in vitro therapeutic screening: a systematic approach to enhance the biological relevance of data obtained. Sci Rep 6:19103. https://doi.org/10.1038/srep19103
19. Piccinini F, Tesei A, Arienti C, Bevilacqua A (2015) Cancer multicellular spheroids: volume

assessment from a single 2D projection. Comput Methods Prog Biomed 118(2):95–106. https://doi.org/10.1016/j.cmpb.2014.12.003

20. Lamprecht MR, Sabatini DM, Carpenter AE (2007) CellProfiler: free, versatile software for automated biological image analysis. BioTechniques 42(1):71–75
21. Schindelin J, Arganda-Carreras I, Frise E, Kaynig V, Longair M, Pietzsch T, Preibisch S, Rueden C, Saalfeld S, Schmid B, Tinevez JY, White DJ, Hartenstein V, Eliceiri K, Tomancak P, Cardona A (2012) Fiji: an open-source platform for biological-image analysis. Nat Methods 9(7):676–682. https://doi.org/10.1038/nmeth.2019
22. Strasters KC, van der Voort HTM, Geusebroek JM, Smeulders AWM (1994) Fast attenuation correction in fluorescence confocal imaging: a recursive approach. Bioimaging 2:78–92
23. Santo VE, Estrada MF, Rebelo SP, Abreu S, Silva I, Pinto C, Veloso SC, Serra AT, Boghaert E, Alves PM, Brito C (2016) Adaptable stirred-tank culture strategies for large scale production of multicellular spheroid-based tumor cell models. J Biotechnol 221:118–129. https://doi.org/10.1016/j.jbiotec.2016.01.031
24. Shaw PJ (2006) Comparison of widefield/deconvolution and confocal microscopy for three-dimensional imaging. In: Pawley JB (ed) Handbook of biological confocal microscopy. SpringerScience+Business Media, New York, pp 453–467

Chapter 2

Phenotypic Screening Using Mouse and Human Stem Cell-Based Models of Neuroinflammation and Gene Expression Analysis to Study Drug Responses

Masin Abo-Rady, Jessica Bellmann, Michael Glatza, Lara Marrone, Lydia Reinhardt, Santiago Tena, and Jared Sterneckert

Abstract

High-throughput phenotypic screening enables the identification of new therapeutic targets even when the molecular mechanism underlying the disease is unknown. In the case of neurodegenerative disease, there is a dire need to identify new targets that can ameliorate, halt, or reverse degeneration. Stem cell-based disease models are particularly powerful tools for phenotypic screening because they use the same cell type affected in patients. Here, we describe the expansion of mouse stem cells and human induced pluripotent stem cells as well as the differentiation of these cells into neural lineages that, when exposed to neuroinflammatory stress, can be used for compound screening followed by hit identification, validation, and target deconvolution.

Key words Pluripotent stem cells, Disease modelling, Phenotypic screening, Neurodegeneration, Neuroinflammation, Neuronal differentiation, Target deconvolution

1 Introduction

A critical issue in drug discovery is the identification and/or selection of a target against which a small molecule can be developed. In the case of neurodegenerative diseases, few therapeutics exist and none can effectively halt, delay, or reverse degeneration. Therefore, new targets are urgently needed. Phenotypic screening is a particularly powerful tool because compounds can be tested in a high-throughput manner without a previous understanding of their molecular mechanism of action. By identifying the target afterward, it is possible to detect new targets and pathways that are effective and druggable. Indeed, phenotypic screening has been more

Masin Abo-Rady, Jessica Bellmann, Michael Glatza, Lara Marrone, and Lydia Reinhardt contributed equally to the chapter and, thus, share first authorship.

Slava Ziegler and Herbert Waldmann (eds.), *Systems Chemical Biology: Methods and Protocols*, Methods in Molecular Biology, vol. 1888, https://doi.org/10.1007/978-1-4939-8891-4_2, © Springer Science+Business Media, LLC, part of Springer Nature 2019

successful at identifying first-in-class compounds compared to target-based approaches, which were more successful at identifying follower drugs [1]. Embryonic stem cells (ESCs) are ideal tools for phenotypic screening. Because ESCs are able to self-renew indefinitely as well as to differentiate into any cell lineage, theoretically limitless numbers of an affected cell type can be generated for phenotypic screening [2–4]. Induced pluripotent stem cells (iPSCs) have distinct advantages over human ESCs because iPSCs avoid the ethical concerns associated with the destruction of embryos and can be generated from patients with observable phenotypes and known or even unknown genotypes in order to generate disease models [5].

Microglial activation is a hallmark of amyotrophic lateral sclerosis as well as many other neurodegenerative diseases. Previously, we reported a mouse-based phenotypic assay for identifying compounds that inhibit motor neurodegeneration induced by microglial activation [6]. Here, we describe a detailed methodology for phenotypic screening that includes the expansion and differentiation of mouse ESCs (mESCs) and human iPSCs into different cellular identities that are subsequently exposed to neuroinflammatory stressors. These assays are used to identify small molecules that protect neurons from degeneration and use microarray analysis to study drug responses.

2 Materials

2.1 Cell Culture

2.1.1 General Materials

1. 6 and 10 cm culture dishes.
2. 10 cm bacterial petri dishes.
3. 6-, 12-, 96-, and 384-well plates.
4. Accutase (1 ml per 6 cm dish).
5. 0.05% trypsin–ethylenediaminetetraacetic acid (*EDTA*) solution.
6. 0.25% trypsin–ethylenediaminetetraacetic acid (*EDTA*) solution.
7. 2 mg/ml collagenase type IV stock solution: diluted in iPSC/hESC medium (*see* Subheading 2.1.4 below).
8. 10 mM purmorphamine (PMA) stock solution: diluted in DMSO.
9. 10 mM all-trans retinoic acid (RA) stock solution: diluted in DMSO.
10. 10 μg/ml activin A stock solution: diluted according to the manufacturer's instructions.
11. 400 mM diethylenetriamine/nitric oxide adduct (DetaNO) stock solution: diluted in DMSO.

12. 20 mM Y-27632 stock solution: diluted in DMSO.
13. Sterile 100 ml containers.
14. Cell strainers.

2.1.2 0.1% Gelatin Solution

1. Warm up 2% gelatin from in a 37 °C warm water bath until gelatin becomes clear and less viscous. Dilute the gelatin in Dulbecco's PBS-1X without Ca^{2+} and Mg^{2+} (DPBS; Lonza) to a final concentration of 0.1% and subsequently sterile filter the freshly prepared solution. The 0.1% gelatin solution can be stored at RT.

2.1.3 Matrigel Aliquots

1. Knockout™ DMEM (KO-DMEM) from Thermo Fisher Scientific.
2. Matrigel solution: Thaw Matrigel overnight at 4 °C on ice. Dilute the Matrigel 1:5 (KO-DMEM–Matrigel) and mix thoroughly by pipetting up and down. Prepare 1 ml aliquots in precooled 15 ml conical tubes and freeze at −20 °C (*see* **Note 1**).

2.1.4 Cell Culture Medium

1. iPSC/hESC medium: Add about 80 ml of DMEM, 20 ml Knockout Serum replacement, 1 ml of 100× penicllin–streptomycin–glutamine solution, 1 ml of nonessential amino acid solution, 0.7 μl of 14.3 M β-mercaptoethanol to a sterile 100 ml container. Store at 4 °C.
2. N2B27 medium: Add about 50 ml of DMEM/F12, about 50 ml of Neurobasal medium, 1 ml of 100× penicllin–streptomycin–glutamine solution, 1 ml of N2 supplement, and 1 ml of B27 supplement to a sterile 100 ml container. Store at 4 °C.
3. Split medium: Add about 100 ml DMEM, and add BSA fraction V to a final concentration of 0.5% to a sterile 100 ml container. Store at 4 °C.
4. Warm all media in a water bath at 37 °C before use.

2.1.5 Induction and Expansion Media

1. iPSC/hESC induction medium: To iPSC/hESC medium, add CHIR99021 to a final concentration of 3 μM, SB-431542 to a final concentration of 10 μM, dorsomorphin to a a final concentration of 1 μM, purmorphamin to a a final concentration of 0.5 μM, ascorbic acid to a final concentration of 200 μM, and Y-27632 to a final concentration of 5 μM in a sterile 100 ml container. Store at 4 °C.
2. N2B27 induction medium: To N2B27 medium, add CHIR99021 to a a final concentration of 3 μM, SB-431542 to a final concentration of 10 μM, dorsomorphin to a final concentration of 1 μM, purmorphamin to a final concentration of 0.5 μM, and ascorbic acid to a final concentration of 200 μM in a sterile 100 ml container. Store at 4 °C.

3. Expansion medium: To N2B27 medium, add CHIR99021 to a final concentration of 3 μM, purmorphamin to a final concentration of 0.5 μM, and ascorbic acid to a final concentration of 200 μM in a sterile 100 ml container. Store at 4 °C.
4. Warm all media in a water bath at 37 °C before use.

2.1.6 Neuronal Differentiation Media

1. Motor neuron patterning medium: To N2B27 medium, add GDNF to a final concentration of 10 ng/ml, add BDNF to a final concentration of 10 ng/ml, add retinoic acid to a final concentration of 1 μM, add Smoothened agonist purmorphamin to a final concentration of 0.5 μM, and ascorbic acid to a final concentration of 200 μM in a sterile 100 ml container. Store at 4 °C.
2. Dopaminergic neuron patterning medium: To N2B27 medium, add GDNF to a final concentration of 10 ng/ml, add BDNF to a final concentration of 10 ng/ml, add Smoothened agonist purmorphamin to a final concentration of 0.5 μM, and ascorbic acid to a final concentration of 200 μM in a sterile 100 ml container. Store at 4 °C.
3. Maturation medium: To N2B27 medium, add GDNF to a final concentration of 10 ng/ml, add BDNF to a final co^ncentration of 10 ng/ml, add TGF-β3 to a final concentration of 1 ng/ml, dbcAMP to a final concentration of 100 μM, and ascorbic acid to a final concentration of 200 μM in a sterile 100 ml container. Store at 4 °C.
4. Warm all media in a water bath at 37 °C before use.

2.1.7 Mouse Stem Cell Media

1. mESC expansion medium: Add about 80 ml Knockout DMEM, 15 ml Knockout Serum Replacement, 5 ml fetal bovine serum, 1 ml of 100× penicllin–streptomycin–glutamine solution, 1 ml of nonessential amino acid solution, 0.7 μl of 14.3 M β-mercaptoethanol, and add Leukemia Inhibitory Factor to a final concentration of 10 ng/ml in a sterile 100 ml container. Store at 4 °C.
2. DK10 medium: Add about 90 ml Knockout DMEM/F12, 10 ml Knockout Serum Replacement, 1 ml of 100× penicllin-–streptomycin–glutamine solution, 1 ml of nonessential amino acid solution, 0.7 μl of 14.3 M β-mercaptoethanol in a sterile 100 ml container. Store at 4 °C.
3. mNSC expansion medium: Add about 90 ml Knockout DMEM/F12, 1 ml N2 supplement, 1 ml of 100× penicllin-–streptomycin–glutamine solution, 7.5 ml of a 7.5% BSA fraction V solution, add Fibroblast growth factor-basic to a final concentration of 10 ng/ml, and add Epidermal growth factor to a final concentration of 10 ng/ml in a sterile 100 ml container. Store at 4 °C.

4. Astrocyte differentiation medium: Add about 90 ml Knockout DMEM/F12, 10% fetal bovine serum, 1 ml N2 supplement, 1 ml of 100× penicllin–streptomycin–glutamine solution, 7.5 ml of a 7.5% BSA fraction V solution, and add Smoothened agonist to a final concentration of 0.5 μM, in a sterile 100 ml container. Store at 4 °C.
5. BV2 medium: Add about 90 ml RMPI, 10% fetal bovine serum, 1 ml of 100× penicllin–streptomycin–glutamine solution in a sterile 100 ml container. Store at 4 °C.
6. Warm all media in a water bath at 37 °C before use.

2.1.8 Plating for Screening Campaign

1. Microplates, 96-well black (TC, μclear, F-bottom, w/Lid).
2. Matrigel.
3. 0.01% poly-L-ornithine (PLO) solution: diluted in water.
4. 1 mg/ml Laminin diluted according to the manufacturer's instructions.

2.1.9 Stressing Neurons for Screening Campaign

1. 1 mg/ml lipopolysaccharide (LPS) stock solution: diluted according to the manufacturer's instructions.
2. 10 μg/ml recombinant human IFNγ (stock solution: diluted according to the manufacturer's instructions.
3. BV2 microglia cells, for example from ATCC (CRL-2467, 2468, or 2469).

2.2 Fixation Buffer

4% paraformaldehyde: Add paraformaldehyde to a final concentration of 4% in PBS. Make fresh before use.

2.3 Compound Libraries (See Note 2)

The exact library to be used depends on the experiment as well as availability and price. Below are examples.

1. Approved drugs: drugs that have become commercially available upon approval of US Food and Drug Administration (FDA), European Medicines Agency (EMA) or other agencies. Such drugs have well-characterized bioactivity, biosafety, and bioavailability, and hence hits belonging to this set of compounds will ensure an accelerated drug development and optimization process. Examples: Prestwick, ApexBio, Enzo, TimTec, Selleck, and BioFocus NIH clinical collection. These compounds are often not very specific, and the targets of these compounds are not fully described, sometimes making identifying the molecular mechanism challenging.
2. Tool compounds: collections of compounds with well described targets, but not necessarily approved drugs. These compounds are ideal for testing the contribution of known and druggable targets in human neurons. Examples: SelleckChem

kinase inhibitor library. Hits from these libraries can be rapidly translated into molecular mechanisms for follow-up studies.

3. Novel compounds: require more development, further characterization and testing, but provide greater possibility for patenting. Examples are MicroSource Spectrum Library, ChemDiv Library. Of particular note, these compounds require target deconvolution.

The concentration used for screening is, to a certain extent, arbitrary. At higher concentrations (e.g., 50 μM or higher), more toxicity and offtarget effects can be observed, but at lower concentrations (lower than 1 μM), fewer hit compounds are identified. One commonly used screening concentration for primary screening is 10 μM.

2.4 Liquid Handling Devices for Compound Screening

1. Dispenser (e.g., Multidrop™ 384 or Matrix WellMate).
2. Washer (e.g., Power Washer 384).

2.5 Microarray Analysis

1. QIAGEN RNeasy columns with on-column DNA digestion.
2. Gene expression microarrays (e.g., Illumina mouse-8 V2 expression BeadChips).
3. Linear amplification kit (Ambion).
4. 2100 Bioanalyzer (Agilent).
5. Microarray analsysis software (e.g., Illumina GenomeStudio).
6. Spreadsheet software (e.g., Micosoft Office—Excel).
7. R, including pvclust package.

3 Methods

3.1 Preparation of Matrigel-Coated Plates

1. Pour 20 ml of ice-cold KO-DMEM in a 50 ml conical tube. Use only ice-cold KO-DMEM.
2. Quickly mix 5 ml taken from the above 20 ml KO-DMEM in the tube of one frozen Matrigel aliquot by pipetting up and down until the Matrigel is dissolved. Do this very quickly.
3. Once fully mixed, take the 5 ml of combined Matrigel and KO-DMEM and add it to the 50 ml conical tube with the remaining KO-DMEM. Pipet up and down to mix.
4. Pipet a thin and uniform coat of diluted Matrigel to the bottom of the plates and distribute by carfully tilting the plate. For a 6 cm dish we recommend using 1.25 ml in each well and for 12 well dishes 500 μl in each well of diluted Matrigel.
5. Plates should be left at room temperature (RT) overnight and can be either used or stored at 4 °C thereafter.

3.2 Cultivation of Mouse Neuronal Cells for Screening

3.2.1 Mouse Embryonic Stem Cell (mESC) Expansion

1. Coat cell culture dishes with 0.1% gelatin for 15–30 min at room temperature or 37 °C. When coating cell culture plates, the volume of 0.1% gelatin needs to cover the surface of the dish; for example, 1.5 ml 0.1% gelatin is used to coat one well of a 6-well plate.
2. Change expansion medium on a daily basis. To stimulate self-renewal of mESC, freshly supplement expansion medium with 10 ng/ml LIF (*see* **Note 3**).
3. To prevent overgrowth and differentiation, passage mESCs every 2–3 days. Typical splitting ratios are between 1:15 and 1:20.
4. Passage mESCs by first washing the cells with Dulbecco's PBS-1X without Ca^{2+} and Mg^{2+}, and then detaching them using 1 ml Accutase per well of a 6 cm cell culture dish. Incubate at 37 °C for up to 5 min.
5. Following the addition of Accutase, generate a single cell suspension by pipetting gently up and down with a P1000 pipette. Transfer a proportion of the cell suspension (depending on your splitting ratio) into a 15 ml conical tube with 5 ml expansion medium (without LIF), and centrifuge the suspension (5 min at 200 × *g*).
6. After centrifugation, remove the supernatant, resuspend the cell pellet in 5 ml expansion medium supplemented with 10 ng/ml LIF, and plate the cell suspension on a gelatinized cell culture dish.

3.2.2 Mouse Neural Stem Cell (NSC) Cultivation and Differentiation into Astrocytes

1. Prepare gelatin-coated dishes as described in Subheading 2.3.
2. NSCs are maintained in NSC expansion medium freshly supplemented with 10 ng/ml of Fibroblast growth factor 2 and Epidermal growth factor on 0.1% gelatin-coated cell culture dishes.
3. Replace the medium daily and passage NSCs every 2–3 days when they reach a confluence of about 80%.
4. After washing once with Dulbecco's PBS-1X without Ca^{2+} and Mg^{2+}, add 1 ml 0.05% trypsin–EDTA and incubate the cells about 2–3 min at room temperature.
5. NSCs are dissociated into single cells by gently pipetting up and down.
6. Trypsin is inactivated by adding 8 ml astrocyte medium. After centrifugation at 200 × *g* for 5 min, the supernatant is discarded and the pellet is resuspended in 10 ml NSC medium containing 10 ng/ml FGF2 and 10 ng/ml EGF.

7. NSCs are plated on a 10 cm cell culture plate freshly coated with gelatin. Typically, the splitting is performed at a ratio of 1:3–1:6.
8. NSCs are differentiated into astrocytes by plating the NSCs at a density of 5×10^5 cells/ml in astrocyte medium on 0.1% gelatin-coated 10 cm cell culture plates.
9. Culture medium during differentiation is changed on days 1 and 3.
10. On day 4, astrocytes are dissociated into single cells using 0.25% trypsin–EDTA and replated on Matrigel-coated 96-well plates at a density of 45,000 cells/100 μl. For a 384-well plate 15,000 cells per well are plated. Astrocytes are matured for 2 weeks and the culture medium is replaced three times a week.

3.2.3 Mouse Motor Neuron Differentiation

mESCs are differentiated into motor neurons (MNs) using embryoid body (EB) formation as described in Wichterle et al. with minor modifications [7]. Mouse motor neurons can be cultured with or without astrocytes (derived using the protocol above), depending on the experiment requirements.

1. Similar to the passaging of mESC (Subheading 2.4, **item 1**), generate a single cell suspension by washing the cells with DPBS (without Ca^{2+}, Mg^{2+}), incubate with Accutase for several min at 37 °C, and gently pipette up and down with a P1000 pipette.
2. The cell suspension is transferred completely to a 15 ml tube with 5 ml DK10 medium and centrifuged down for 5 min at $200 \times g$. Subsequently, the supernatant is removed to get rid of the Accutase and the pellet is resuspended in 5–8 ml DK10 medium.
3. Differentiation is induced by plating at a density of 2×10^5 cell/ml in DK10 medium in 10 cm petri dishes (to avoid attachment to the bottom surface).
4. For the neural induction, culture medium is supplemented on days 2 and 4 with 1 μM PMA and 1 μM RA to specify MN differentiation.
5. To exchange the medium, EBs are transferred to a 15 ml tube and centrifuged for 5 min at $200 \times g$ and the medium discarded.
6. Prepare 21 ml DK10 medium freshly supplemented with 1 μM RA and 1 μM PMA. Take 2 × 10 cm petri dishes and pipet 5 ml DK10 containing RA and PMA into each petri dish and resuspend the settled EBs gently in 10 ml DK10. Split the suspension between the two 10 cm petri dishes.

7. On day 4 of the differentiation, medium is exchanged again using DK10 freshly supplemented with RA and PMA.
8. Because MN formation should be completed by day 6, exchange the medium on day 6, using DK10 without any supplements.
9. On day 7 of the differentiation, EBs are harvested in 15 ml tubes, the supernatant is aspirated, and the EBs are disaggregated by incubation for 45 min at 37 °C (preferably in a water bath) in 2 ml Accutase.
10. Subsequently, a single cell suspension is generated by pipetting gently up and down. The Accutase-containing cell suspension is further diluted with 7 ml DK10 medium, and centrifuged for 5 min at 200 × *g*.
11. When cells are further processed without FACS-sorting (*see* **Note 4**), determine the exact cell number and plate a mixture of MNs and other neural cell types on a astrocyte monolayer at a density of 45,000 cells. Medium to be used is 100 μl DK10 medium supplemented with 10 ng/ml GDNF.
12. When the cells are FACS-sorted, the cell pellet is resuspended with 5 ml DK10, filtered through a cell strainer, and again centrifuged for 5 min at 200 × *g*.
13. The cell pellet is resuspended in 500–1000 μl DK10 medium and placed on ice. Simultaneously, a 15 ml conical tube with 5 ml DK10 supplemented with 10 ng/ml GDNF is prepared and also placed on ice.
14. Next, the MNs are enriched by FACS sorting using the Hb9-GFP reporter.
15. Finally, plate the MNs at a density of 20,000 MNs in 100 μl on a 96 well plate in DK10 medium supplemented with 10 ng/ml GDNF (*see* **Note 4**).

3.3 Human Neurons for Screening

3.3.1 Small Molecule Neural Precursor Cell (smNPC) Derivation [8]

1. Grow induced pluripotent stem cells (iPSCs) in iPSC/hESC medium on a 6-well plate until nearly confluent (*see* **Note 5**).
2. Using a cannula, colonies are cut in smaller pieces by scratching over the plate in a checkered pattern (Fig. 1).
3. Remove medium, and add 2 mg/ml collagenase in hiPSC/hESC medium and incubate until borders of the colonies start detaching.
4. Wash plate with PBS or split medium and carefully detach colonies using a cell scraper.
5. Collect colony pieces in split medium and let them settle by gravity.
6. Discard supernatant and resuspend the colonies in iPSC/hESC induction medium and plate them on a petri dish. For 6 cm

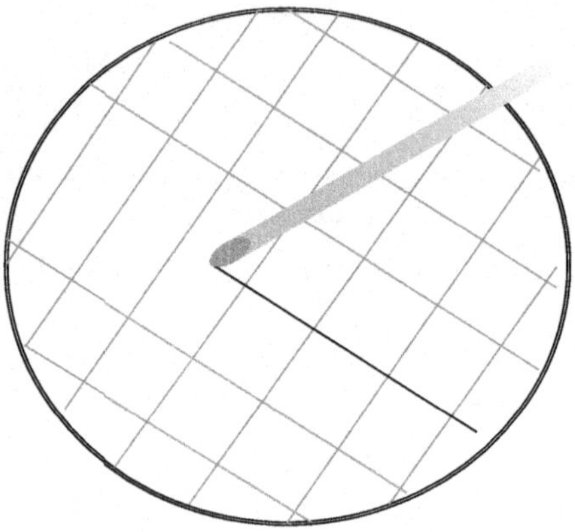

Fig. 1 Colonies are divided into smaller pieces

petri dishes, use 5–6 ml, and for 10 cm dishes, use 10–12 ml medium.

7. Two days following plating of the colonies, remove iPSC/hESC induction medium and replace with N2B27 induction medium. To exchange the medium, tilt the dish and carefully suck off the old medium, leaving as many EBs as possible in the dish. It is not possible to completely remove the old medium, a tiny amount will remain in the dish. Add the N2B27 induction medium. For 6 cm petri dishes, use 5–6 ml, and for 10 cm dishes, use 10–12 ml medium.
8. After another 2 days, exchange the induction medium for expansion medium.
9. After another 2 days, plate 10–20 EBs, preferably with elongated outgrowths (Fig. 2) on 12-well Matrigel-coated dishes (*see* Subheading 2.3 for how to coat dishes with Matrigel) and disaggregate them into smaller fragments using trituration with a P1000 pipette.
10. 4–6 days after plating, subculture cells and seed 4×10^5–6×10^5 cells per 12-well plate. Higher split ratios are used to select for the desired smNPCs.
11. smNPCs are fed every 2 days, depending on the rate of medium consumption. Splitting is performed, usually every 5–6 days, at a ratio of 1:10 to 1:20, depending on the confluence of the cultures.

3.3.2 smNPC Passaging

smNPC lines are maintained on Matrigel-coated cell culture dishes. To prevent overgrowth and differentiation, colonies are passaged. Cells are split at a 1:5 to 1:20 ratio, depending on their density.

1. Prewarm Accutase in a 37 °C water bath.
2. Remove old medium and incubate cells with Accutase for 5–10 min at 37 °C. For a 12-well plate apply 500 μl Accutase.

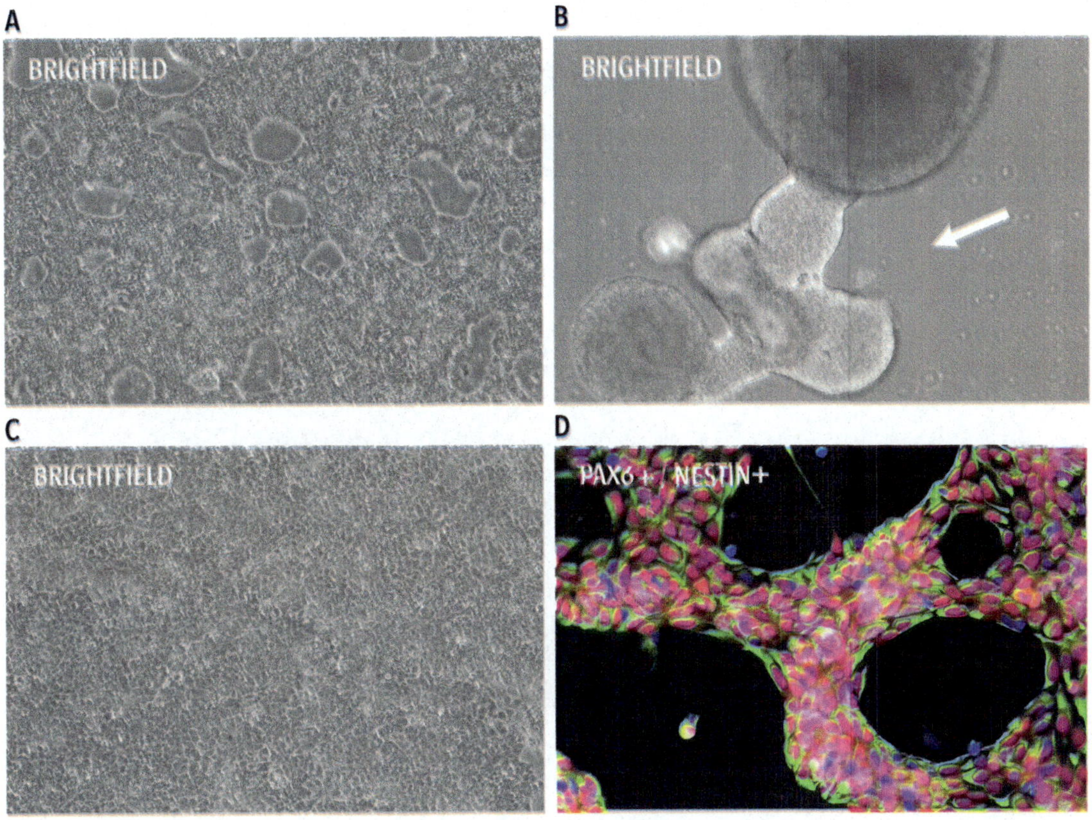

Fig. 2 smNPC Derivation from iPSCs (**a**) iPSCs (**b**) EBs with elongated outgrowth (**c**) smNPCs (**d**) Immunostaining of smNPCs can be performed for quality control. MNs smNPCs are positive for PAX6 (red) and nestin (green)

3. Cells are disaggregated and the mixture of Accutase and cells is diluted 1:10 in split medium. For a passaging ratio of 1:10, take 50 μl of the cell suspension.
4. Centrifuge cells at 200 × *g* for 5 min and remove the supernatant.
5. The cell pellet is resuspended in fresh expansion medium and plated on Matrigel-coated cell culture dishes.
6. From passage 3–5, the cell culture should be free of nonneural cells. Store frozen aliquots beginning from passage 4. To do this, ¼–½ of one well of a 12-well plate with smNPCs are resuspended in cold N2B27 with 10% DMSO. Use standard freezing conditions.

3.3.3 Differentiation of smNPCs into Motor Neurons (Fig. 3)

1. 3×10^5 smNPCs are plated on a 12-well plate in expansion medium.
2. On the next day, day 2, medium is exchanged to MN patterning medium.

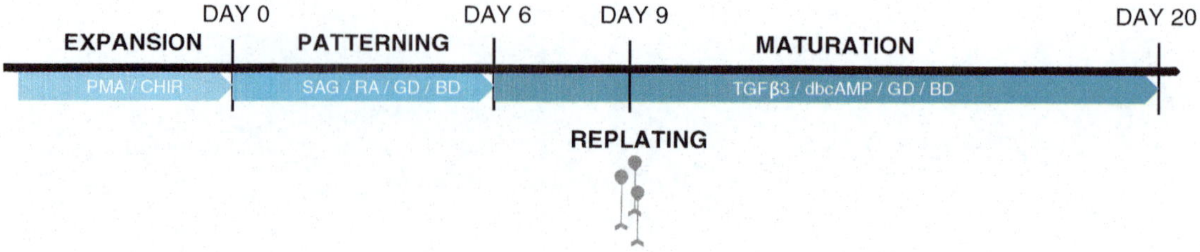

Fig. 3 Summary of motor neuron differentiation

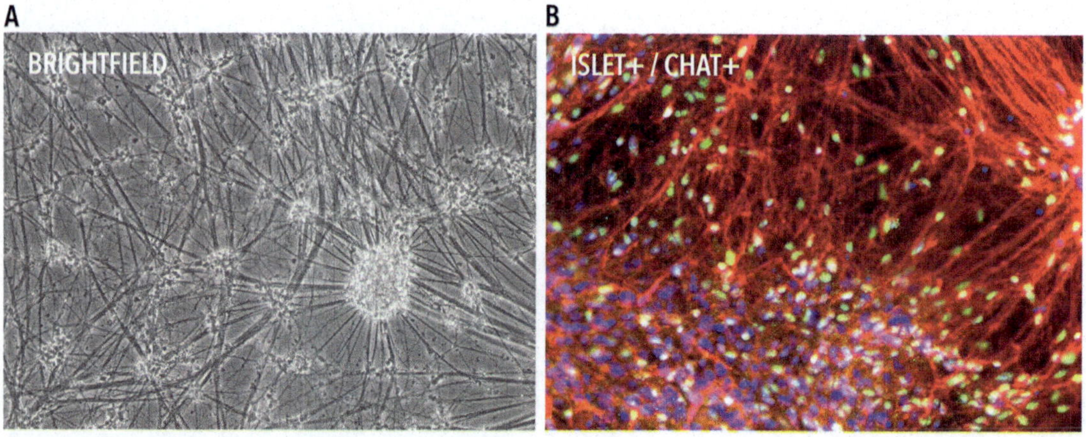

Fig. 4 Phase contrast micrograph of mature motor neurons differentiated from smNPCs. A) MNs B) Immunostaining of MNs can be performed for quality control and to determine the differentiation efficiency. MNs are positive for ISLET1 (green) and CHAT (red)

3. After 3 days in MN patterning medium, on day 5, cells can be replated in the desired format. For reseeding, incubate premature neurons 15–20 min in prewarmed Accutase.
4. Cells are cultured for at least 14 days in maturation medium (Fig. 4).

3.3.4 Differentiation of smNPCs into Dopaminergic Neurons (Fig. 5)

1. 3×10^5 smNPCs are plated on a 12-well plate in expansion medium.
2. The next day, day 2, medium is exchanged to dopaminergic neuron patterning medium.
3. During days 6–8, cells are transferred to maturation medium containing 5 ng/ml Activin A. On the following days, cells are cultured in maturation medium supplemented with 2 ng/ml Activin A.
4. After 3 days in maturation medium, cells can be replated in the desired format. For reseeding, incubate premature neurons 15–20 min in prewarmed Accutase.
5. Cells are cultured for at least 14 days in maturation medium until they are mature (Fig. 6).

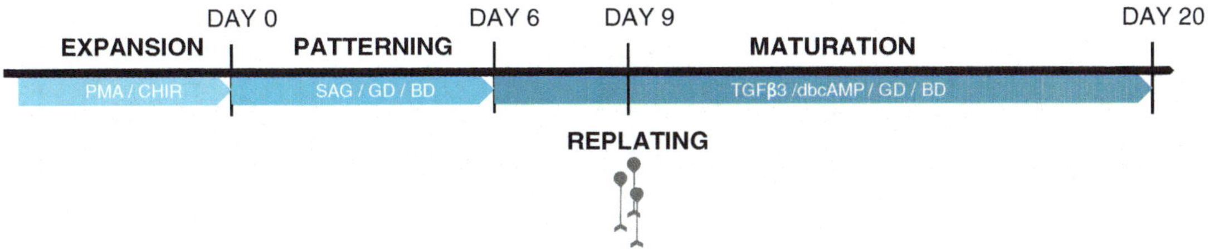

Fig. 5 Summary of dopaminergic neuron differentiation

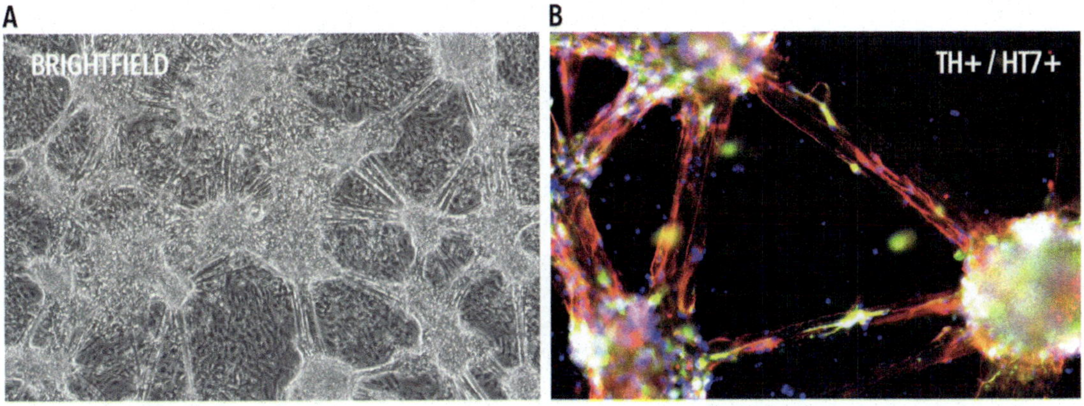

Fig. 6 Phase contrast micrograph of mature dopaminergic neurons differentiated from smNPCs (**a**) Dopaminergic neurons (DANs) (**b**) Immunostaining of DANs can be performed for quality control and to determine the differentiation efficiency. DANs are positive for tyrosine hydroxylase (TH; green) and Tau (antibody HT7; red)

3.4 Plating Human Neurons for a Screening Campaign

1. Differentiate and reseed MNs as previously described (Subheading 2.4, **item 3**).
2. Coat 96-well plates with Matrigel (final dilution 1:100 in KO-DMEM), with 50–100 μl per well 1 day prior to cell seeding (*see* **Note 6**).
3. Discard the remaining liquid of the wells and seed 40,000 cells/well of a 96-well plate in a final volume of 100 μl per well. Make sure to distribute the cells evenly (*see* **Note 7**).
4. The time point of a compound screening strongly depends on cell type and the underlying questions to be answered. We recommend a maturation time of 2 weeks for human smNPCs differentiated into MNs before screening. A longer time period might be beneficial (e.g., when studying models of late-onset diseases, to facilitate maturation and the appearance of age-related phenotypes).
5. Example 96-well plate layouts are shown in Fig. 7 (*see* **Note 8**). To measure the impact and ability of a compound to rescue

A

	1	2	3	4	5	6	7	8	9	10	11	12
A	cp 1	cp 2	cp 3	cp 4	cp 5	cp 6	cp 7	cp 8	cp 9	cp 10	cp 11	L1
B	cp 12	cp 13	cp 14	cp 15	cp 16	cp 17	cp 18	cp 19	cp 20	cp 21	cp 22	L2
C	cp 23	cp 24	cp 25	cp 26	cp 27	cp 28	cp 29	cp 30	cp 31	cp 32	cp 33	L3
D	cp 34	cp 35	cp 36	cp 37	cp 38	cp 39	cp 40	cp 41	cp 42	cp 43	cp 44	L4
E	cp 45	cp 46	cp 47	cp 48	cp 49	cp 50	cp 51	cp 52	cp 53	cp 54	cp 55	H1
F	cp 56	cp 57	cp 58	cp 59	cp 60	cp 61	cp 62	cp 63	cp 64	cp 65	cp 66	H2
G	cp 67	cp 68	cp 69	cp 70	cp 71	cp 72	cp 73	cp 74	cp 75	cp 76	cp 77	H3
H	cp 78	cp 79	cp 80	cp 81	cp 82	cp 83	cp 84	cp 85	cp 86	cp 87	cp 88	H4

B

	1	2	3	4	5	6	7	8	9	10	11	12
A						empty						
B		cp 1	cp 2	cp 3	cp 4	cp 5	cp 6	cp 7	cp 8	cp 9	L2	
C		cp 10	cp 11	cp 12	cp 13	cp 14	cp 15	cp 16	cp 17	cp 18	L3	
D		cp 19	cp 20	cp 21	cp 22	cp 23	cp 24	cp 25	cp 26	cp 27	L4	
E		cp 28	cp 29	cp 30	cp 31	cp 32	cp 33	cp 34	cp 35	cp 36	H1	
F		cp 37	cp 38	cp 39	cp 40	cp 41	cp 42	cp 43	cp 44	cp 45	H2	
G		cp 46	cp 47	cp 48	cp 49	cp 50	cp 51	cp 52	cp 53	cp 54	H3	
H						empty						

Fig. 7 Plate layout. (**a**) An example 96-well plate layout showing low survival (L1-L4) and high survival (H1-H4) controls. (**b**) Because of evaporation at the edges, the outer wells can be left out of the analysis. An example plate layout is shown

cells from dying, each compound needs to be compared to control cells. Every plate should contain the following internal controls: Low control (L; 0% survival), that is, highest stress concentration; only DMSO; High control (H; 100% survival), that is, unstressed cells (Fig. 8a).

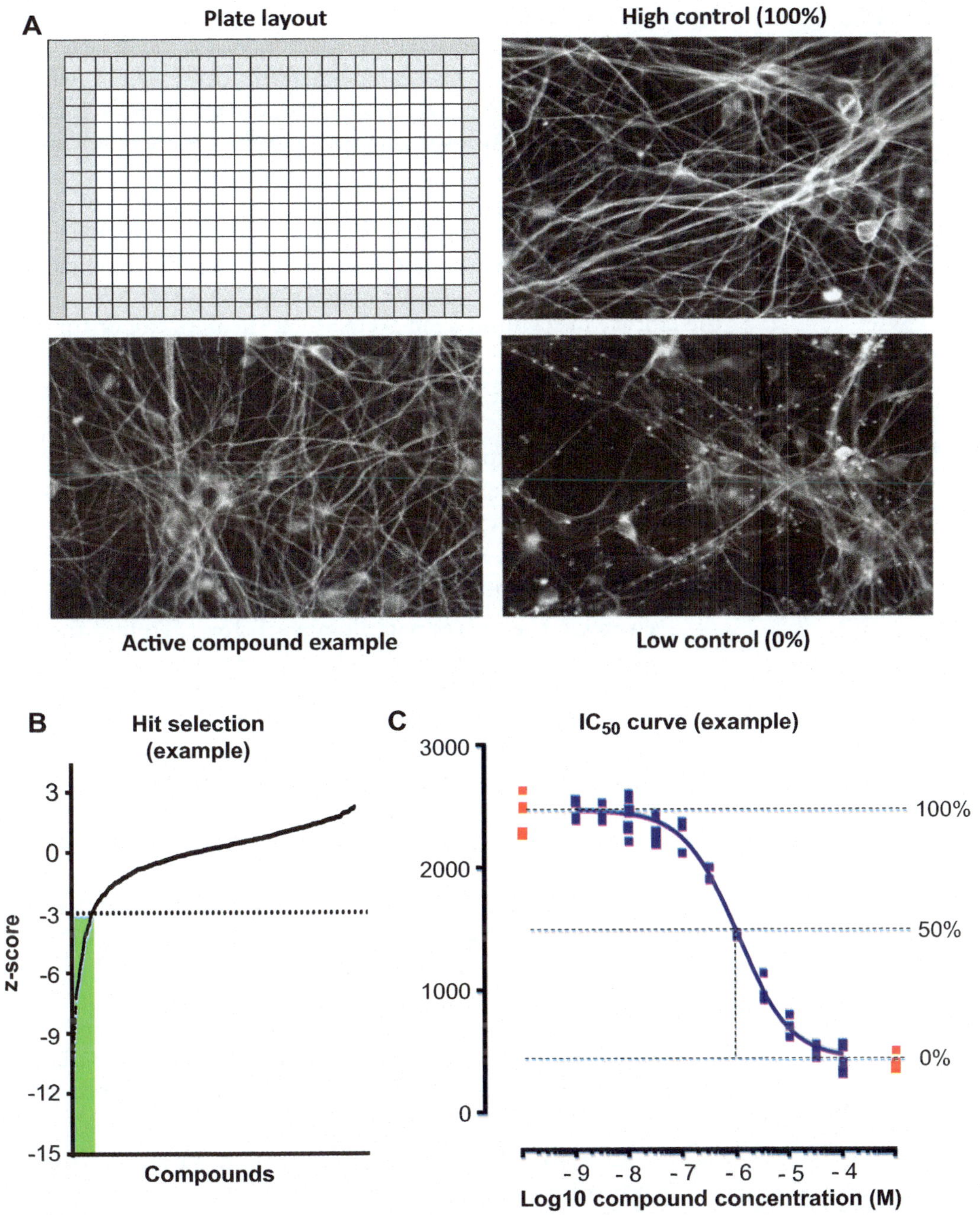

Fig. 8 Example of compound screening workflow. (**a**) Plate layout for 384-well format experiment (upper left). Fluorescent micrographs representing the high control (upper right), low control (lower right), and a potential active compound able to protect neurons from oxidative stress. (**b**) Determination of hits as compounds with a calculated *z*-score < -3 (green region highlights selected hits). (**c**) Graphical representation of an IC_{50} curve

3.5 Stressing Neurons for a Screening Campaign

Microglia act as immune cells for the central nervous system. In optimal conditions, microglia protect neurons. However, when abnormally activated, they generate nitric oxide (NO), which is neurotoxic. Therefore, compounds that protect neurons, including dopaminergic and motor neurons, from activated microglia-induced degeneration could be effective therapeutics. However, it is important to determine whether a compound directly protects neurons from degeneration, or whether the observed effect is due to a selective toxicity to microglia cells. This can be investigated by comparing the ability to rescue neurons from activated BV2 cells to the efficacy in rescuing from DetaNO, a chemical donor for NO.

3.5.1 Stressing Neurons Using Activated BV2 Microglial Cells

1. Plate BV2 cells on an uncoated plastic plate and grow in microglia medium.
2. For microglia activation, detach cells from the plate, resuspend in maturation medium including 1 μg/ml LPS and 10 ng/ml IFN-γ.
3. Add maturation medium, including cells, to assay plate 1 day after plating neurons.
4. Determine the appropriate dosage of stress. To find the right parameters for a microglia induced stress assay, several activators of BV2 cells need to be tested. A combination of interferon gamma (IFNγ) and LPS leads to a robust generation of nitric oxide (NO). First, a series of concentrations is tested to determine at which dose the activators are toxic for BV2 cells using, for example, propidium iodide staining after 24–36 h. Second, a neuronal kill curve is performed, to set the parameters of time and seeding cell number. Neuronal reporter lines, such as Hb9-GFP for MNs, or immunostaining with TuJ1 antibody can be used to evaluate neuronal viability using high-content imaging (*see* Subheading 3.6 below).

3.5.2 Stressing Neurons Using DetaNO

1. To determine the optimal amount of DetaNO to use for a stress assay, a series of concentrations should be tested, for example, 600 μM, 400 μM, 200 μM, 100 μM, 50 μM for a certain amount of time, for example 30 h.
2. In our assays, 400 μM DetaNO is used for smNPC-derived neurons.

3.6 Phenotypic Assessment

The observed phenotype should be quantifiable and expressed as a measurement. One example is based on using reporter lines, such as Hb9-GFP, or immunostaining for a neuronal marker such as TUBB3 (*see* Subheading 3.7.2). Neuronal survival can be defined using parameters such as as total neurite length per well using bioinformatics platforms for image analysis. The open-source software for cell image analysis, CellProfiler combined with KNIME, as an analytic platform, represent powerful tools for high-content image analysis. A detailed description of these tools, however, goes beyond the scope of this chapter [9].

The *Z*-prime factor (Z') statistic is widely used to estimate assay quality by taking into account data variability information of both the positive (high) and the negative (low) control, as well as the dynamic range information. Values between 0.5 and 1 correspond to excellent quality for cell-based assays; values between 0.3 and 0.5 are still considered acceptable; values below 0.3 indicate the assay is not suitable. When possible, calculate Z' with entire plates using at least two or more independent experiments.

$$Z'\text{factor} = 1 - \frac{3(\sigma p + \sigma n)}{|\mu p - \mu n|}$$

σ_p standard deviation of the positive control
σ_n standard deviation of the negative control
μ_p mean of the positive control
μ_n mean of the negative control.

3.7 Compound Screening

Screening campaigns use libraries of small molecules, which are collections of products usually consisting of structurally diverse and/or bioactive compounds having either narrow or broad-spectrum activities (or even unknown activities). Quality libraries should include information regarding structure, purity, quantity, and physiochemical properties.

3.7.1 Example of Liquid Handling for Screening Using a Immunofluorescence Readout

1. The example provided here is for a 384-well plate. Insert the 384-well source plate (compound library) in its holder within the dispenser device. Note that each well of a 384-well plate already contains cells in adhesion in 40 μl medium. Immunostaining can also be performed on 96-well plates with the aid of robotics or by hand. In the latter case, the working volume per well is 100 μl.
2. Define the plate layout by assigning each component of the source plate to the destination plate. Set a dispensing volume of 40 nl/well for a final concentration of 10 μM of each compound from 10 mM stocks.

   Upon completion, incubate cells at 37 °C according to the preestablished conditions.
3. Proceed to cell fixation by means of further robotics to ensure a rapid, homogeneous, and accurate processing of plates.
4. Perform washings and staining by dispensing liquids in five sequential steps of 20 μl aspiration followed by dispensing of 20 μl of either PBS or staining solutions (*see* **Notes 9** and **10**).

3.7.2 Hit Identification

Hits are defined according to their *z*-score, a statistical measure that judges whether the response to a particular treatment is large enough to warrant further attention. The z-score is a measure of how many standard deviations below or above the population mean a single readout value lies.

1. Calculate *z*-score using the equation (*see* **Note 11**):

$$z = \frac{x - \mu}{\sigma}$$

   σ standard deviation of the population
   μ mean of the population

2. Determining the threshold used to define hit compounds is in an assay-specific manner. Usually, a score of ± 3 indicates that a value is much higher/lower than average, suggesting that the compound under consideration exerts a substantial effect (Fig. 8b) (*see* **Note 12**).

3.7.3 Hit Validation

1. Retest hit compounds from the library to ensure validity of the results produced in the primary screening. If the effect is confirmed, proceed to the next step.
2. Order fresh compound stocks from commercial vendors. This is crucial to make sure that the observed effect is reproducible and not caused by an artifact of the library (such as compound degradation). It is suggested to first repeat the screen at 10 μM to reconfirm the effectiveness of repurchased compounds.
3. Test each hit at increasing concentrations, for example 10 nM to 100 μM. By constructing such a dose–response curve, EC_{50} (the concentration of a compound that gives half-maximal response = upward sloping curve) or IC_{50} (the concentration of a compound where the response is reduced by half = downward sloping curve) can be determined. In an ideal situation, single measurements distribute evenly between the control values, which define 100% and 0% of the effect (Fig. 8c). Incomplete dose–response curves (without a top or a bottom plateau) have undefined IC_{50} and determining the right concentration at which the compound should be used for further experiments becomes impracticable (*see* **Note 13**).

3.8 Strategies for Target Deconvolution

Molecular target identification is an essential step for progression from a phenotypic screen hit to a lead compound because it enables improved drug design and chemical optimization, and target specific toxicity and side effects can be addressed more easily. During this step, the compound's mechanism of action is resolved and target proteins responsible for the phenotypic response are identified. The affinity and functional activity between the compound and its putative target protein are assessed and confirmed.

A variety of different technologies can be used to identify molecular targets that are responsible for the observed phenotypic effect: microarray analysis (*see* Subheading 3.8.1 below), structural similarly search (*see* **Note 14**), affinity chromatography-based methods (*see* **Note 15**), and Cellular thermal shift assay (CETSA; *see* **Note 16** and Chapter 4). The choice for a specific method has to be made individually for each project and compound based on its properties.

3.8.1 Microarray Analysis

Whole-genome expression analysis enables the unbiased observation of compound effects on the transcriptome of treated cells. A popular approach is the use of gene expression microarrays that allow for cheap, fast, and standardized sample processing. Using this technique one can analyze the expression of known genes and transcripts for several samples in parallel and evaluate the cellular response to the drug.

1. Seed microglial cells or astrocytes at a density of 1.5×10^5 cells per cm^2. NSC-derived astrocytes are plated on day 4 of differentiation and matured for 14 days. Cells are then exposed to 1 μg/ml LPS, 10 ng/ml IFNγ, and the indicated hit compound for 4 h. Hit compound concentration depends on the EC_{50} or IC_{50}.
2. Extract RNA according to the manufacturer's instructions (i.e., using QIAGEN RNeasy columns with on-column DNA digestion).
3. For hybridization on Illumina mouse-8 V2 expression BeadChips (*see* **Note 17**), samples are processed using the linear amplification kit (Ambion), generating biotin-labeled cRNA, according to manufacturer's instructions. For quality-checking, cRNA is analyzed using a 2100 Bioanalyzer (Agilent) and then subsequently hybridized as recommended, using reagents provided by the manufacturer.
4. In Illumina GenomeStudio, raw data is background-subtracted and normalized using the cubic spline algorithm. Differential gene expression is determined based on thresholds for both expression ratios and statistical significance employing the Illumina custom algorithm considering standard deviations from replicate beads within each array.
5. Export data as a full table for import into Microsoft Excel or other spreadsheet software.
6. Determine signal threshold based on p-value stringency. Signal intensity values with a detection of p value >0.01 are arbitrarily set to the lowest signal intensity value with a detection of p value $= 0.01$. This procedure underestimates expression changes for genes undetectable in the reference sample (or vice versa) but avoids nonsense ratios, such as negative ones.

7. Hierarchical clustering of genes is performed using the one minus, the sample correlation metric and the Unweighted Pair-Group Method using Average (UPGMA) linkage method using the pvclust package in R (*see* **Note 18**).

4 Notes

1. Matrigel should not reach temperatures over 0 °C, as it will start polymerizing.
2. A full description of storing a compound library and performing routine quality control is beyond the scope of this chapter. Instead, we recommend partnering or using a professionally run facility. The concentration used for screening is, to a certain extent, arbitrary. At higher concentrations (e.g., 50 μM or higher), more toxicity and off-target effects can be observed, but at lower concentrations (lower than 1 μM), fewer hit compounds are identified. One commonly used screening concentration for primary screening is 10 μM.
3. Usually, mESCs grow in round-shaped colonies. However when the density of cells in the dish becomes confluent, mESC colonies fuse and spontaneously differentiate.
4. The HGD2 mESCs should carry a GFP reporter under the control of the *Hb9* promoter. The *Hb9* gene encodes a transcription factor (TF) that is specifically expressed in postmitotic motor neurons (MNs). As a result, after differentiation into MNs, GFP is expressed, which can be used for identifying MNs by fluorescent microscopy or to enrich a pure population of MNs by fluorescent activated cell sorting (FACS). Survival of MNs can be further increased by supplementing the medium with 5 μM Y-27632).
5. The materials and protocols for culturing iPSCs are not described here because many diverse methods exist and have been described previously. Culturing iPSCs in Essential 8 Medium (E8) is recommended. Culture medium is changed daily. After each split 5 μM Y-27632 is added for 24 h in the culture medium. For smNPC derivation, iPSC are transferred on feeder cultures.
6. If an immunofluorescent read out is required, Matrigel might not be the optimal coating, because Matrigel increases background noise and some cell types and lines plated on Matrigel tend to form dense clusters while aging, complicating the separation of single cells for image analysis. Alternatively, a coating with poly-L-ornithine and laminin can be used to prevent these effects. For this, coat 96-well plates with 30% PLO in PBS (40 μl/well) and incubate overnight at 37 °C. Suck off

the supernatant and add 40 μl/well laminin (1:50 in N2/B27 medium, stock 1 mg/ml) and incubate at least 3 h at 37 °C.

7. Cell numbers should be adjusted, according to the read out; for example, for subsequent immunofluorescent staining and image analysis, a discrimination of single cells is desired, and thus a lower cell density is favorable (25,000/well). For other parameters like cytotoxicity, measured by the release of cytosolic LDH by dead and dying cells using the Cytotoxicity Detection Kit a higher cell number is beneficial (50,000–60,000/well). Feed cells every 2–3 days with maturation medium (100 μl/well).

8. Some high-content imaging systems are not able to capture the outer wells of a 96-well plate.

9. Note that the deepest vertical position of aspiration and dispensing needs to be carefully set according to the type of plate and cells.

10. Neuronal networks may be washed away by strong rinsing, thus causing sample loss. Calibrate with good care the conditions that best suit your system.

11. A *z*-score of 1 is 1 standard deviation above the mean. A *z*-score of 2 is 2 standard deviations above the mean. A *z*-score of −2 is 2 standard deviations below the mean.

12. Various online tools can convert *z*-scores to *p*-values; for example, a *z*-score of 3 corresponds to a 2-tailed *p*-value of about 0.0027.

13. Hits are considered validated when they produce a reliable IC_{50}, demonstrating that the effect is concentration-dependent and reproducible. The IC_{50} defines the optimal concentration at which the compound should be used for further investigation in the defined experimental settings. Several tools are available to calculate the IC_{50}. For instance, GraphPad Prism can easily fit a dose–response curve for IC_{50} determination.

14. Computational tools to search chemical structure databases are essential to finding leads early in a drug discovery project. Structurally similar molecules tend to have similar properties. Once a lead compound has been discovered from a chemical screen, a series of structural analogues with improved pharmaceutical properties can be designed. The goal is to detect common functional groups, structural motifs, or substructures and to identify molecules that have similar properties to known entities.

    - SciFinder is a comprehensive database for chemical literature, journal articles and patent records. Furthermore, the database enables the search for chemical substances and reactions.

- PubChem provides information on the biological activities of small molecules.
- ChemDB interactively deconstructs target compounds into component precursors and reconstructs similar building-blocks into combinatorial libraries representing the *virtual chemical space* near the target compound.

15. Affinity chromatography can be used to identify target proteins that interact with small molecules. In that approach, small molecules are structurally modified by introducing a functional group coupled to a spacer, referred to as a linker. This linker is used to immobilize the ligand to a solid support such as a bead or matrix. After incubation with cell extracts, unbound proteins are removed by several washing steps. Bound proteins are then eluted by using buffer conditions that disrupt the protein–ligand interaction, and subsequently identified by mass spectrometry. To minimize the identification of nonspecifically bound proteins, the same procedure is carried out in parallel using an inactive analogue with a similar structure.

    The linker variant and position needs to be determined for each compound individually. Steric hindrance, or the creation of a hydrophobic environment, can result in nonspecific binding or interfere with the activity of the small molecule. Many obstacles can be avoided by adding linkers with polar function groups. Most frequently, biotin and polyethylene glycol in combination with agarose beads have been used successfully for target deconvolution applications.

    After the identification of biological targets that interact with the small molecule, validate if the initially detected phenotypic effect can be recapitulated by modulation of the identified biological target. Functional studies using genetic overexpression of the target protein can mimic the effects of small molecules acting as agonists. For small molecules acting as inhibitors or antagonists, genetic knockdowns/knockouts using small interfering RNA (siRNA) or gene editing tools like CRISPR/Cas9 of the identified biological target can be used to functionally validate phenotypes.

16. The CETSA for evaluating drug target interactions in cells relies on the thermal stabilization of proteins upon ligand binding (*see* Chapter 4). Cells are treated with the compound of interest and subsequently lysed. After separation of cell debris and aggregates from a soluble protein fraction, unbound proteins denature and precipitate at elevated temperatures, while ligand-bound proteins remain in solution. After heat treatment, protein levels are detected using quantitative

western blotting or two target-directed antibodies using AlphaScreen technology.

17. Illumina stopped providing mouse gene expression BeadChips. However, similar gene expression microarrays can be found at, for example, Affymetrix or Agilent. Alternatively, RNA-seq can be used to analyze the transcriptome.
18. Validation of gene expression changes are performed by qRT-PCR. Relative expression levels are calculated using the delta delta Ct method, normalized to biological reference samples and using GAPDH and ACTB as housekeeping genes.

Acknowledgments

This work was supported by the Deutsche Forschungsgemeinschaft, EU Joint Programme–Neurodegenerative Disease Research, the Bundesministerium für Bildung und Forschung, and the Max Planck Society. J.B., M.G., and L.M. are supported by fellowships from the Hans and Ilse Breuer Stiftung. We thank Tanya Levin for editing the manuscript.

References

1. Swinney DC, Anthony J (2011) How were new medicines discovered? Nat Rev Drug Discov 10 (7):507–519. https://doi.org/10.1038/nrd3480 nrd3480 [pii]
2. Martin GR (1981) Isolation of a pluripotent cell line from early mouse embryos cultured in medium conditioned by teratocarcinoma stem cells. Proc Natl Acad Sci U S A 78(12):7634–7638
3. Evans MJ, Kaufman MH (1981) Establishment in culture of pluripotential cells from mouse embryos. Nature 292(5819):154–156
4. Thomson JA, Itskovitz-Eldor J, Shapiro SS, Waknitz MA, Swiergiel JJ, Marshall VS, Jones JM (1998) Embryonic stem cell lines derived from human blastocysts. Science 282 (5391):1145–1147
5. Sterneckert JL, Reinhardt P, Scholer HR (2014) Investigating human disease using stem cell models. Nat Rev Genet 15(9):625–639. https://doi.org/10.1038/nrg3764
6. Höing S, Rudhard Y, Reinhardt P, Glatza M, Stehling M, Wu G, Peiker C, Bocker A, Parga JA, Bunk E, Schwamborn JC, Slack M, Sterneckert J, Scholer HR (2012) Discovery of inhibitors of microglial neurotoxicity acting through multiple mechanisms using a stem-cell-based phenotypic assay. Cell Stem Cell 11 (5):620–632. https://doi.org/10.1016/j.stem.2012.07.005
7. Wichterle H, Lieberam I, Porter JA, Jessell TM (2002) Directed differentiation of embryonic stem cells into motor neurons. Cell 110 (3):385–397. https://doi.org/10.1016/S0092-8674(02)00835-8
8. Reinhardt P, Glatza M, Hemmer K, Tsytsyura Y, Thiel CS, Hoing S, Moritz S, Parga JA, Wagner L, Bruder JM, Wu GM, Schmid B, Ropke A, Klingauf J, Schwamborn JC, Gasser T, Scholer HR, Sterneckert J (2013) Derivation and expansion using only small molecules of human neural progenitors for neurodegenerative disease modeling. PLoS One 8(3):e59252. https://doi.org/10.1371/journal.pone.0059252
9. Stoter M, Niederlein A, Barsacchi R, Meyenhofer F, Brandl H, Bickle M (2013) CellProfiler and KNIME: open source tools for high content screening. Methods Mol Biol 986:105–122. https://doi.org/10.1007/978-1-62703-311-4_8

Chapter 3

Quantitative, Real-Time Measurements of Intracellular Target Engagement Using Energy Transfer

Matthew B. Robers, James D. Vasta, Cesear R. Corona, Rachel Friedman Ohana, Robin Hurst, Manisha A. Jhala, Kenneth M. Comess, and Keith V. Wood

Abstract

Intracellular target affinity and residence time are fundamental aspects of pharmacological mechanism (Lu and Tonge, Curr Opin Chem Biol 14:467–474, 2010). Although various robust biochemical approaches exist to measure these binding characteristics, analysis of compound binding with isolated targets may not accurately reflect engagement in the milieu of living cells. To realize the influence of cellular context, methods are needed that are capable of quantifying affinity and residence time in the presence of the intracellular factors that may impact target engagement. Bioluminescence resonance energy transfer (BRET) offers a solution for intracellular target engagement when quantitative metrics or kinetic analyses are required.

Key words Target engagement, Residence time, Intracellular affinity, NanoBRET(TM), BRET

1 Introduction

While quantitation of compound binding to isolated targets is readily achieved, similar biophysical assessments in living systems represent a challenge. Cell-based assays of target engagement are most commonly achieved by assessing the dynamics of cell signaling. Although these methods allow for interrogation of compound potency using live cells, pathway analysis methods generally integrate the activities of multiple protein targets. Consequently, the molecular targets of drug molecules can be ambiguous when analyzed within the framework of a pathway. The ideal target engagement format provides a direct, biophysical assessment of compound binding at the selected intracellular target, therefore providing a bridge between reductionist biochemical assays and traditional cellular pathway analysis readouts.

Slava Ziegler and Herbert Waldmann (eds.), *Systems Chemical Biology: Methods and Protocols*, Methods in Molecular Biology, vol. 1888, https://doi.org/10.1007/978-1-4939-8891-4_3, © The Author(s) 2019

A number of novel techniques have been recently developed that allow for the interrogation of targets derived from cells or cell extracts. Each technique offers certain advantages and disadvantages that should be evaluated according to the requirements of the study. The cellular thermal shift assay (CETSA) represents a significant advancement for cellular analysis, as it queries compound binding of endogenously expressed targets, using intact cells [1]. However, CETSA is an indirect measure of compound engagement, and is reportedly prone to noteworthy false negative and false positive results for clinically relevant probe molecules [1]. Activity-based proteomics reagents offer a more direct assessment of compound binding via mass spectrometry (MS) [2]. However, the majority of such MS techniques involve dissolution of the cell membrane, and consequently a dilution of key cellular factors that are known to impact inhibitor potency. While these methods enable analysis of endogenously expressed targets, each has a significant limitation that should be considered. Activity-based proteomics methods do not assess target engagement under a thermodynamic equilibrium with the unlabeled drug molecule in living cells, thus limiting their ability to measure intracellular affinity. Acquisition of CETSA data to estimate IC_{50} values can be labor-intensive and does not necessarily correlate directly to thermodynamic affinity. Such approaches also fail to offer real-time analysis of target engagement inside intact cells. When quantitation of affinity and residence time is desired, alternate approaches should be considered.

A recently described technique using Bioluminescence Resonance Energy Transfer (BRET) offers a quantitative analysis of target engagement in living cells under both equilibrium and non-equilibrium conditions [3, 4]. This method, termed NanoBRET (TM), is a highly sensitive measure of molecular proximity and offers a direct and real-time capability for measuring molecular interactions in live cells. Briefly, reporter complexes are formed inside live cells, when a luciferase (e.g., NanoLuc(R)) -tagged target protein reaches dynamic equilibrium with cell-permeable fluorescent probe molecules (tracers). Upon binding of an unlabeled compound, the complex is disrupted, leading to a loss of BRET that can easily be monitored and scaled in a microplate format (Fig. 1). In many cases, the NanoBRET tracers can be readily designed from drug or tool compounds, generally offering a quantitative analysis of apparent affinity and a real-time analysis of residence time [3, 5–7].

However, this approach has technical requirements that must first be considered and carefully followed in order to achieve quantitation in live cells. The following chapter outlines: (1) considerations for development of novel cell-permeable tracers, (2) optimization of conditions for intracellular BRET, and (3) protocols for quantitative and real-time analysis of target engagement in live cells in the context of HTS work flows.

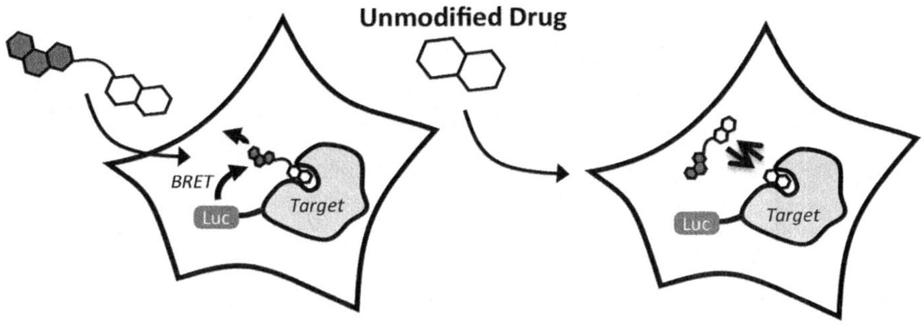

Fig. 1 Illustration of target engagement via energy transfer

2 Materials

2.1 Development and Characterization of NanoBRET™ Tracers with Optimal Cell Permeability

1. Functionalized small molecule precursor with a primary amine group.
2. 5 mL amber vials.
3. 30 mL plastic bottle.
4. Anhydrous dimethylformamide (DMF).
5. *N,N*-Diisopropylethylamine.
6. NanoBRET™ 590 SE and NanoBRET™ 618-TFP.
7. Analytical HPLC.
8. Reaction quenching solution: 1:1:0.01 (v/v/v) water–acetonitrile–trifluoroacetic acid (TFA).Combine 5 mL water and 5 mL acetonitrile in a 30 mL plastic bottle and add 0.1 mL trifluoroacetic acid (TFA).
9. Trifluoroacetic acid (TFA) HPLC Buffer: 0.1% TFA in water. Add 4 mL TFA to a 4 L bottle of HPLC grade water and mix.

2.2 Verification of Tracer Permeability and Optimization of BRET

1. Desired cell type (e.g., HEK293).
2. White, nonbinding surface (NBS) 96-well plates, appropriately coated to avoid adsorptive properties of the fluorescent tracers.
3. NanoBRET Tracer Dilution Buffer (Promega).
4. Polypropylene plastic ware (buffer troughs, serial dilution troughs, etc.).
5. Miscellaneous tissue culture reagents and plasticware, including, standard cell culture growth medium.
6. Assay medium: Opti-MEM without phenol red (Gibco).
7. Transfection reagent.
8. Promoterless DNA (Transfection Carrier DNA, Promega).
9. Extracellular NanoLuc(R) Inhibitor (Promega).

10. Luciferase substrate (NanoBRET(TM) Nano-Glo(R) Substrate, Promega).
11. Dimethylsulfoxide (DMSO).
12. Tracer: 100–400 μM in DMSO (stored at −80 °C; avoid freeze/thaw cycles).
13. Plasmid DNA encoding NanoLuc(R) fusion under control of a mammalian promoter (e.g., in a pFN31K or pFN32K vector, Promega).
14. 250 μg/mL digitonin in water (or 50 mg/mL digitonin in DMSO).
15. 5× Passive Lysis Buffer (Promega).
16. A BRET-compatible luminometer equipped with 450 nm (bandpass) and 600 nm (longpass) filters. (e.g., Glomax Discover, PerkinElmer EnVision, or BMG Clariostar).

2.3 Considerations for Quantitation of Compound Affinity by Competitive Displacement of Tracers

1. Same reagents as described in Subheading 2.2.

2.4 Liquid Handling Requirements and Considerations for High-Throughput Screening and Lead Optimization

1. Same reagents as described in Subheading 2.2.
2. Acoustic liquid handling technology capable of nanoliter dispensing of test and control compounds in DMSO solvent, e.g., Labcyte Echo 550 Liquid Handler.
3. Electronic 16-channel repeat pipettors, e.g., Thermo Fisher E1-ClipTip™ 384 Multichannel Electronic Pipette and tips.
4. Agilent PlateLoc Thermal Microplate Sealer and Agilent peelable aluminum seals.
5. White, 384-well nonbinding surface (NBS) plates (Corning).

2.5 Optimization of Energy Transfer for Residence Time Studies in Live Cells

1. Same reagents described in Subheading 2.2.
2. Assay plates: white, opaque, tissue cultured-treated (TC) 96-well plates.
3. Vacuum aspirator with eight-channel adapter.
4. A BRET-compatible luminometer, equipped as described earlier (*see* Subheading 2.2), and with repeat read capabilities over at least 2 h intervals.

3 Methods

3.1 Development and Characterization of NanoBRET™ Tracers with Optimal Cell Permeability

To achieve quantitation in living cells, the BRET technique exploits permeable fluorescent reporter probes (or tracers), derived from chemical matter with appropriate binding characteristics for the select target protein or protein family. Tracers may be derived from compounds with high selectivity for the target protein, but in many cases, broad-spectrum inhibitors or published tool compounds can be utilized to achieve greater coverage if desired [4].

The chemical synthesis of functional tracers involves the covalent conjugation of a biologically active small molecule to a fluorescent organic dye. Certain organic dyes have proven optimal in maintaining cell permeability with spectral properties required for NanoBRET(TM) (Fig. 2a). These dyes are readily coupled to base compounds containing pendant amine groups via a one-step reaction of a highly reactive, activated ester group on the dye to produce a stable amide linkage.

Tracers must permeate live cells and function in a complex intracellular environment. Thus, there are several physical requirements that must be considered during the design phase. First, high affinity compounds may yield the most useful intracellular tracers. Within the milieu of cells, strong affinity may be required to overcome competition from endogenous metabolites or enzymatic

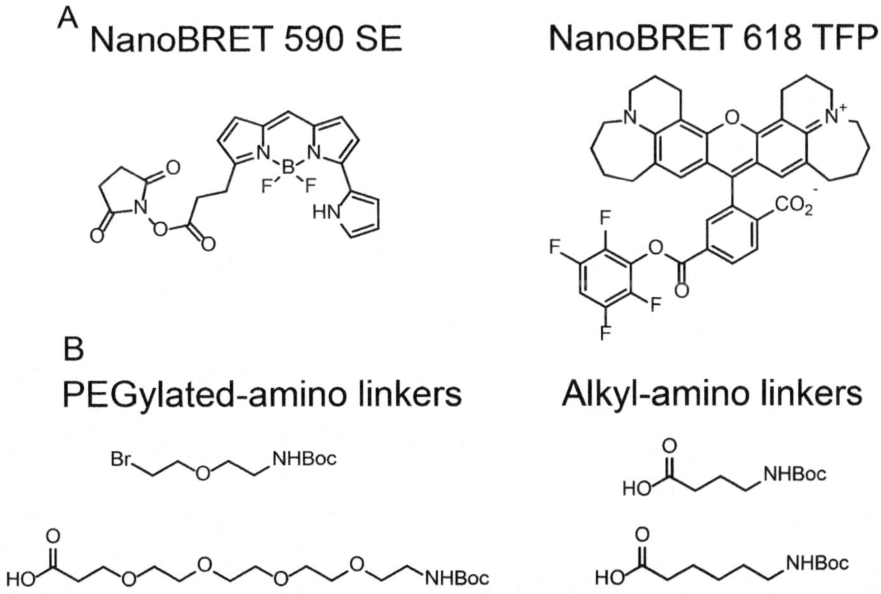

Fig. 2 The structures of NanoBRET(TM) activated ester dyes (**a**) and example linker building blocks (**b**). NanoBRET(TM) dyes balance permeability and energy transfer efficiency, resulting in higher signal, stability and reactivity. Dyes directly coupled to the inhibitor yield the tracers, though it may be necessary to install a linker group to provide separation between the core and the fluorescent tag. In general, C3 to C6 alkyl, as well as 01 to 04 PEGs are suitable as starting points

substrates that may compete directly with the tracer. Second, the intrinsic binding kinetics of tracer for the target may also be important for consideration. In general, compounds with rapid equilibration kinetics also offer better capabilities for real-time analysis of target engagement (i.e., residence time analysis). Finally, the base scaffold must have a solvent exposed reactive group for conjugation. This site can be determined through cocrystal structure, SAR analysis, or homology models. Although a variety of synthetic approaches can be employed, NanoBRET(TM) dyes prepared as activated esters can be readily conjugated to the scaffold via a primary or secondary aliphatic amine. These amine groups have been shown to react quickly and quantitatively with activated ester dyes and amino acid linkers.

A variety of dye classes have been evaluated in BRET assay formats, and dyes representing two fluorophore classes (Fig. 2a) have been identified that demonstrate consistently robust cell permeability and optimal spectral overlap for energy transfer. Derivatives containing varied linkers and dyes may perform differently in the assay (examples provided in Fig. 2). Therefore, as a starting point for assay development, it is initially recommended that at least four derivatives be prepared and evaluated in the BRET configuration: Minimal/no linker derivatives conjugated to NanoBRET™ 618 and NanoBRET™ 590 dyes and extended linkers (e.g., multi-PEG or alkyl derivatives) conjugated to NanoBRET™ 618 and NanoBRET™ 590 dyes. This panel of tracer candidates can be created from common intermediates and will provide structural diversity required to ensure the highest probability of success, and if required, guide further optimization. To serve as examples, these principles have been employed to create a panel of candidate cell-permeable tracers based on crizotinib, a multikinase inhibitor scaffold (Fig. 3).

3.1.1 Example Synthesis of Amine-Linked Tracers

1. Charge the small molecule precursor, functionalized with an amine, (9.9 μmol) into a 5 mL amber vial.
2. Add 0.5 mL anhydrous dimethylformamide (DMF) and stir.
3. Add 8.6 μL *N,N*-Diisopropylethylamine, allowing the mixture to stir for 10 min.
4. Add 5.0 mg NanoBRET™ 590 SE. Cap and allow to react in the dark for 1 h.
5. Monitor the reaction for consumption of the starting amine by analytical HPLC. The reaction is judged complete when this material has been consumed. The reaction progresses quickly and is often near completion after 30 min (*see* **Note 1**).
6. Dilute the crude mixture with the reaction quenching solution.
7. Purify the compound using reverse-phase preparative HPLC. Most tracers can be purified using standard HPLC methods.

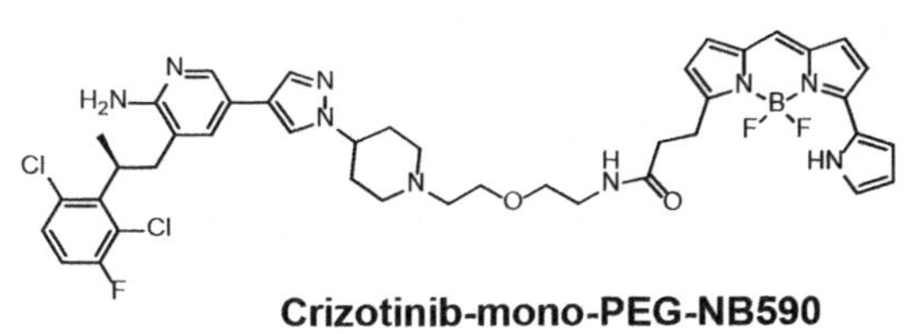

Crizotinib-mono-PEG-NB590

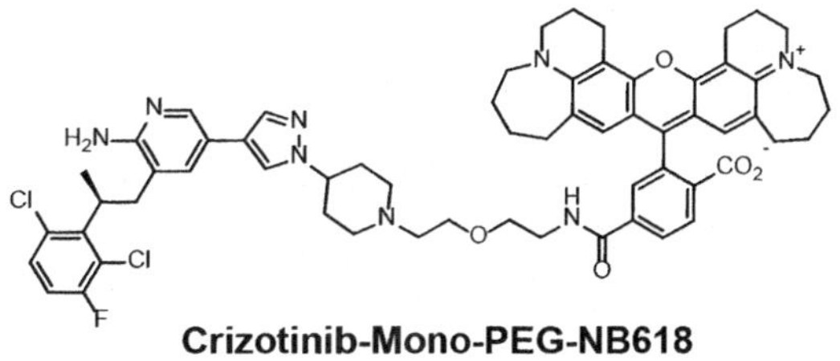

Crizotinib-Mono-PEG-NB618

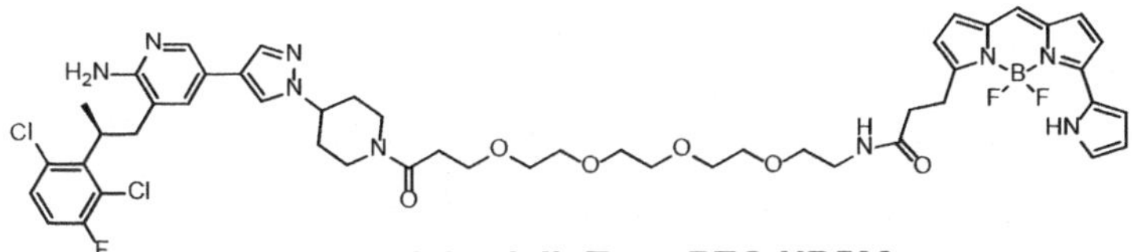

Crizotinib-Tetra-PEG-NB590

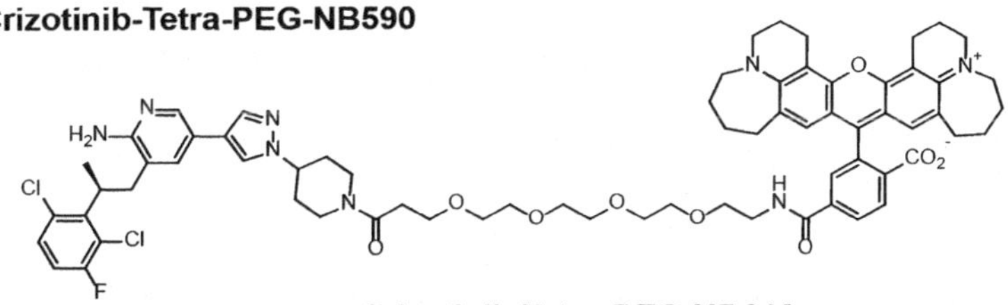

Crizotinib-Tetra-PEG-NB618

Fig. 3 Chemical structures of four candidate tracer derivatives based upon the crizotinib scaffold. Similar dye/linker variants can be created for a wide variety of base molecules by conjugation to solvent exposed primary or secondary amines

Use a gradient of trifluoroacetic acid (TFA) HPLC Buffer (or 0.1% formic acid in water) in acetonitrile as an eluent.

8. Pool product containing fractions and concentrate to dryness under reduced pressure.
9. Dissolve the resulting film in approximately 10 mL acetonitrile and concentrate to dryness a total of three times (*see* **Note 2**).
10. Dry the resulting film for a minimum of 1 h to remove trace volatile solvents.
11. Dissolve the resulting solid in an appropriate volume of DMSO to give a 2.0 mM solution.
12. Calculate the concentration of the solution using the following molar extinction coefficients: $\varepsilon_{\text{NanoBRET™ 590}} = 83{,}000\ \text{M}^{-1}\ \text{cm}^{-1}$, $\varepsilon_{\text{NanoBRET™ 618}} = 96{,}000\ \text{M}^{-1}\ \text{cm}^{-1}$.

3.2 Verification of Tracer Permeability and Optimization of BRET

BRET is highly sensitive to molecular distance and orientation between the donor (luciferase) and acceptor (fluorescent tracer) partners. This feature enables a bona fide measurement of direct molecular interactions inside intact cells. However, optimization may be warranted to identify the conditions producing optimal BRET between the donor/acceptor partners. This can generally

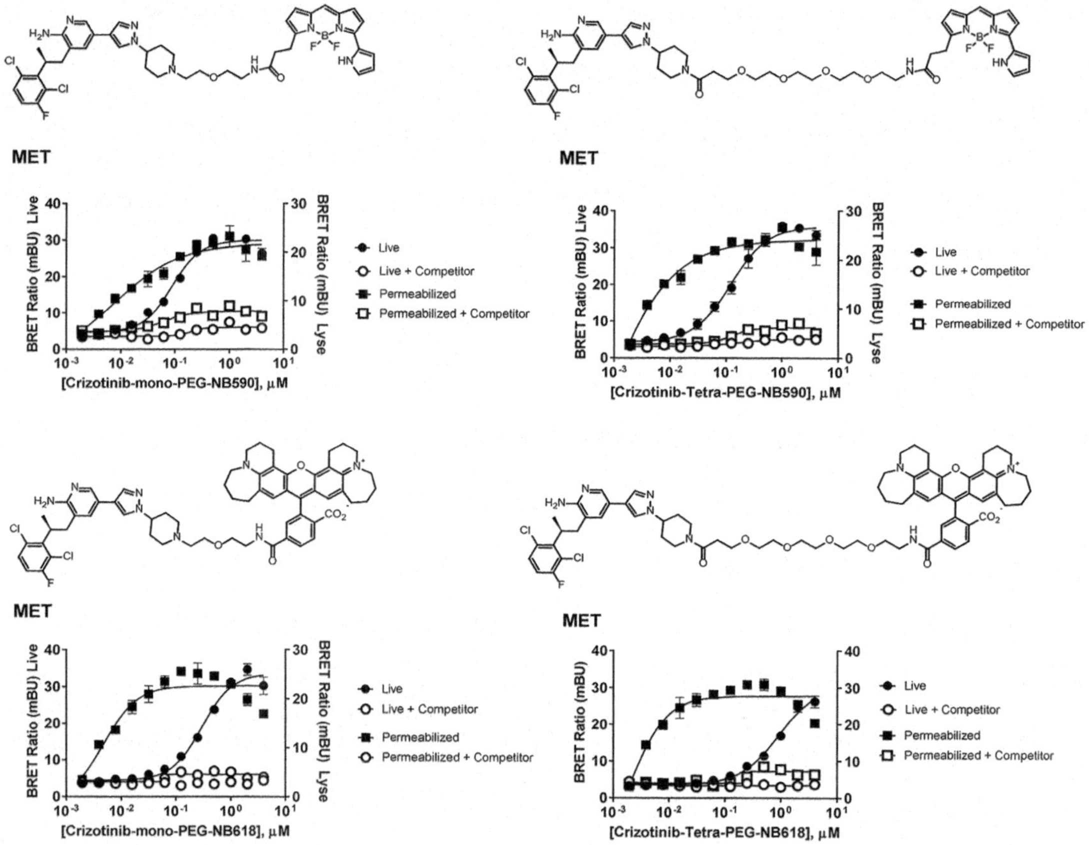

Fig. 4 Verification of tracer permeability and affinity. HEK293 cells transfected with NanoLuc(R) fusion plasmids were titrated with tracers. Cells treated with 50 μg/mL digitonin were included as controls for engagement of poorly permeable tracers. Each tracer demonstrated specific BRET with MET kinase in live cells and permeabilized cells. Both of the NanoBRET(TM) 590 tracers yielded the strongest cellular engagement and proved to be optimal for MET kinase. In certain cases, alternate dye/linker combinations may prove optimal

be achieved without a laborious effort, via transient transfection of plasmids encoding target/NanoLuc(R) fusion proteins (*see* **Note 3**) [4]. The following basic protocol can be used to determine the optimal orientation of NanoLuc(R) luciferase for target engagement analysis via BRET. It is generally recommended that novel NanoBRET(TM) tracers are evaluated in both live cells, as well as permeabilized cells (via addition of digitonin or Passive Lysis buffer), to assess tracer affinity in the absence of interfering cellular factors. Figure 4 provides an example where the protocol is used to determine the optimal tracer for a kinase target, and Figs. 5 and 6 provide examples wherein the protocol is used to determine optimal NanoLuc(R) fusion orientation or expression level, respectively.

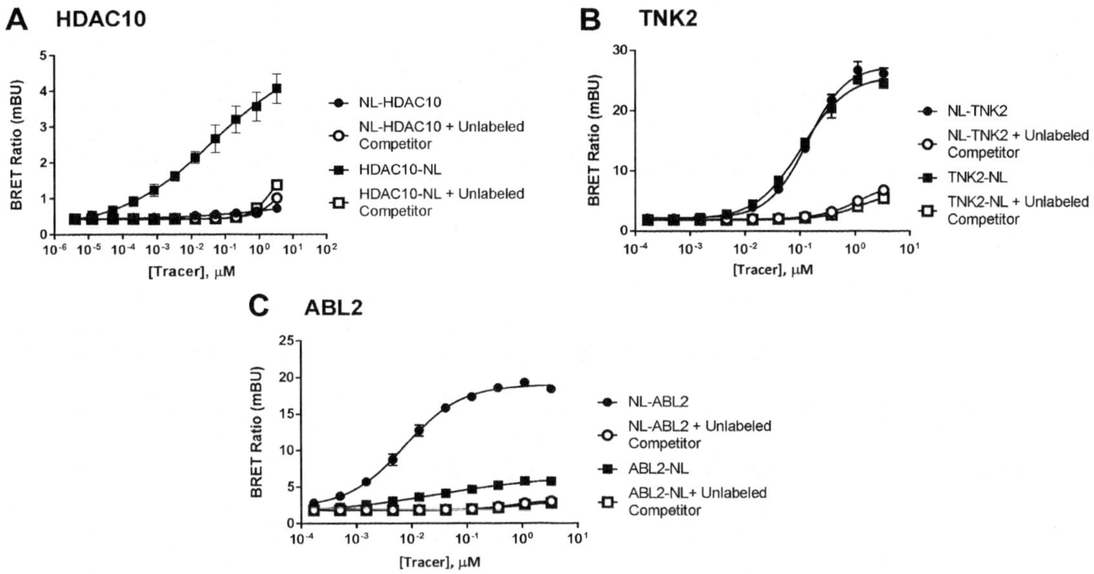

Fig. 5 Optimization of NanoLuc(R) orientation for BRET. (**a**) For HDAC10, BRET was optimal with NanoLuc(R) on the *C*-terminus of the target protein. For (**b**) TNK2 both *N*- and *C*-terminal NanoLuc(R) fusions performed optimally in BRET experiments. (**c**) For ABL2, both *N*- and *C*-terminal NanoLuc(R) fusions performed in BRET experiments, but the *N*-terminal placement of NanoLuc(R) performed optimally. *N*- and *C*-terminal orientations should be tested when possible

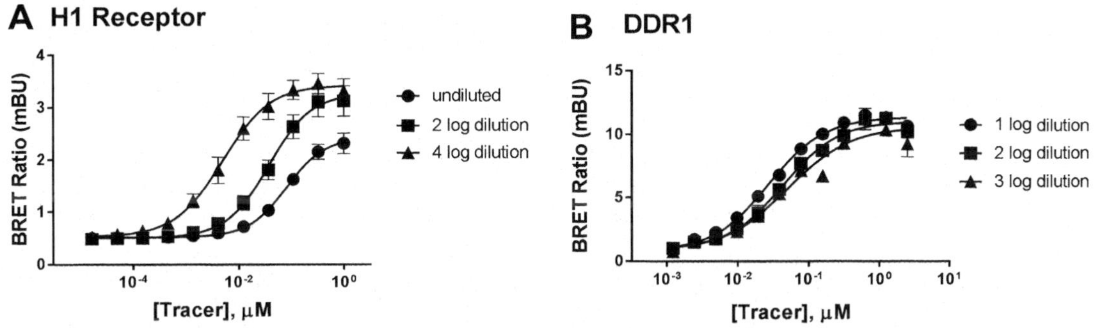

Fig. 6 Optimization of expression levels for BRET studies. Expression levels can be easily optimized by dilution of the expression plasmids into a promoterless carrier DNA (Transfection Carrier DNA). This maintains a constant amount of DNA for transfection, but allows for lower expression levels across the population of transfected cells. (**a**) For the GPCR example (Histamine H1 receptor), BRET signal and tracer engagement was sensitive to expression levels, supporting use of lower expression levels for the receptor of interest. (**b**) For DDR1 kinase, BRET and tracer engagement was unaffected by expression levels. In general, it is recommended that low expression levels be used to ensure physiological conditions and avoid ligand depletion

3.2.1 [Day 1]: Transient Transfection of HEK293 Cells with NanoLuc(R) Fusions

1. Cultivate desired cells (e.g., HEK293) appropriately prior to assay and resuspend cells into a single-cell suspension using complete cell culture medium.
2. Adjust the cell density to 2×10^5/mL in cell culture medium in a sterile, conical tube.

3. Prepare plasmid DNAs for transfection. For first-time experiments, use a 1:10 dilution (mass:mass) of expression construct into a promoterless DNA (Transfection Carrier DNA). At this stage it may be useful to perform a titration of expression, by diluting the expression construct into Transfection Carrier DNA. Examples of extending this titration are shown in Fig. 6.
4. Prepare transfection complexes using the manufacturer's protocol.
5. Mix one part (e.g., 1 mL) of transfection complex with 20 parts (e.g., 20 mL) of HEK293 cells in suspension at 2×10^5/mL. Mix gently by inversion five times in a sterile, conical tube.
6. Dispense cells/lipid–DNA complex into a sterile tissue culture flask, and incubate at least 20 h to allow expression to occur. We recommend a cell density of approximately 55,000–80,000 cells/cm^2 during the transfection (for example use approximately 4–6 million cells for a T75 flask). Larger or smaller bulk transfection should be scaled accordingly, using this ratio.

3.2.2 [Day 2]: Preparation of Cells with 1× Tracer for BRET Assay

1. Harvest cells via trypsinization, and resuspend in growth medium including serum. Centrifuge the cells and resuspend the pellet in Opti-MEM without serum or phenol red.
2. Adjust the cell density to 2×10^5/mL in Opti-MEM without serum or phenol red.
3. Prepare cells for target engagement analysis in live or permeabilized cells. For live cell target engagement experiments, dispense 85 μL per well of cell suspension into white, 96-well NBS plate. Periodically mix cells appropriately to avoid settling of the cell suspension. For live cell analysis, proceed to **step 4**. For permeabilized cell experiments dispense 68 μL cell suspension per well into a white, 96-well plate. Add 17 μL of lysis reagent (either 5× Passive Lysis Buffer or 250 μg/mL digitonin).
4. Prepare the complete 20× tracer. For characterization of tracer affinity, perform a serial dilution of the tracer in DMSO at a 100× concentration (*see* **Note 4**). For characterization of unlabeled drug affinity (if the optimal concentration of tracer has already been determined) prepare a 100× concentration of the tracer in pure DMSO.
5. Dilute the 100× tracers to 20× with Tracer Dilution Buffer according to the manufacturer's instructions. Add 4 parts Tracer Dilution Buffer to 1 part of 100× tracer (from DMSO) to prepare the Complete 20× tracer reagent.
6. Mix gently several times to ensure that the DMSO solution is mixed with the tracer dilution buffer.
7. Add 5 μL of complete 20× tracer reagent per well to cells in suspension.

8. Mix on orbital shaker for 15 s at 900 rpm. Mixing may vary between orbital shakers and should be optimized for each individual unit accordingly. Due to the viscosity of the Tracer Dilution Buffer, it is necessary to dispense this solution slowly.

3.2.3 [Day 2]: Addition of Test Compounds for Competition Analysis and Determination of Specific BRET

1. Prepare a suitable unlabeled test compound at 1000× final concentration in 100% DMSO. Typically, the 1000× concentration of compound is in the 10–100 mM range. Then dilute to a 10× final concentration in Opti-MEM without serum or phenol red.
2. The characterization of BRET tracers can be performed as a single concentration of unlabeled compound for tracer dose–response curves. It is generally recommended that a 20–100× molar excess of parental drug is coincubated with the tracer to determine the nonspecific BRET signal.
3. For characterization of unlabeled drug IC_{50}, create a serial dilution of test compound for drug IC_{50} analysis.
4. Add 10 μL per well of 10× inhibitor/test compound to the 96-well plates containing cells with 1× tracer. Mix on orbital shaker for 15 s at 900 rpm.
5. Incubate the plate at 37 °C and 5% CO_2 for 2 h. Allow plate to cool to room temperature for approximately 15 min, then proceed to the next section below.
6. For permeabilized cell assays, incubate the plate at room temperature, protected from light for 1–2 h (depending on the characteristics of the test compound). Then proceed with the next step (*see* Subheading 3.2.4).

3.2.4 Detection in Microplate Luminometer Equipped with Appropriate BRET Filters

1. Immediately prior to BRET measurements, prepare 3× Complete NanoBRET(TM) Nano-Glo(R) Substrate.
2. For live cell analysis, add Extracellular NanoLuc(R) Inhibitor in Opti-MEM and add to cells according to the manufacturer's instructions (Promega) (*see* **Note 5**).
3. For Permeabilized cell analysis, prepare 3× NanoBRET(TM) Nano-Glo(R) Substrate in Opti-MEM without serum or phenol red, and omit the Extracellular NanoLuc(R) Inhibitor. This solution consists of only a 1:166 dilution of NanoBRET(TM) Nano-Glo(R) Substrate in Opti-MEM.
4. Following addition of 50 μL of NanoBRET(TM) Nano-Glo(R) Substrate, measure donor emission (e.g., 450 nm) and acceptor emission (e.g., 610 nm or 630 nm) using a NanoBRET(TM)-compatible luminometer.
5. To generate raw BRET ratio values, divide the acceptor emission value (e.g., 610 nm) by the donor emission value (e.g., 450 nm) for each sample. Convert raw BRET units to

milliBRET units (mBU) by multiplying each raw BRET value by 1000.

6. (Optional) If a background correction is desired, use the Nano-BRET Equation (Eq. 1).

$$\left(\frac{\text{Acceptor}_{\text{sample}}}{\text{Donor}_{\text{sample}}} - \frac{\text{Acceptor}_{\text{no tracer control}}}{\text{Donor}_{\text{no tracer control}}}\right) \times 1000 \quad (1)$$

3.3 Considerations for Quantitation of Compound Affinity by Competitive Displacement of Tracers

As described above (*see* Subheading 3.2), the basic protocol details the measurement of the intracellular engagement potency (IC_{50} of BRET attenuation) of a test compound at a single concentration of tracer. In general, compound affinity is measured under steady-state conditions between the test compound, the tracer, and the intracellular target protein. For the simple determination of compound rank-order affinity, it is recommended to use the tracer at a fixed concentration approximating 50–80% occupancy at the target. However, tracers can be exploited to provide a more quantitative apparent affinity value, as engagement potency is often dependent on the concentration and affinity of the tracer used in the assay. As the tracer and test compound bind to the target in a competitive and mutually exclusive manner, the relationship between the apparent IC_{50} of the test compound and the concentration of tracer used in the assay is governed by the Cheng–Prusoff Equation [8] (Eq. 2).

$$K_{\text{i, app}} = \frac{IC_{50}}{1 + \left(\frac{[\text{tracer}]}{K_{\text{d, app}}}\right)} \quad (2)$$

Equation 2 provides the quantitative capability to extract an apparent dissociation constant ($K_{i,app}$) for the test compound. This analysis yields a value that is insensitive to tracer concentration. Considering the limits of Eq. 2 with respect to tracer concentration, in situations where the concentration of tracer is much less than its affinity ($K_{d,app}$), the measured IC_{50} for the test compound will generally be within twofold of the $K_{i,app}$ (i.e., if $[\text{tracer}] << K_{d,app}$, then $IC_{50} \cong K_{i,app}$). Conversely, when the concentration of tracer increases above the $K_{d,app}$, the apparent IC_{50} value of the test compound shifts to higher concentrations.

The Cheng–Prusoff relationship can be used to determine the $K_{i,app}$ for a test compound in multiple ways. For example, if the apparent affinity of the tracer ($K_{d,app}$) has already been measured, Eq. 2 can be used to convert an IC_{50} value measured at a single, known concentration of tracer to the $K_{i,app}$ for the test compound. Alternatively, Eq. 2 can be algebraically linearized (Eq. 3) and used experimentally to extract the $K_{i,app}$, a method known as linearized Cheng–Prusoff analysis [9]. In practice, the apparent IC_{50} value for the test compound is measured at multiple tracer concentrations,

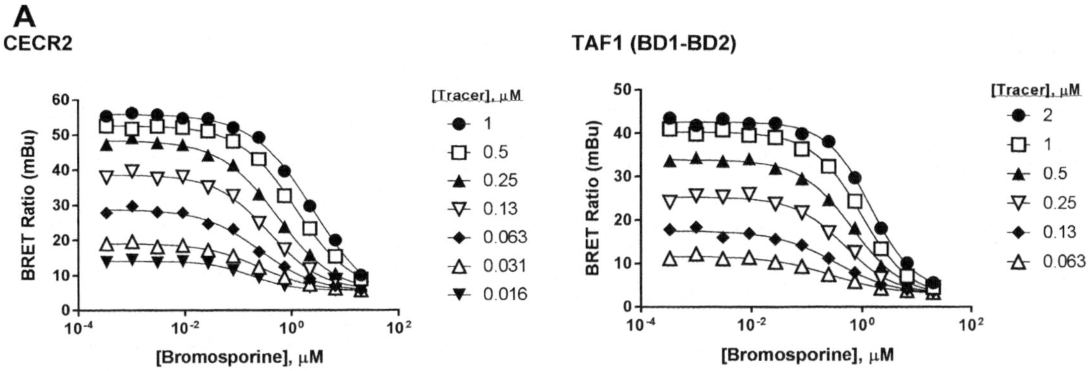

	1	0.5	0.25	0.13	0.063	0.031	0.016
IC50	2.6	1.2	0.53	0.38	0.25	0.26	0.18

	2	1	0.5	0.25	0.13	0.063
IC50	1.4	0.95	0.64	0.47	0.33	0.28

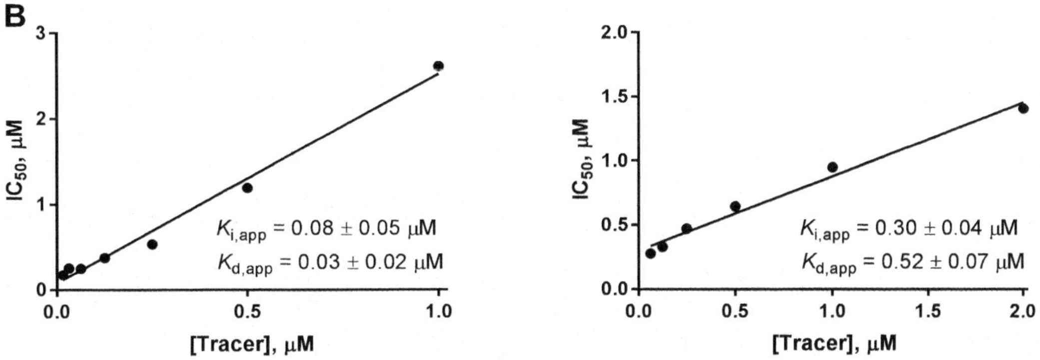

Fig. 7 Linearized Cheng–Prusoff analysis to determine bromosporine affinity to CECR2 Bromodomain (BD) and TAF1 BD1-BD2. HEK293 cells transiently expressing NanoLuc(R) fusions to bromodomains were seeded into 96-well plates and then treated with a dilution series of bromosporine at increasing concentrations of tracer. Panel A depicts dose–response curves for bromosporine at each individual tracer concentration. Individual curves were fitted with the sigmoidal dose–response curve to determine the apparent IC_{50}. Replots of the IC_{50} values as a function of tracer concentration (panel B) show linearity and allowed calculation of the $K_{i,app}$ for bromosporine and $K_{d,app}$ of the tracer using Eq. 3

after which the IC_{50} values are replotted as a function of tracer concentration. Regression analysis yields a linear plot with a y-intercept equal to the $K_{i,app}$, and slope equal to $K_{i,app}/K_{d,app}$. Figure 7 provides examples in which linearized Cheng–Prusoff analysis was used to quantify the $K_{i,app}$ for the binding of the broad spectrum bromodomain inhibitor bromosporine to NanoLuc-bromodomain fusions.

$$IC_{50} = K_{i,app}\left(1 + \frac{[\text{tracer}]}{K_{d,app}}\right) \quad (3)$$

The linearized Cheng–Prusoff analysis is generally recommended for determinations of $K_{i,app}$ for test compounds, particularly if binding of the tracer is saturable. For targets where tracer binding is weak and not saturable, but signal window (SW) is still adequate for IC_{50} determination (in practice, SW $\geq$ 1.6), the measured IC_{50} will often be a good approximation for the $K_{i,app}$ (generally within twofold, as discussed above) (*see* **Note 6**).

3.4 Liquid Handling Requirements and Considerations for High-Throughput Screening (HTS) and Lead Optimization

Cellular target engagement assays can provide critical information for hit prioritization at both the HTS and lead optimization stages, particularly when the biochemical and cell-based assays fail to correlate. Accordingly, compounds that are similarly active in biochemical, cellular target engagement, and cellular activity assays can be assigned highest priority, while those that are very weak or inactive in a cellular target engagement assay, yet active in the other two formats, should be binned as suspicious. Such compounds are likely to trigger cellular activity assay inhibition through indirect effects. A sensitive cellular target engagement assay may also show correlative activity with the biochemical assay for certain compounds, yet fail to demonstrate cellular activity. In this case, the compound or compound series may be valid, yet require chemical optimization to allow cellular activity. It would therefore be valuable to apply the NanoBRET™ target engagement assay for HTS and as a key aid for dissection of mechanism in lead optimization.

It is possible to configure the steps and equipment to higher throughput, in order to enable efficient intracellular target engagement-based HTS and lead optimization. In order to afford the throughput required for HTS and lead optimization, several steps to streamline the NanoBRET™ process are suggested. Strategies to maximize efficiency include switching from 96-well to 384-well format to reduce the number of plates required and the plate "real estate" occupied by controls, reducing the number of steps by using assay-ready plates, and implementation of robotic liquid handling protocols. Preparation of assay-ready plates requires acoustic droplet liquid dispensing technology. While relatively expensive, it is becoming increasingly available in core facilities within both academic and industrial research centers.

It is recommended that each 384-well plate contain a set of controls occupying one entire row of 24 wells and comprising a dose response with a 12-point concentration range. Be sure to include appropriate controls including a zero BRET control (i.e., no tracer) as well as a "BRET specificity control" (i.e., a high concentration of a control inhibitor in the presence of tracer) (*see* **Notes** 7 and **8**).

Table 1 summarizes the results of target engagement analysis for a panel of full-length intracellular kinases in a scaled format (i.e., automated, 384-well format). As a measure of statistical robustness, each assay yielded a Z' above 0.4, supporting suitability in HTS work flows. As a future direction, scaling the BRET technique to higher density (e.g., 1536-well plates) may be possible.

Table 1
Data and statistical calculations for various kinase enzyme assays in high-throughput format

	Target									
	MAP4K1	**BTK**	**LTK**	**ITK**	**MAP4K2**	**TBK1**	**DDR1**	**CDK2**	**TEC**	**TBK1**
Plus tracer signal (mBRET)	92.69	174.6	104.7	78.46	55.41	124.3	54.82	36.22	134.6	55.77
Plus tracer signal stdev	3.85	3.12	3.87	1.57	2.38	5.41	3.08	2.60	13.1	4.01
No tracer signal (mBRET)	20.08	21.86	20.36	20.80	19.85	20.41	19.80	17.40	24.56	20.90
No tracer signal stdev	0.54	3.15	0.61	0.91	0.55	2.13	2.17	0.80	3.51	2.71
Z′	0.82	0.88	0.84	0.87	0.75	0.78	0.55	0.46	0.55	0.42
S:B	4.62	7.99	5.14	3.77	2.79	6.09	2.77	2.08	5.48	2.67
Tracer K_d (M)	1.50E-06	6.00E-07	1.70E-06	9.70E-08	5.40E-07	4.90E-07	9.80E-07	2.70E-06	3.20E-07	1.70E-06

3.4.1 [Day 1]: Predispense of Test Compounds to Assay Plate

1. Aliquot test and control compound stocks in 100% DMSO at high concentration (typically either 5 or 10 mM depending on desired final assay concentration of 10 or 20 μM, respectively) to acoustic dispense qualified mother plates. Depending on equipment, one or more additional dilutions likely are required.
2. Adjust instrument to dispense 80 nL of highest concentration, then 1:3 dilutions using 100% DMSO as backfill in order to create a 12-point dilution series of 80 nL total volume in each well. Since the lowest dispense amount may be 5 nL, several intermediate dilutions of the initial 5 or 10 mM stock likely are required to achieve the required dispenses.
3. Seal plates tightly and store at 4 °C until day 2 of assay.

3.4.2 [Day 2]: Preparation of Cells with 1× Tracer for BRET Assay

1. Harvest cells as described in Subheading 3.2.2 and adjust the cell density to 2×10^5/mL in Opti-MEM without serum or phenol red.
2. Dispense 38 μL per well of cell suspension/extracellular Nano-Luc Inhibitor into white, 384-well plate containing previously dispensed, serially diluted test and control compounds.
3. Prepare the BRET tracer at the lowest concentration determined to yield a useable signal:background and Z' greater than 0.5, optimally 50–80% of full target occupancy.
4. Add 2 μL of Complete 20× tracer per well to cells in suspension. Mix thoroughly using appropriate instrumentation.
5. Incubate plates in a humidified, 37 °C/5% CO_2 incubator for 2 h.
6. Add 20 μL per well of 3× Complete NanoBRET(TM) Nano-Glo(R) Substrate for a 384-well plate. Incubate 2–3 min at room temperature prior to first read; readings are stable at least 60 min.
7. Measure BRET as described in Subheading 3.2.4.

3.5 Optimization of Energy Transfer for Residence Time Studies in Live Cells

Although equilibrium-based analysis is predictive of engagement in a closed system (i.e., conditions of static drug concentrations), nonequilibrium analysis may better predict target occupancy in open systems, where drug concentrations are dynamic. The compound residence time (i.e., k_{off}^{-1}) may better dictate target occupancy in the dynamic regime of a living organism [10, 11]. The energy transfer technique can be readily configured to measure real-time binding kinetics in live cells, thus providing an approach to assess drug residence in an artificial open system. The open system can be simulated by applying an acute treatment of test compound to the cells, followed by removal of unbound compound and a real-time measurement of residence time using energy transfer (Fig. 8).

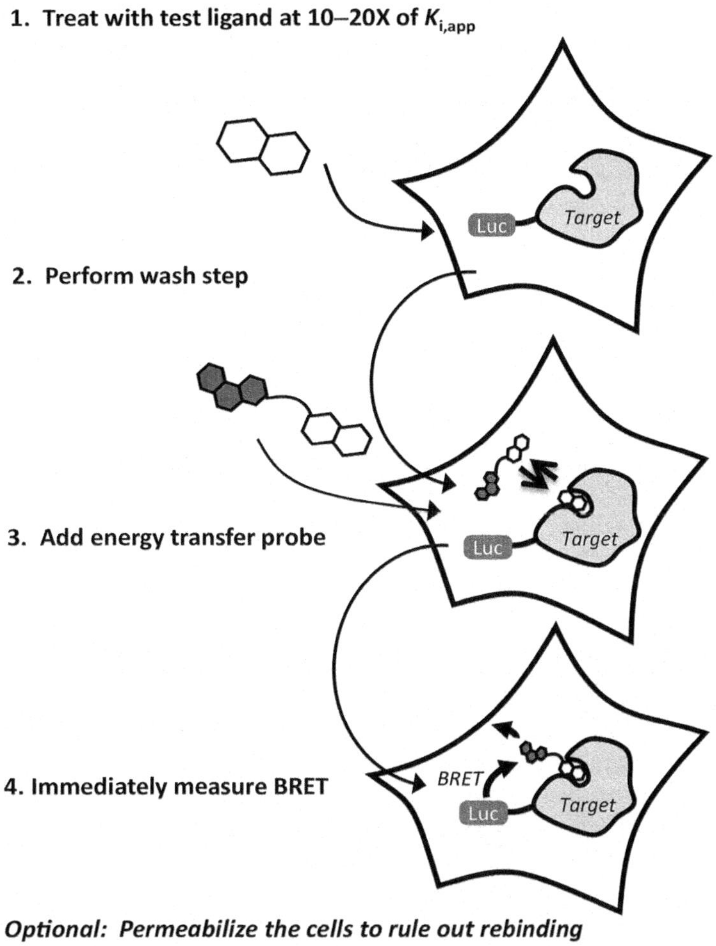

Fig. 8 Illustration of BRET for intracellular residence time analysis

Prior to performing a residence time analysis, it is important to determine a few key parameters. First, it is important to select an optimal tracer and determine a working concentration of the tracer for BRET measurement after compound washout. In general, it is recommended to use a saturating concentration of tracer that has fast association and dissociation kinetics with the target of interest. While this may be achievable in practice, the binding of the tracer might not be saturable for some targets. Thus, it is important to optimize the association kinetics at the highest tracer concentration possible. For new tracers and/or new target combinations, it is recommended to initially evaluate the kinetic parameters in a dose-responsive manner to ensure that an optimal tracer and working concentration are selected for the downstream residence time analysis. This is accomplished via observed rate constant (k_{obs}) analysis, wherein the establishment of BRET at the target is measured kinetically over time at increasing tracer concentrations.

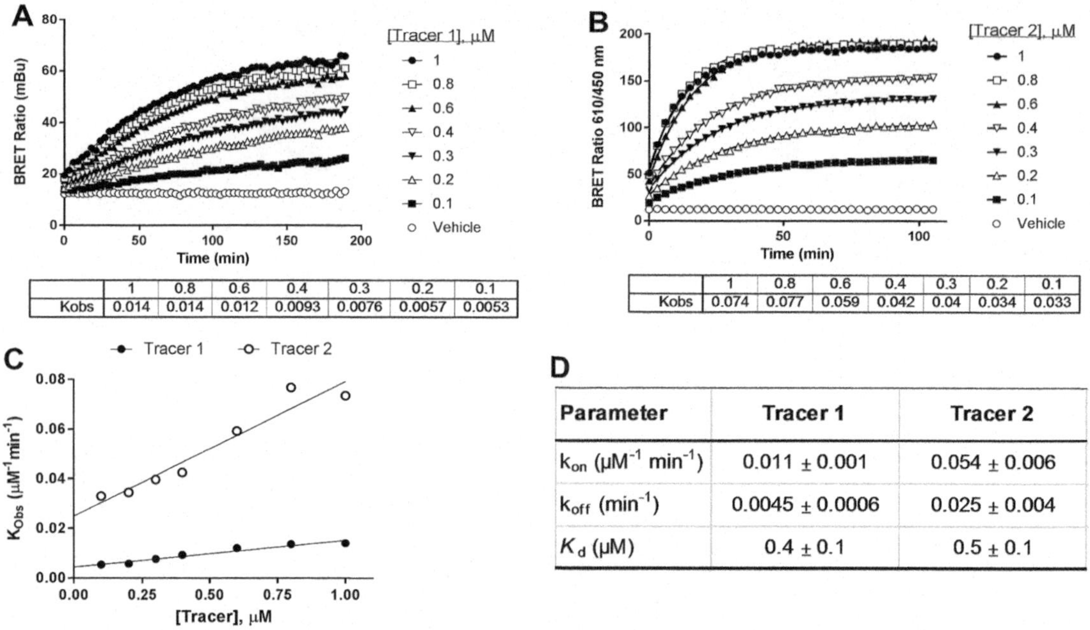

	1	0.8	0.6	0.4	0.3	0.2	0.1
Kobs	0.014	0.014	0.012	0.0093	0.0076	0.0057	0.0053

	1	0.8	0.6	0.4	0.3	0.2	0.1
Kobs	0.074	0.077	0.059	0.042	0.04	0.034	0.033

Parameter	Tracer 1	Tracer 2
k_{on} (μM^{-1} min^{-1})	0.011 ± 0.001	0.054 ± 0.006
k_{off} (min^{-1})	0.0045 ± 0.0006	0.025 ± 0.004
K_d (μM)	0.4 ± 0.1	0.5 ± 0.1

Fig. 9 Kinetic optimization of bromodomain tracers. HEK293 cells transiently expressing NanoLuc(R)-BRD9 bromodomain were seeded into 96-well plates, after which a dilution series of tracer was added and the BRET measured kinetically. Panels A and B depict tracer association curves for two different bromodomain tracers. Individual points represent the mean of three technical replicates. Nonlinear regression analysis using the one-phase association equation (Eq. 4) was performed to extract an observed rate constant (k_{obs}) at each tracer concentration. Observed rate constants were replotted with respect to tracer concentration (panel C), yielding linear relationships. Kinetic parameters extracted by linear regression using Eq. 5 are tabulated in panel D. Tracer 2 gave faster association and dissociation kinetics compared to tracer 1 and was chosen for downstream residence time analyses at a working concentration of 1 μM

Subsequently, the k_{obs} is determined by nonlinear regression analysis of the kinetic curves for each tracer concentration using the one-phase association equation (Eq. 4), assuming pseudo-first order association [12]. The k_{obs} values are replotted with respect to tracer concentration, which yields a linear relationship. Fitting Eq. 5 to the data allows dissection of the k_{on} (from the slope) and k_{off} (from the y-intercept) of the tracer. Figure 9 provides an example in which a k_{obs} analysis is performed to choose between two different tracers for kinetic analysis of the BRD9 Bromodomain.

$$Y = Y_o + (\text{Plateau} - Y_o) \times \left(1 - e^{-k_{obs} * t}\right) \quad (4)$$

$$k_{obs} = k_{on} \times [\text{tracer}] + k_{off} \quad (5)$$

Generally, higher values of both k_{on} and k_{off} are desired for residence time analysis to ensure that the tracer binding is at equilibrium with the target at any given time once the test compound dissociates. However, in many cases, a simple comparison of the

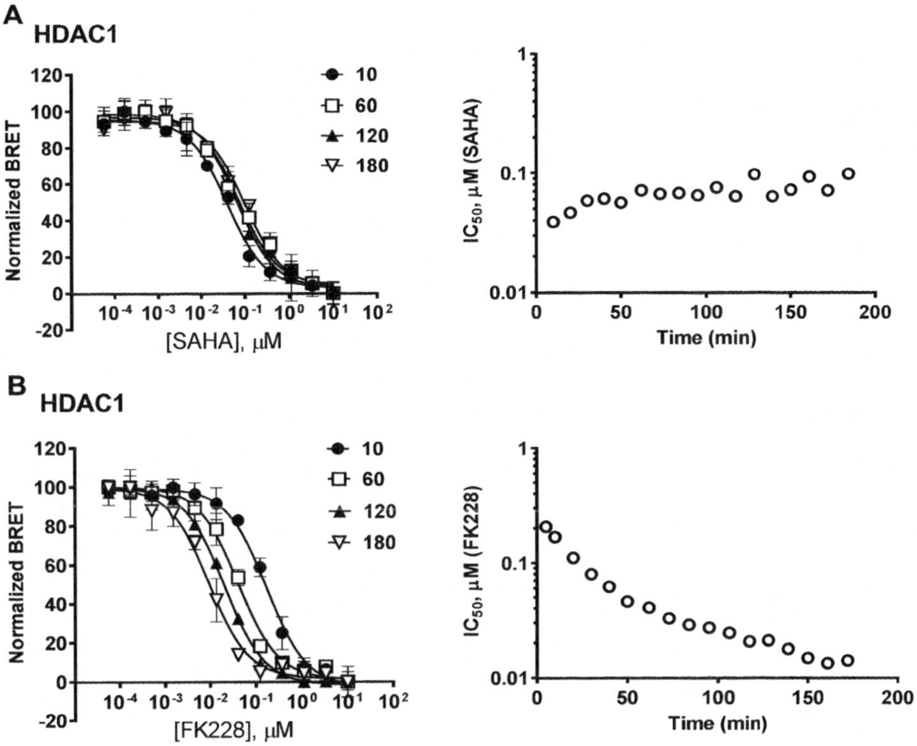

Fig. 10 Determination of optimal inhibitor concentrations for compound treatment prior to residence time analysis. BRET analysis was performed in real-time to determine an appropriate equilibration time and inhibitor concentration for cellular washout experiments. In general, an inhibitor concentration approximating 90% occupancy (e.g., 10–20× $K_{i,app}$) is optimal prior to washout studies. (**a**) For HDAC1, equilibration with SAHA was proven to be rapid, indicating that a concentration of 1 μM SAHA is suitable following a short incubation time. (**b**) For FK228, equilibration kinetics were slower, indicating that a dose of 100 nM is appropriate, and 3 h of equilibration is required to achieve HDAC1 occupancy

relative k_{obs} value itself may be sufficient to select an optimal tracer among a few lead candidates. Though equilibration for tracers may vary greatly (nearly instantaneous to hours), equilibration of the tracer at the target within 20–30 min is sufficient for a qualitative assessment (i.e., rank-order assessment) of test compound residence time.

After the tracer and working concentrations are established, it is recommended to determine an appropriate acute dose of the test compound, as well as the length of time required to achieve equilibration with the target in the cellular context (*see* **Note 9**). Figure 10 provides an example for the determination of the acute dose and equilibration time for the binding of two test compounds to the histone deacetylase, HDAC1. Figures 11 and 12 provide example residence time analyses for the binding of the same two test compounds at HDAC1, with cell permeabilization controls to evaluate the effect of intracellular trapping, respectively.

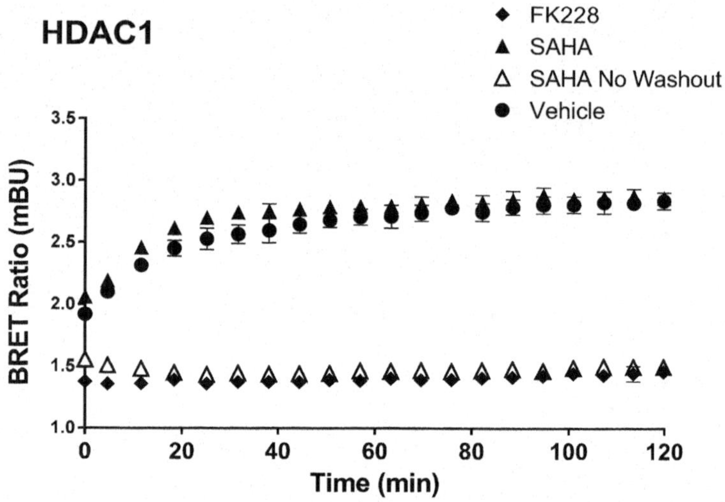

Fig. 11 Residence time analysis of HDAC1 with SAHA and FK228 in live cells. SAHA shows rapid dissociation, as expected. The occupancy of HDAC1 with FK228 following washout was equivalent to the full occupancy control (SAHA, no washout), thus indicating durable binding. Thus, FK228 shows protracted residence time at HDAC1, consistent with previous reports

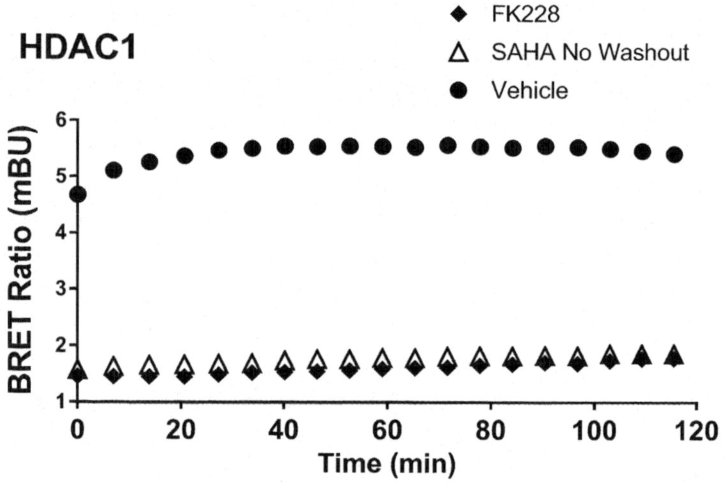

Fig. 12 Control residence time experiments in permeabilized cells to ensure that kinetic profiles are not influenced by intracellular trapping of the test compound. FK228 maintains a long residence time after permeabilization of the cells, thus supporting previous results in live cells of durable binding

3.5.1 Adding Test Compound to Cells

1. After transfecting cells (as described in Subheading 3.2.1), resuspend cells using prewarmed assay medium (e.g., Opti-MEM without serum or phenol red).
2. Adjust the density to 2×10^5 cells/mL in assay medium in a 15 mL sterile polypropylene conical tube. We recommend a minimum volume of 1 mL of cells, ideally 5 mL of cells.

3. Prepare test compound at 10× final concentration in appropriate solvent (e.g., 1% DMSO).
4. To determine the optimal concentration of test compound for residence time analysis, it is recommended to first determine the compound IC_{50} using the equilibrium-based Target Engagement Protocol (*see* Subheading 3.2.2). Introduce the test compound at a final concentration of 10–20× $K_{i,app}$ determined under equilibrium conditions. This insures that adequate washout conditions can be achieved prior to off-rate measurements. Add 100 μL of 10× test compound to each 1 mL of cells in assay medium and mix gently.

   (Optional) Include a positive control treatment of a compound at a high and saturating dose as a full-occupancy control, as well as a negative control treatment with compound vehicle.
5. Place the conical tube in a rack, and incubate the tube with the cap loosened at 37 °C, 5% CO_2 for a minimum of 2 h or until sample is expected to reach equilibrium.

3.5.2 Preparing BRET Reagents

1. To prepare Complete 20× tracer reagent, first prepare a 100× solution of BRET tracer in 100% DMSO. Add 4 parts Tracer Dilution Buffer to 1 part of 100× tracer to generate Complete 20× tracer reagent.
2. For analysis of residence time in live cells, prepare a 2× solution of NanoBRET(TM) Nano-Glo(R) Substrate plus Extracellular NanoLuc Inhibitor by diluting the NanoBRET(TM) Nano-Glo (R) Substrate 1:250 and the Extracellular NanoLuc(R) Inhibitor 1:750 in a conical tube with Opti-MEM without serum or phenol red. For example, for a 96-well plate, mix 40 μL of NanoBRET(TM) Nano-Glo(R) Substrate, 13.3 μL of Extracellular NanoLuc(R) Inhibitor and 9946 μL of assay medium to produce 10 mL of 2× NanoBRET™ Nano-Glo(R) Substrate plus Extracellular NanoLuc(R) Inhibitor Solution. Mix gently by inversion 5–10 times. For analysis of residence time in permeabilized cells, include 100 μg/mL digitonin and omit the Extracellular NanoLuc(R) inhibitor in the 2× solution.

3.5.3 Preparing Cells and Dispensing Reagents for the Residence Time Assay

1. After incubation with test compound is complete, centrifuge cells at 200 × *g* for 5 min.
2. Optional Wash: To remove unbound test compound, resuspend cells in prewarmed assay medium, centrifuge cells and remove medium. Maintain the concentration of the positive control compound for the full-occupancy control.
3. Remove medium, resuspend cells in prewarmed assay medium, and adjust cell density to 2×10^5 cells/mL.

4. Dispense 90 μL per well of cell suspension with test compound or controls into white, 96-well plates. Periodically mix cells to avoid cell settling in the tube.

   (Optional) For background correction, dispense 100 μL of cell suspension in triplicate as no-tracer control samples.

5. Add 100 μL of 2× NanoBRET(TM) Nano-Glo(R) Substrate prepared above (*see* Subheading 3.5.2).
6. Dispense 10 μL of Complete 20× NanoBRET™ Tracer Reagent per well to cells. Mix the 96-well plate on an orbital shaker for 15 s at 900 rpm.

   (Optional) Prepare a separate set of samples without tracer for optional background correction steps.

7. Immediately proceed to the next step (*see* Subheading 3.5.4).

3.5.4 BRET Detection in Kinetic Mode

1. Set a NanoBRET™ Assay-compatible instrument to perform repeat measurements for the desired time interval. As a starting point, use 5-min intervals between each measurement, and be sure to include a sufficient number of measurements to determine compound dissociation. Set the instrument at 25 °C to minimize sample evaporation. However, as temperature may impact residence time profiles, it may be valuable to evaluate physiological temperatures.
2. Measure donor emission wavelength (e.g., 450 nm) and acceptor emission wavelength (e.g., 600 nm), and determine the BRET ratio as described earlier (*see* Subheading 3.2.4).

3.5.5 Adapting the Residence Time Protocol to Adherent (ADH) Format

In addition to the residence time protocol described above, residence time analysis can be performed in adherent (ADH) format in which the cells are adhered to a tissue culture-treated assay plate and the compound incubations are performed directly in that assay plate. Adherent format has both benefits and drawbacks compared to the conventional format in the nonbinding surface assay plate (NBS format) described above that include the ability to test significantly more experimental conditions simultaneously (test compound concentrations and/or targets), and improved workflow for washout that reduces time between the initial washout and the tracer addition step and preserves kinetic information at early time points. However, in some cases the kinetics of the BRET signal may not achieve a constant value under steady state conditions, or may appear biphasic compared to those assessed in NBS-treated plates.

Users should consider their specific goals for the residence time analysis to help select between the assay formats. Figure 13 depicts an example residence time determination for dasatinib and ibrutinib at BTK kinase in both NBS and ADH formats for comparison.

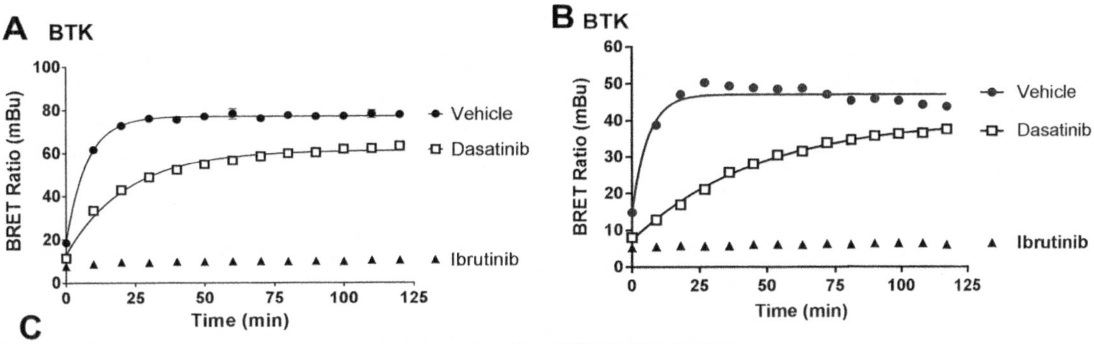

C

Compound	NBS k_{obs} (min^{-1})	Adherent k_{obs} (min^{-1})
Vehicle	0.13 ± 0.01	0.17 ± 0.04
Dasatinib	0.045 ± 0.002	0.021 ± 0.002

Fig. 13 Comparison of residence time analysis in NBS vs adherent (ADH) formats. HEK293 cells expressing BTK-NanoLuc(R) kinase fusion were treated with DMSO vehicle, 250 nM dasatinib, or 1 μM ibrutinib for 2 h. The cells were washed with Opti-MEM to remove unbound drug, after which 1 μM kinase tracer 4 was added and NanoBRET™ measurements were recorded in kinetic mode. For kinase tracer 4, qualitative residence time analysis for dasatinib and ibrutinib revealed comparable apparent dissociation rates in both NBS format (panel **a**) and ADH format (panel **b**). In both formats, dasatinib showed protracted residence time at BTK, while ibrutinib completely prevents tracer association due to covalent binding. Observed rate constants (k_{obs}) for tracer association after washout of vehicle or dasatinib are tabulated in panel C for comparison

1. Mix desired cells with lipid–DNA complexes as described in Subheading 3.2.1.
2. Dispense 100 μL cells and the lipid–DNA complex into a sterile, tissue-culture treated 96-well plate, and incubate at least 20 h to allow expression. We recommend a cell density of approximately 55,000–80,000 cells/cm^2 during the transfection (for example, use approximately 20,000 cells/well for a 96-well Corning #3917 assay plate).
3. On day 2, gently remove medium from the assay plate with transfected HEK293 cells via aspiration.
4. Add back 90 μL per well of prewarmed assay medium. Include suitable controls as indicated in Subheadings 3.5.1 and 3.5.3.
5. Add 10 μL of 10× test compound to each assay well. For full-occupancy control, include a sample with a saturating dose of a positive control compound. For zero-occupancy control, add only DMSO. If the optional "No tracer control samples" were added to the experiment, do not add test compound.
6. Mix on orbital shaker for 15 s at 900 rpm.
7. Incubate the assay plate at 37 °C, 5% CO_2 for a minimum of 2 h or until sample is expected to reach equilibrium. Shorter or longer incubation times may be required depending on the characteristic of the test compound.

8. After incubation with test compound is complete, gently aspirate to remove the medium and test compound from the assay plate. For the full-occupancy control, do not perform this step. Test compound should remain in the sample.
9. To remove residual unbound test compound, gently add 150 μL of assay medium to each well and incubate at room temperature for 30 s. For the full-occupancy control, do not perform this step. Test compound should remain in the sample.
10. Gently aspirate the medium and dispense 90 μL per well of assay medium.
11. Add 100 μL of 2× NanoBRET(TM) Nano-Glo(R) Substrate plus Extracellular NanoLuc(R) Inhibitor solution to each well of the 96-well plate (prepared as described in Subheading 3.5.2 in the protocol for residence time in NBS format).
12. Dispense 10 μL of Complete 20× NanoBRET(TM) Tracer Reagent per well. Mix the 96-well plate on an orbital shaker for 15 s at 900 rpm.
13. Immediately measure BRET in kinetic mode as described in Subheading 3.5.4 in the protocol for residence time in NBS format.
14. Process data as described above in the protocol for residence time in NBS format (*see* Subheading 3.5.4).

4 Notes

1. Progress of the reaction can be monitored by analyzing a small aliquot every 15 min. Additional time may be required to achieve satisfactory results. Slow reaction can often be remedied by adding an additional 8.6 μL portion of diisopropylethyl amine. Overnight reaction is not recommended.
2. It is recommended that fraction pools be concentrated to dryness under reduced pressure. Coevaporation with acetonitrile will help remove spurious residual TFA that has been shown to affect the stability of some tracers. Lyophilization has been shown to enhance this effect during freeze-drying and is not recommended.
3. When possible, placement of the luciferase reporter at both termini of the target protein is recommended to identify the optimal donor orientation. Installation of a 4–10 residue linker between the target and luciferase reporter is also recommended. Standardized poly Gly/Ser linkers (e.g., Gly-Ser-Ser-Gly) are commonly used to ensure some level of flexibility between the target and the reporter. Although a variety of vectors can be configured for such purposes, common mammalian expression vectors such as pFN-31K or pFC-32K are predesigned for NanoBRET(TM).

4. It is recommended that the maximum concentration of tracer used at this stage is 0.4 mM so that the maximum concentration on cells does not exceed 4 μM. This is typically the solubility limit of cell-permeable NanoBRET(TM) tracers, so higher concentrations in culture medium should be avoided.
5. The Extracellular NanoLuc(R) inhibitor is recommended for live cell analysis as a means to inhibit any Nluc signal arising from targets in cell culture debris. For experiments with cell lysates, it should be omitted.
6. When planning the linearized Cheng–Prusoff analysis, it is important to consider the concentration range of both tracer and the test compound. Ideally, both will be used at or below the associated solubility threshold. When these thresholds are surpassed, deviations in linearity in the regression analysis are commonly observed, thus complicating quantitative analysis with NanoBRET(TM). When choosing tracer concentrations, it is ideal to include concentrations both above and below the $K_{d,app}$, typically in the range of 0.25–20× $K_{d,app}$. This ensures adequate competition and establishment of the linear right-shifting of test compound IC_{50}. As the analysis requires linear regression, linear spacing of the tracer concentration is ideal, although not always practical. Hence, a twofold dilution series of the tracer is often used to bracket the $K_{d,app}$. In the case of the test compound, generally an 11-point, threefold dilution series is most appropriate for capturing the IC_{50} values, often starting at 50× the apparent IC_{50} at a particular tracer concentration, or 1000–2000× the $K_{i,app}$ of the test compound for a full analysis. In situations where a new test compound/tracer combination is being evaluated, it is often practical to perform a pilot analysis with a twofold dilution series of tracer starting in the range of 1–4 μM and a threefold dilution series of the test compound starting in the range of 10–20 μM, after which the concentration ranges can be adjusted as needed.
7. When test compounds are applied to the plate in 12-point concentration dilution, as many as 30 IC_{50} curves can be determined on a single 384-well plate. If duplicate dose–response curves are determined on a single plate and a duplicate control compound dilution series also is run on every plate for interplate and interday statistical comparisons, then a reasonable throughput of 14 compounds per plate can be achieved. The NanoBRET™ signal is stable for at least 60 min; therefore at least 15–20 plates can be filled and scanned efficiently using appropriate bulk dispensing equipment without loss of signal between reading the first and last plates.
8. Adapting the competitive displacement format to high throughput requires additional tactics to assure consistency plate-to-plate and day-to-day. These include normalizing the

amount of DMSO in every well and freezing many vials of cells at the same passage number. The robustness of this format should be assessed by repeating tracer dose response and dose response for a selected control test compound on multiple plates, on different days, with different transient transfections, and using cells with different passage numbers or different vials of frozen cells. Test-retest Minimum Significant Ratio (MSR) and moving MSR values should be within threefold in each repeat assay [13]. As is typical for assay development and daily evaluation, assay behavior also should be evaluated using Z' values. With appropriate assay development, the high-throughput format behaves essentially identical to the 96-well standard format. Tracer dose response $K_{d,app}$, signal–background ratio, and Z' values have been found to be comparable between 96- and 384-well plate formats for multiple different kinase NanoBRET(TM) assays studied, for example. Table 1 illustrates data for ten archetypal enzymes across multiple kinase classes. In each case, the 384-well assay was developed, and the statistical effect size (measured as Z') is acceptable for HTS and lead optimization. There is no need to first develop and optimize an assay in 96-well format prior to adapting 384-well format.

9. The acute dose can be determined by measuring the $K_{i,app}$ or IC_{50} of the test drug according to the protocols described above in Subheadings 2 and 3. Once determined, we recommend an acute dose of 10–20× $K_{i,app}$ or IC_{50} for the residence time analysis, which balances both meaningful binding of the test compound to the target of interest (approximately 90% occupancy) as well as the ability to wash away the compound after equilibration. To determine an appropriate equilibration time, it is recommended to monitor the affinity of the test compound to the target over time, and select a sufficient amount of time for stabilization of the IC_{50} value (indicative of equilibration). The binding of many test compounds will reach equilibrium with the target within 2–3 h in a cellular context, although the incubation time might require adjustment for poorly permeable compounds.

References

1. Savitski MM, Reinhard FB, Franken H, Werner T, Savitski MF, Eberhard D, Martinez Molina D, Jafari R, Dovega RB, Klaeger S, Kuster B, Nordlund P, Bantscheff M, Drewes G (2014) Proteomics. Tracking cancer drugs in living cells by thermal profiling of the proteome. Science 346(6205):1255784. https://doi.org/10.1126/science.1255784
2. Bantscheff M, Hopf C, Savitski MM, Dittmann A, Grandi P, Michon AM, Schlegl J, Abraham Y, Becher I, Bergamini G, Boesche M, Delling M, Dumpelfeld B, Eberhard D, Huthmacher C, Mathieson T, Poeckel D, Reader V, Strunk K, Sweetman G, Kruse U, Neubauer G, Ramsden NG, Drewes G (2011) Chemoproteomics profiling of HDAC inhibitors reveals selective targeting of HDAC complexes. Nat Biotechnol 29 (3):255–265. https://doi.org/10.1038/nbt.1759

3. Robers MB, Dart ML, Woodroofe CC, Zimprich CA, Kirkland TA, Machleidt T, Kupcho KR, Levin S, Hartnett JR, Zimmerman K, Niles AL, Ohana RF, Daniels DL, Slater M, Wood MG, Cong M, Cheng YQ, Wood KV (2015) Target engagement and drug residence time can be observed in living cells with BRET. Nat Commun 6:10091. https://doi.org/10.1038/ncomms10091
4. Vasta JD, Corona CR, Wilkinson J, Zimprich CA, Hartnett JR, Ingold MR, Zimmerman K, Machleidt T, Kirkland TA, Huwiler KG, Ohana RF, Slater M, Otto P, Cong M, Wells CI, Berger BT, Hanke T, Glas C, Ding K, Drewry DH, Huber KVM, Willson TM, Knapp S, Müller S, Meisenheimer PL, Fan F, Wood KV, Robers MB (2018) Quantitative, wide-spectrum kinase profiling in live cells for assessing the effect of cellular ATP on target engagement. Cell Chem Biol 25(2):206-214.e11. doi: https://doi.org/10.1016/j.chembiol.2017.10.010
5. Forster M, Chaikuad A, Bauer SM, Holstein J, Robers MB, Corona CR, Gehringer M, Pfaffenrot E, Ghoreschi K, Knapp S, Laufer SA (2016) Selective JAK3 inhibitors with a covalent reversible binding mode targeting a new induced fit binding pocket. Cell Chem Biol 23(11):1335–1340. https://doi.org/10.1016/j.chembiol.2016.10.008
6. Stoddart LA, Johnstone EK, Wheal AJ, Goulding J, Robers MB, Machleidt T, Wood KV, Hill SJ, Pfleger KD (2015) Application of BRET to monitor ligand binding to GPCRs. Nat Methods 12(7):661–663. https://doi.org/10.1038/nmeth.3398
7. Waring MJ, Chen H, Rabow AA, Walker G, Bobby R, Boiko S, Bradbury RH, Callis R, Clark E, Dale I, Daniels DL, Dulak A, Flavell L, Holdgate G, Jowitt TA, Kikhney A, McAlister M, Mendez J, Ogg D, Patel J, Petteruti P, Robb GR, Robers MB, Saif S, Stratton N, Svergun DI, Wang W, Whittaker D, Wilson DM, Yao Y (2016) Potent and selective bivalent inhibitors of BET bromodomains. Nat Chem Biol 12(12):1097–1104. https://doi.org/10.1038/nchembio.2210
8. Cheng Y, Prusoff WH (1973) Relationship between the inhibition constant (K1) and the concentration of inhibitor which causes 50 per cent inhibition (I50) of an enzymatic reaction. Biochem Pharmacol 22(23):3099–3108
9. Newton P, Harrison P, Clulow S (2008) A novel method for determination of the affinity of protein: protein interactions in homogeneous assays. J Biomol Screen 13(7):674–682. https://doi.org/10.1177/1087057108321086
10. Lu H, Tonge PJ (2010) Drug-target residence time: critical information for lead optimization. Curr Opin Chem Biol 14(4):467–474. https://doi.org/10.1016/j.cbpa.2010.06.176
11. Copeland RA, Pompliano DL, Meek TD (2006) Drug-target residence time and its implications for lead optimization. Nat Rev Drug Discov 5(9):730–739. https://doi.org/10.1038/nrd2082
12. Edwards PR, Gill A, Pollard-Knight DV, Hoare M, Buckle PE, Lowe PA, Leatherbarrow RJ (1995) Kinetics of protein-protein interactions at the surface of an optical biosensor. Anal Biochem 231(1):210–217
13. Haas JV, Eastwood BJ, Sittampalam GS, Devanaryan V, Iversen PW, Weidner JR (2004) Minimum significant ratio–a statistic to assess assay variability. In: Sittampalam GS, Coussens NP, Brimacombe K et al (eds) Assay guidance manual. Eli Lilly & Company and the National Center for Advancing Translational Sciences, Bethesda (MD)

Chapter 4

Target Engagement of Small Molecules: Thermal Profiling Approaches on Different Levels

Elena S. Reckzeh, Andreas Brockmeyer, Malte Metz, Herbert Waldmann, and Petra Janning

Abstract

The identification of protein targets and the elucidation of the **m**olecular **m**echanism **o**f **a**ction (MMoA) of bioactive small molecules are central goals of chemical biology. Many different techniques for target identification and engagement are developed, but none of them is generic. Here we describe one of these techniques—the **ce**llular **t**hermal **s**hift **a**ssay (CETSA). The assay works without any labeling of proteins or small molecules, which allows the investigation of the unaltered interaction between the interaction partners. Briefly, the influence of small molecules on the thermal stability of proteins within whole cell lysates is investigated. We describe this approach in two variants: the conventional immunoblot-based approach (CETSA), as well as an unbiased approach based on a proteome-wide mass spectrometric analysis (**t**hermal **p**roteome **p**rofiling, TPP). The CETSA approach requires preknowledge about possible target proteins and can only detect a few proteins at once. Although TPP is technically more demanding, it allows for the identification of multiple (off)targets without any preknowledge.

Key words Proteomics, Target identification, Target engagement, (Off)target, Cellular thermal shift assay (CETSA), Thermal proteome profiling (TPP), Immunoblot, Nano-LC-MS/MS

1 Introduction

Identifying the protein targets of bioactive small molecules is one of the most challenging tasks in chemical biology research. Even though a lot of different approaches based on proteomic, genetic, and computational methods have been developed (for reviews see for example [1–4]), none of them is generic. Well-established methods are based on affinity enrichment of proteins on a solid support with the bioactive small molecule immobilized. Several versions of this type of target identification have been described [5], but most of them have the drawback that the small molecule has to be labeled in advance. For these approaches it is necessary to know a position of the molecule that can be altered without loss of biological activity. This knowledge can be gained by studying the

Slava Ziegler and Herbert Waldmann (eds.), *Systems Chemical Biology: Methods and Protocols*, Methods in Molecular Biology, vol. 1888, https://doi.org/10.1007/978-1-4939-8891-4_4, © Springer Science+Business Media, LLC, part of Springer Nature 2019

structure–activity relationship (SAR). This requires a library of structurally similar compounds and a complete set of biological tests, i.e., a major workload in chemical synthesis and biological testing. Kanoh et al. [6] describe an approach where the small molecule is bound via photocrosslinking in different directions to the solid support. For this approach it is not necessary to know which position of the small molecule can be altered without loss of biological function, but just a part of the molecules will be able to bind into the binding pocket. To avoid these drawbacks, methods without any labeling were developed [1, 7, 8]. One of these methods—the cellular thermal shift assay (CETSA)—is described in this protocol in two variations: an immunoblot-based version which can be performed in each biochemistry lab without special equipment, but preknowledge of one or a few dedicated target proteins is required, and a technically more demanding mass spectrometry-based version—also named as thermal proteome profiling (TPP)—which allows to identify target and offtarget proteins without any preknowledge. Martinez et al. [9] described first the successful transfer of a thermal shift assay from an isolated protein [10] to whole cell lysate. The assay measures the melting behaviour of known proteins in a cell lysate with and without small molecule by means of immunoblot readout. The method needs no labeling but is limited to the readout of a small number of proteins, for which an antibody is available and binding of the small molecule is expected. Very soon after this first publication Savitzki et al. [8] described a mass spectrometry-based readout which allows the measurement of the thermal stability of proteins on a proteome wide scale without any preknowledge about the target or offtarget proteins in cell lysates and in intact cells. The first method they published [8, 11] was restricted to cytosolic proteins but very soon an optimization was described which allows the additional detection of membrane-bound proteins [12, 13]. We adapted the TPP approach [8] using whole cell lysates in our lab and implemented useful modifications, for example, we replaced the described inGel fractionation and digestion by a much easier to handle inSolution digestion and exchanged the R script mediated data evaluation by an easy to handle Excel-based macro. Furthermore, we combined the western blot and the mass spectrometry-based readouts to achieve a dual readout using one sample preparation. The principle of the method is visualised in Fig. 1. A compound-treated lysate and a vehicle control lysate are prepared in parallel. First, each of the samples is split into 10 aliquots which are each subjected to one out of ten temperatures between 37 and 67 °C. After sedimentation of the precipitated proteins resulting from the heat treatment each aliquot is split again: the main part is used for the mass spectrometry and the smaller one for the western blot-based readout. In case of western blot-based readout parallel SDS page separation of each fraction is performed, followed by immunoblotting and

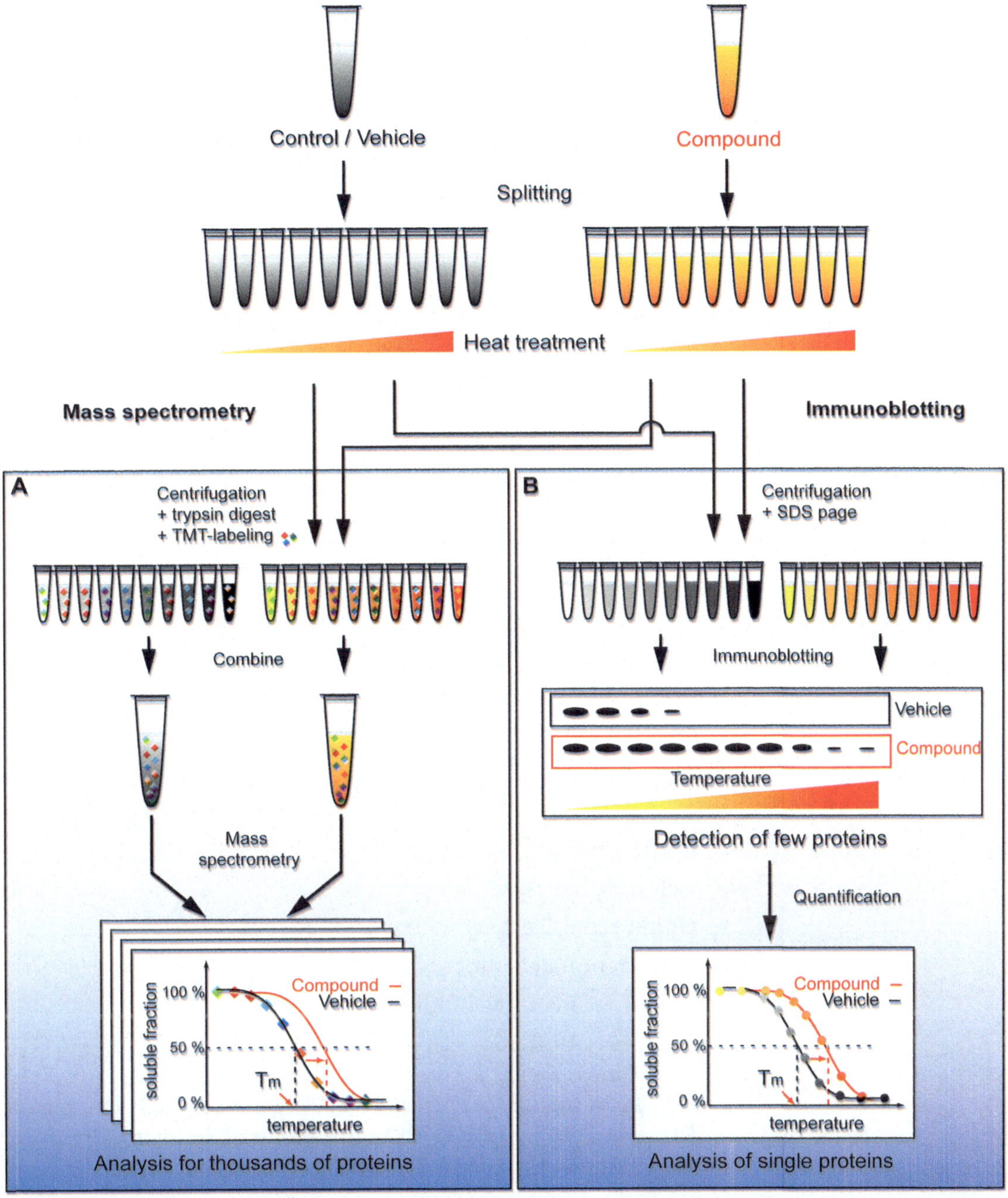

Fig. 1 Overview on thermal proteome profiling (**a**) and cellular thermal shift assay (**b**). After incubation of cell lysate with compound or vehicle, samples are split up in ten aliquots. Each aliquot is treated with one out of ten temperatures within a temperature gradient. The soluble protein fraction can be analyzed by means of (**a**) mass spectrometry (thermal proteome profiling, TPP) or (**b**) immunoblotting (cellular thermal shift assay, CETSA) to generate the corresponding melting curves for multiple proteins or single proteins, respectively

quantitative readout of the blot signals. In case of mass spectrometry-based readout, the fractions are reduced, alkylated, digested using trypsin, and labeled using TMT10plex reagent (one label corresponding to one temperature). Afterward, all compound-treated fractions and all control fractions are combined respectively, followed by offline RP-HPLC prefactionation of peptides using high-pH conditions and low pH conditions for online RP-HPLC-MS/MS coupling. If no target or offtarget of a small molecule is known, usually the mass spectrometry-based readout is performed first. In a second step the potential hits resulting from this readout are rechecked in the immunoblot readout as first validation step. In this chapter the protocol including sample preparation, detection and data evaluation is described in detail.

2 Materials

All reagents are prepared using ultrapure water. Unless otherwise indicated, it is not necessary to use autoclaved water.

2.1 Cell Culture and Harvesting

1. Jurkat cells (*see* **Note 1**).
2. Cell culture medium: Commercial RPMI 1640 media, containing 2 mM L-glutamine, 1 mM sodium pyruvate, 4.5 g/L glucose, 10 mM 2-(4-(2-hydroxyethyl)-1-piperazineethanesulfonic acid (HEPES), and 1.5 g/L $NaHCO_3$, supplemented with 10% (v/v) fetal bovine serum filtered through a 0.2 μm sterile filter, 100 units penicillin/mL and 0.1 mg/mL streptomycin.
3. Cell culture flasks (T75).
4. Commercial 0.4% (w/v) trypan blue stain.
5. Cell counting chamber (disposable or reusable).
6. 1.5 mL protein low binding tubes.
7. Phosphate-buffered saline (PBS): 8.1 mM Na_2HPO_4, 136.9 mM NaCl, 2.7 mM KCl, 1.5 mM KH_2PO_4, pH 7.4. Add 1150 mg of water-free Na_2HPO_4, 8 g of NaCl, 200 mg of KCl and 200 mg of KH_2PO_4 and fill up to 1 L with ultrapure water. Adjust the pH to 7.4 with 1 M HCl and autoclave.
8. For adherent cells: (*see* **Note 1**): Trypsin solution containing 0.25% (w/v) trypsin and 0.02% (w/v) EDTA in PBS or enzyme-free cell dissociation buffer.
9. Lysis buffer: 0.4% (v/v) NP-40 alternative diluted in PBS (*see* **Note 2**).

2.2 Lysate Preparation

1. Lysis buffer (*see* Subheading 2.1, **item 9**).
2. 70% (v/v) Ethanol (to clean the polycarbonates tubes mentioned in this section, *see* **Note 3**).

3. Polycarbonate tubes (0.5 mL polycarbonate tubes, e.g., Beckman Coulter, #343776, *see* **Note 3**).
4. Protein low-binding tubes (1.5 mL).
5. Phosphate-buffered saline (PBS) (*see* Subheading 2.1, **item** 7).
6. Ultracentrifuge.

2.3 Determination of Protein Concentration

1. Bovine serum albumin (BSA) standard solutions: 2 mg/mL, 1 mg/mL, 0.5 mg/mL, 0.25 mg/mL, 0.125 mg/mL BSA. Dissolve 2 mg BSA in 1 mL lysis buffer. Prepare four serial 1:1 dilutions with lysis buffer (e.g., 50 μL lysis buffer plus 50 μL of the respective higher concentration of BSA standard solution).
2. Commercial Bradford stock solution. Prepare Bradford solution as recommended by the manufacturer.
3. Photospectrometer.
4. Phosphate-buffered saline (PBS) (*see* Subheading 2.1, **item** 7).
5. Lysis buffer (*see* Subheading 2.1, **item 9**).
6. GraphPad Prism or Excel (for linear regression for BSA standard curve).

2.4 Thermal Shift

1. Compound: e.g., 10 mM stock solution in dimethylsulfoxid (DMSO).
2. PCR tubes (0.2 mL).
3. Protein low-binding tubes (1.5 mL).
4. Protein low-binding pipette tips.
5. Polycarbonate tubes (0.5 mL, Beckman Coulter, *see* **Note 3**).
6. Dimethylsulfoxide (DMSO, biology grade).
7. 70% (v/v) ethanol (*see* **Note 3**).
8. Phosphate-buffered saline (PBS) (*see* Subheading 2.1, **item** 7).
9. Lysis buffer (*see* Subheading 2.1, **item 9**).
10. PCR mastercycler (temperature gradient necessary, *see* **Note 4**).
11. Ultracentrifuge.

2.5 Splitting of Samples

1. Commercial 1 M triethylammonium bicarbonate (TEAB) buffer (pH 8.5) (*see* **Note 5**).
2. Protein low-binding tubes (1.5 mL).
3. Protein low-binding pipette tips.
4. Autoclaved ultrapure water.
5. Dilution buffer: 100 mM TEAB buffer. Dilute 500 μL of 1 M TEAB buffer with 4.5 mL of autoclaved H_2O. Prepare freshly under laminar flow hood before use.

2.6 Mass Spectrometry-Based Detection

All solutions should be prepared under the laminar flow hood to avoid protein contamination.

2.6.1 Tryptic Digest

1. Protein low-binding tubes (1.5 mL).
2. Protein low-binding pipette tips.
3. Acetone (HPLC-grade, prechilled to −20 °C).
4. 10 mM HCl: Add 100 μL of 1 M HCl to 9900 μL autoclaved, ultrapure water.
5. Dilution buffer: 100 mM TEAB buffer. Add 500 μL of commercial 1 M TEAB buffer (pH 8.5, *see* **Note 5**) to 4.5 mL of autoclaved ultrapure water. Prepare freshly before use.
6. Reducing reagent solution: 200 mM tris(2-carboxyethyl) phosphine (TCEP) in dilution buffer. Add 28.7 mg of TCEP–HCl to 200 μL of autoclaved ultrapure water to prepare a 0.5 M solution. To prepare the reducing reagent solution add 140 μL of the 0.5 M TCEP solution to 140 μL of autoclaved ultrapure water and add 70 μL of commercial 1 M TEAB buffer (pH 8.5, *see* **Note 5**). Prepare freshly before use.
7. Alkylating reagent: 375 mM iodacetamide in 200 mM TEAB buffer. Add 26.0 mg of iodacetamide to 300 μL of autoclaved ultrapure water and add 75 μL of commercial 1 M TEAB buffer (pH 8.5, *see* **Note 5**). Prepare freshly, shortly before use.
8. Trypsin stock solution: 0.4 μg/mL trypsin solution. Add 100 μg of trypsin (proteomics grade) to 250 μL of 10 mM HCl. Prepare freshly, shortly before use.
9. Tryptic digest solution: 28 ng/mL trypsin solution. Add 165 μL of trypsin stock solution to 2200 μL of freshly prepared dilution buffer.

2.6.2 TMT Labeling

All solutions should be prepared under the laminar flow hood to avoid protein contamination.

1. Isotopic label (isobaric tandem mass tag 10-plex (TMT10plex) label, ThermoFisher Scientific).
2. Syringe (gas-tight, 100 μL, e.g., Hamilton).
3. Labeling solution: Equilibrate the TMT label reagent at room temperature. Dissolve each 0.8 mg label by adding 82 μL of anhydrous acetonitrile and vortex occasionally over a 5 min time period.
4. Quenching solution: 5% (w/v) hydroxylamine. Dissolve 5 mg of hydroxylamine in 100 mL of autoclaved water.
5. Vacuum concentrator.

2.6.3 High pH Reversed-Phase Fractionation of TMT-Labeled Peptides

1. Eluent A: 20 mM ammonium formate solution, pH 11. Add 1.26 g of LC-MS grade ammonium formate to about 750 mL of LC-MS grade water. Adjust the pH to 11 with commercial LC-MS grade ammonium hydroxide solution >25% (w/w) and LC-MS grade formic acid. Fill up to 1000 mL with LC-MS grade water.
2. Eluent B: 40% eluent A / 60% acetonitrile (v/v).
3. Ultrasonic bath.
4. Vortexer.
5. Vacuum concentrator.
6. Autosampler vials with septum caps (e.g., 250 μL polypropylene vials, caps for 250 μL polypropylene vials).
7. Protein low-binding tubes for collection (1.5 mL).
8. HPLC: Micro-HPLC system capable for flow-rates of 50 μL/min and meandering fractionation (e.g., UltiMate™ 3000 system with Fraction Collector, Dionex) equipped with a HPLC column with extended pH range stability, 1 mm ID, C18 material, (e.g., Xbridge Peptide BEH C18, 130 Å, 3.5 μm, 1 mm ID × 15 cm, Waters).

2.6.4 LC-MS/MS of Fractionated TMT-Labeled Peptides

1. Eluent C: 0.1% LC-MS grade formic acid in LC-MS grade water (v/v).
2. Eluent D: 0.1% LC-MS grade formic acid in LC-MS grade acetonitrile (v/v).
3. Eluent E: 0.1% LC-MS grade trifluoroacetic acid in LC-MS grade water (v/v).
4. Ultrasonic bath.
5. Autosampler vials with septum caps (e.g., 250 μL polypropylene vials, caps for 250 μL polypropylene vials).
6. HPLC: Nano-(U)HPLC system capable for flow-rates of 300 nL/min and precolumn desalting (e.g., UltiMate™ 3000 RSLCnano system, Dionex) equipped with a nano-HPLC column, 75 μm ID, C18 material, (e.g., Acclaim PepMap RSLC C18, 2 μm, 100 Å, 75 μm ID × 50 cm, nanoViper, Dionex), precolumn cartridges (e.g., Acclaim PepMap100 C18, 5 μm, 100 Å, 300 μm ID × 5 mm, Dionex) and nano-spray emitter (e.g., Standard Coated SilicaTip™ Emitter, 360 μm OD, 20 μm ID, 10 μm Tip ID, 10.5 cm, New Objective).
7. Mass spectrometer: High performance mass spectrometer capable for high resolution measurements of precursor peptide masses and MS/MS fragments (e.g., Q Exactive™ HF Hybrid Quadrupole-Orbitrap Mass Spectrometer, Thermo Scientific) equipped with a nano-spray source (e.g., Nanospray Flex Ion Source, Thermo Scientific) (*see* **Note 6**).

2.6.5 Database Search

1. System requirements for MaxQuant: Intel Pentium III/800 MHz or higher, 2 GB RAM, Windows Vista SP2, 64 bit system, .NET Framework 4.5 or higher, but for a smooth analysis we suggest a much stronger system; e.g., system: Intel Xeon E5-2698 2,30 GHz (2 processors), 128 GB RAM, Windows Server 2012 (64 bit).
2. Software: MaxQuant [14].

2.6.6 Excel Macro "TPP_Makro_1.0"

1. Microsoft Excel 2010 or higher.
2. Excel-Macro: TPP_Makro_1.0.xlsm (freely available at http://www.mpi-dortmund.mpg.de/forschung/chemische-biologie/janning).

2.7 Immunoblot Detection

All devices and reagents listed are exemplary for sodium dodecyl sulfate (SDS)–polyacrylamide gel electrophoresis, semidry blot transfer, and detection by means of fluorescent western blot imager system. The transfer and the detection system can be exchanged for another method according to the experimenter's needs.

1. Prestained protein marker.
2. Blotting paper.
3. Polyvinylidene difluoride (PVDF) membrane.
4. Blocking solution (dependent on the detection method suitable blocking solution, e.g., Odyssey Blocking Buffer, LI-COR).
5. Primary antibody solution (antibody against protein of interest, 5 mL per blot): Dilute the primary antibody as recommended by the supplier in blocking solution.
6. Secondary antibody solution (antibody against species of the primary antibody host, 5 mL per blot): Dilute the secondary antibody as recommended by the supplier in blocking solution.
7. Semidry blotting device (for semidry blotting).
8. Cooling chamber.
9. Isopropanol (HPLC grade).
10. Methanol (laboratory grade).
11. 10% (w/v) ammonium persulfate (APS): Add 500 mg of APS to 5 mL of water, store aliquoted at −20 °C.
12. 10% (w/v) sodium dodecyl sulfate (SDS) solution: Add 10 mg SDS and fill up to 100 mL with water.
13. Commercial Ponceau S solution.
14. Phosphate-buffered saline (PBS) (*see* Subheading 2.1, **item 7**).
15. Lysis buffer (*see* Subheading 2.1, **item 9**).

16. 0.5 M Tris(hydroxymethyl)aminomethane (Tris)–HCl (pH 6.8): Add 30.29 g of Tris to about 400 mL water. Adjust the pH to 6.8 with 1 M commercial HCl solution and fill up to 500 mL with water.
17. 5× SDS sample buffer: Add 2 g SDS, 1.55 g dithiotreitol (DTT), 5 mg bromophenolblue to 10 mL of 0.5 M Tris–HCl (pH 6.8) and 10 mL glycerol. The buffer can be stored aliquoted at −20 °C.
18. 1× SDS sample buffer: Add 30 μL 5× SDS sample buffer to 120 μL lysis buffer.
19. Running buffer: 0.25 M Tris, 2 M glycine, 10% (w/v) SDS. Add 30.285 g of Tris, 150.15 g of glycine, 10 g of SDS and fill up to 1 L with water.
20. 1 M Tris–HCl (pH 6.8): Add 60.57 g of Tris to about 400 mL of water. Adjust the pH to 6.8 with 1 M HCl and fill up to 500 mL with water.
21. 1.5 M Tris–HCl (pH 8.8): Add 90.855 g of Tris to about 400 mL of water. Adjust the pH to 8.8 with 1 M HCl and fill up to 500 mL with water.
22. Transfer buffer: 25 mM Tris, 19.2 mM glycine, 20% (v/v) methanol. Add 3.03 g of Tris and 14.41 g of glycine to 200 mL of methanol and fill up to 1 L with water.
23. Tris-buffered saline (TBS buffer): 5 mM Tris, 15 mM NaCl, pH 7.5. Add 6.06 g of Tris and 8.77 g of NaCl to about 900 mL of water. Adjust the pH to 7.5 with 1 M HCl and fill up to 1 L with water.
24. 5% Stacking gel (3 mL, for two gels): Add 0.5 mL of commercial 30% (w/w) acrylamide solution, 0.38 mL of 1 M Tris–HCl (pH 6.8), 30 μL of 10% (w/v) SDS solution and 30 μL of 10% (w/v) APS solution to 2.1 mL of water. Add 3 μL of tetramethylethylendiamine (TEMED) shortly before usage. Prepare freshly before use.
25. 10% Running gel (10 mL, for two gels): Add 3.3 mL of 30% acrylamide, 2.5 mL of 1.5 M Tris–HCl (pH 8.8), 100 μL of 10% (w/v) SDS solution, and 100 μL of 10% (w/v) APS solution to 4.0 mL of water. Add 4 μL of TEMED shortly before usage. Prepare freshly before use.
26. Software for densitometric analysis (e.g., Image Studio, LI-COR).
27. Software for graphical representation (e.g., GraphPad Prism).

3 Methods

3.1 Cell Culture and Harvesting

The choice of cell line depends on the cellular protein abundance of the proposed target protein(s). In case no target protein is hypothesized, suspension cell lines are recommended due to easier handling without an additional detachment step (*see* Fig. 2; *see* **Note 1** for information on adherent cells) [11].

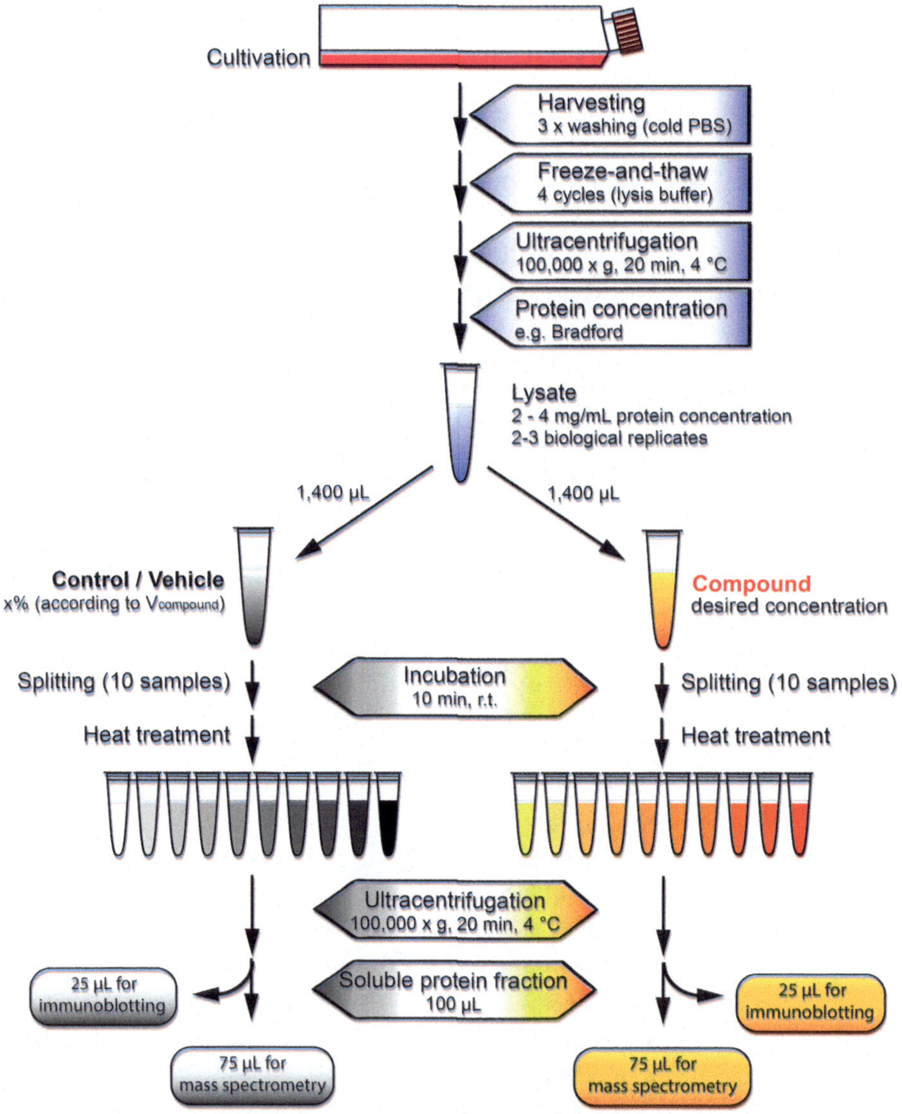

Fig. 2 Overview on lysate generation and thermal shift. Cells are harvested when desired cell density is reached. After washing, cells are resuspended in lysis buffer and subjected to four freeze-and-thaw cycles and ultracentrifugation. The determination of the protein concentration can be achieved by means of a Bradford-based technique. The lysate is treated with compound or vehicle and subsequently subjected to a temperature gradient. An additional centrifugation step yields the soluble protein fraction that can be analyzed by means of immunoblot and/or mass spectrometry

1. Grow Jurkat cells in cell culture medium using T75 cell culture flasks until a cell density of 1.5–2.0 × 10^6 cells/mL is reached (*see* **Note 1**). To determine the cell density transfer the suspension cells into a 50 mL conical centrifuge tube and centrifuge at 350 × *g* for 3 min. Discard the supernatant and resuspend cells in 10 mL RPMI 1640 media. Mix 10 μL cell suspension and 10 μL trypan blue solution in a 1.5 mL tube and transfer 10 μL of this mixture on the cell counting chamber (disposable or reusable) to count the cells automatically or under the microscope. Calculate the cell density in case you manually count the cells.
2. Incubate the cells on ice for 2 min.
3. Centrifuge at 350 × *g* for 3 min at room temperature.
4. Discard the supernatant with a serological pipette and resuspend the cells in 25 mL ice-cold phosphate-buffered saline (PBS).
5. Harvest the cells by centrifugation at 350 × *g* at room temperature for 2 min.
6. Repeat **steps 4** and **5** two more times.
7. Discard the supernatant and resuspend the cells in 1.5 mL ice-cold lysis buffer (*see* **Note 2**).
8. Transfer the cell suspension into a 1.5 mL protein-low binding tube. The cells can now be stored at −80 °C after snap freezing them using liquid nitrogen. In case the cell suspension is used directly for lysate generation, keep the cell suspension on ice and proceed with Subheading 3.2, **step 2**.

3.2 Lysate Preparation

The lysate is generated by means of freeze-thaw method. It is important to apply the recommended temperatures described below since too high temperatures or too fast thawing can damage the cellular proteins (*see* Fig. 2). It is important to work on ice the whole time to avoid proteolytic cleavage of the proteins by cellular proteases (*see* **Note 7**).

1. Thaw the frozen cell suspension from the previous harvesting step to 60–80% (visual judgement) in a thermoblock at 23 °C. Thaw the residual 20–40% on ice.
2. Snap freeze the cell suspension in liquid nitrogen.
3. Repeat **steps 1** and **2** three more times.
4. Transfer the cell lysate into 0.5 mL polycarbonate tubes (*see* **Note 3**) and keep the samples on ice.
5. Separate the insoluble membrane fraction using an ultracentrifuge. Spin the samples at 100,000 × *g* at 4 °C for 20 min.

6. Gently transfer the resulting supernatant into protein low-binding tubes. Avoid touching the pellet with the pipette tip.
7. The lysates can now be stored at −80 °C after snap freezing using liquid nitrogen. In case the protein concentration of the prepared lysates is determined directly, keep the lysates on ice and proceed with Subheading 3.3, **step 2**. In case the protein concentration should be determined timely independent from the thermal shift assay, freeze a 5 μL aliquot separately and proceed with Subheading 3.3, **step 1** using this aliquot.

3.3 Determination of Protein Concentration

The protein concentration of the lysate is essential for proteomic-based thermal shift assays. Therefore, the protein concentration should be identical in all biological replicates to maintain the same compound to protein ratio. Here we describe the determination of lysate protein concentration using the Bradford approach.

1. Unfreeze the lysates at 23 °C in a thermoblock until 60–80% (visual judgement) are unfrozen. Unfreeze the residual 20–40% on ice.
2. Determine a standard curve: Add 1 μL of each concentration of BSA standard solution to 999 μL Bradford solution in duplicates. Use 1 μL lysis buffer in 999 μL Bradford solution as a blank value. Incubate the samples at least 5 min (not more than 1 h) and measure the absorbance with a photospectrometer at 595 nm. Generate a standard curve by plotting the absorbance against the BSA protein concentration. Determine the equation of your standard curve by generating a linear regression curve.
3. Prepare four cuvettes with 999 μL Bradford solution each.
4. Dilute three times 1 μL lysate in one of the four cuvettes to generate technical triplicates.
5. Use 1 μL of pure lysis buffer as a blank value.
6. Incubate the samples at least 5 min (not more than 1 h) before you measure the absorbance of the Bradford solutions at 595 nm in the photospectrometer. The absorbance should be in the range between 0 and 1 (*see* **Note 8**).
7. Calculate your protein concentration using the equation of your linear regression curve and the mean value of all three technical replicates.
8. The lysates can now be stored at −80 °C after snap freezing them using liquid nitrogen. In case the thermal shift assay is performed directly, keep the lysates on ice and proceed with Subheading 3.4, **step 2**.

3.4 Thermal Shift

Briefly, the lysate is treated with either compound or vehicle and afterward split in ten different portions, each treated with one out of ten different temperatures in a certain temperature range (*see* **Note 4** and Fig. 2). Depending on the individual protein stability and the (de)stabilization of a protein by binding to a small molecule, the proteins will denature at different temperatures. In general, a thermal shift can also be performed using whole cells [9, 15]. Please consider the mentioned literature for this. It is important to keep the samples on ice the whole time (unless stated otherwise) to avoid proteolytic cleavage of the proteins by cellular proteases (*see* **Note 7**).

1. Unfreeze the lysates at 23 °C in a thermoblock until 60–80% (visual judgement) are unfrozen. Unfreeze the residual 20–40% on ice.
2. Create a program with an appropriate temperature gradient for the PCR mastercycler (*see* **Note 4**). The temperature range should be between 37 °C and 67 °C and should contain ten different temperatures within (e.g., 36.9, 40.1, 44.1, 47.9, 51.0, 54.4, 58.0, 61.3, 64.5, and 67.0). The temperature should be held for 3 min, followed by an infinite cooling step to 4 °C (*see* **Note 4**).
3. The final protein concentration of the lysate should be between 2 and 4 mg/mL. Dilute the cell lysate if necessary with lysis buffer (*see* **Note 2**) to generate at least 3 mL of lysate featuring the desired protein concentration in a 15 mL conical centrifuge tube.
4. Prepare two protein low-binding 1.5 mL tubes and fill 1.4 mL lysate in each tube.
5. Add the appropriate amount of compound to the lysate of one tube to achieve the desired final compound concentration (e.g., 1 μM). If necessary, predilute the compound stock with the vehicle solvent, e.g., dimethylsulfoxide (DMSO) (*see* **Note 9**).
6. Add the same volume of vehicle, e.g., DMSO, to the other lysate (vehicle control).
7. Invert both tubes three times and incubate the samples for 10 min at room temperature.
8. In the meantime, prepare 20 0.2 mL PCR tubes (ten tubes for each sample). We recommend to label the tubes with C1–C10 (for the compound-treated) and V1–V10 (for the vehicle-treated), respectively.
9. Transfer 120 μL vehicle-treated lysates into each of the V1–V10 labeled PCR tubes and 120 μL compound-treated lysate into each of the C1–C10 labeled PCR tubes, respectively. Close the lid properly.

10. Choose the right program in the PCR mastercycler (*see* **step 2**). Start the program as soon as the incubation time in **item** 7 is over.
11. Put the samples on ice after the program is finished (*see* **Note** 7).
12. Transfer the content of each PCR tube in one clean, labeled 0.5 mL polycarbonate tube (*see* **Note 3**). Keep the samples on ice at all times (*see* **Note** 7).
13. Sediment the denatured proteins by means of ultracentrifugation at 100,000 × *g* at 4 °C for 20 min. In case not all 20 samples can be centrifuged at the same time, make sure that the compound sample and its respective vehicle control is processed in the same run (C1 and V1, C2 and V2, C3 and V3, ...).
14. Carefully pipette 100 μL of each supernatant into separate new 1.5 mL protein low-binding tubes. Strictly avoid touching the pellet or near surrounding with the pipette tip to avoid false results. Keep the supernatant on ice at all times (*see* **Note** 7).

3.5 Splitting of Samples

The thermal shift can be detected via mass spectrometry-based readout (*see* Subheading 3.6) and via immunoblot-based (*see* Subheading 3.7) readout. Therefore, samples have to be split at this stage.

1. For the mass spectrometry-based readout, label ten 1.5 mL protein low-binding tubes with V1-V10 and another ten tubes with C1-C10, pipette 75 μL of dilution buffer (100 mM TEAB buffer) in each tube and add 75 μL supernatant from Subheading 3.4, **step 14** to the tubes according to their label. Proceed with Subheading 3.6.1, **step 1**.
2. For the immunoblot-based readout, take the remaining 25 μL from Subheading 3.4, **step 14**. To store the samples, snap freeze in liquid nitrogen and store them at −80 °C until further usage. In case the immunoblot-based readout is performed directly, keep the samples on ice and proceed with Subheading 3.7, **step 2**.

3.6 Mass Spectrometry-Based Detection

This section will describe the determination of protein-specific thermal profiles by quantitative mass spectrometry.

3.6.1 Tryptic Digest

This step is necessary to process all proteins into their tryptic fragments. These fragments will later be identified by means of mass spectrometry. Conduct the preparation of the soluble protein fractions and the following tryptic digest in a laminar flow hood to avoid any kind of contamination. Figure 3a shows an overview on this part of the experimental procedure.

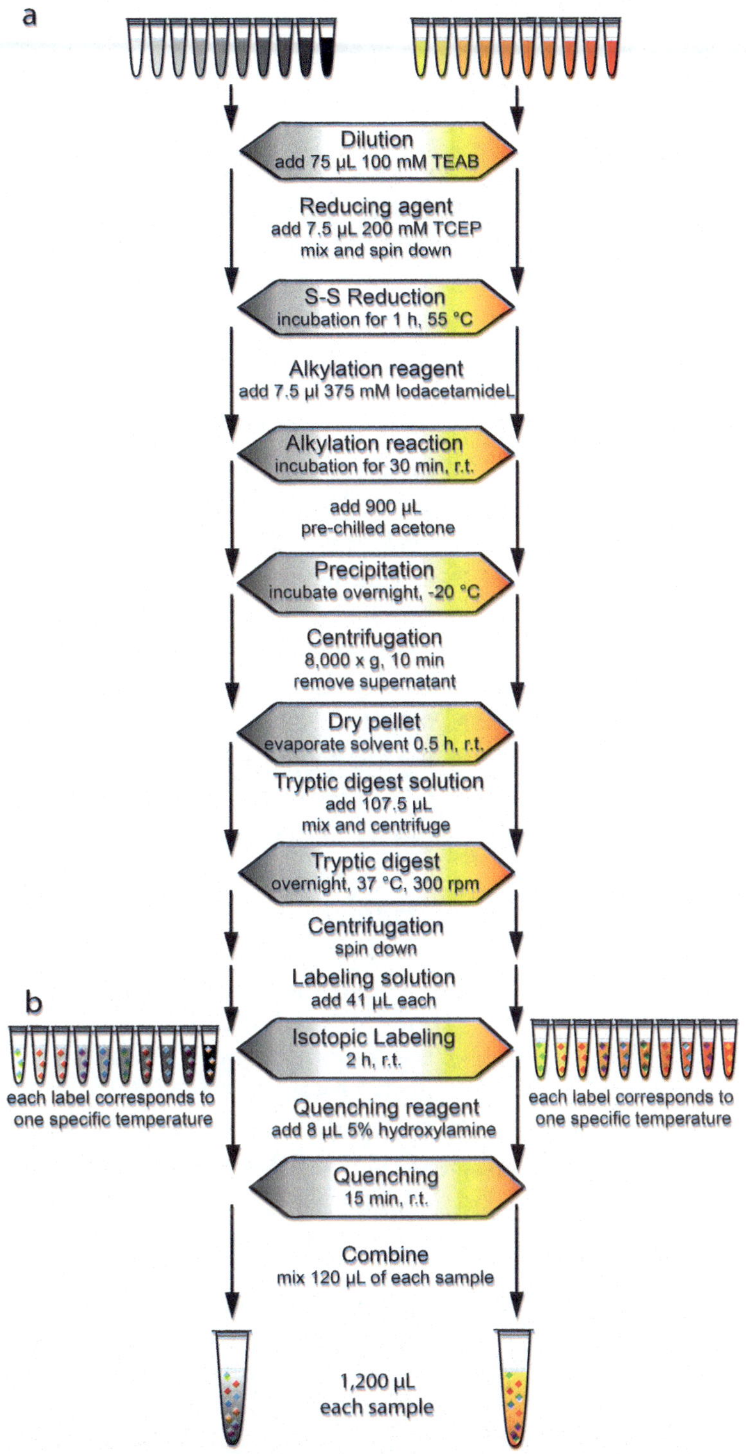

Fig. 3 Overview on sample preparation for mass spectrometry-based detection. (**a**) The main steps in sample preparation include the dilution, reduction of disulfide bridges, alkylation of free thiols, and precipitation. The resulting protein pellet is digested using trypsin. (**b**) The generated peptides can be labeled using TMT10plex isotopic label. After combining the labeled compound- and vehicle-treated lysates, the samples can be measured by means of MS/MS and further analyzed

1. Add 7.5 μL of reducing reagent solution to each sample from Subheading 3.5, **step 1** to reduce the intramolecular disulfide bridges. Subsequently invert the sample three times.
2. Briefly vortex samples and spin them down by shortly centrifuging the samples at 10,000 × *g* to collect all liquid.
3. Incubate the complete sample at 55 °C for 1 h using the thermoblock.
4. Add 7.5 μL of alkylating reagent to the samples to alkylate the reduced thiol groups and incubate for 30 min protected from light at room temperature.
5. Precipitate the proteins by adding 900 μL prechilled acetone. Store the samples over night at −20 °C to ensure complete protein precipitation.
6. Centrifuge the samples at 8000 × *g* for 10 min at 4 °C.
7. Put the samples on ice. Carefully remove about 90% of the supernatant without touching the pellet with the pipette tip. Do so for one sample at a time and keep all other samples on ice.
8. Let the pellets dry in the open tubes under the laminar flow hood for 30–45 min at room temperature.
9. Resuspend the protein pellets in 107.5 μL tryptic digest solution. Vigorously vortex the samples for 20 s.
10. Centrifuge the samples a few seconds at 10,000 × *g* to collect all liquid on the bottom of the tube.
11. Incubate the complete samples at 37 °C in a thermoblock while shaking at about 300 rpm for 2–3 h. Cover the lid of the thermoblock with aluminum foil to assure even heat distribution.
12. Vortex the samples for 20 s and centrifuge the samples a few seconds at 10,000 × *g* to collect all liquid.
13. Continue incubation of the complete samples at 37 °C overnight while shaking at 300 rpm.

3.6.2 TMT Labeling

The peptides of the trypsin-digested proteins are specifically labeled according to their temperature treatment using isotopic tags. The purpose of this step is to ensure the definite assignment of each sample by mass spectrometry to the respective temperature treatment. After specific labeling, all compound samples will be combined in one and all vehicle samples in another tube, respectively. All steps should be conducted using a laminar flow hood to avoid any kind of contamination. Consider Fig. 3b for an overview on this method.

1. Spin down the samples for a few seconds at 10,000 × *g* to collect all liquid on the bottom of the tube.
2. Prepare 20× 0.5 mL aliquots of anhydrous acetonitrile using 1.5 mL tubes for syringe washing.
3. Add half of the labeling solution (41 μL) to one sample (e.g., C1) using a gas-tight, 100 μL Hamilton syringe (*see* **Note 10**). Vortex sample briefly after addition.
4. Wash the syringe using one of the 20 prepared anhydrous acetonitrile aliquots to avoid a remainder of labeling reagent in the syringe.
5. Use a pipette to transfer 100 μL of the respective control (e.g., V1) into the tube containing the remaining labeling solution resulting from **step 3** of this section. Vortex the sample briefly after addition.
6. To avoid cross-contamination between different labels, wash the syringe intensively using one of the 20 prepared anhydrous acetonitrile aliquots.
7. After preparation of all samples, each sample is vortexed again briefly and centrifuged afterward for a few seconds at 10,000 × *g* at room temperature to collect all liquid on the bottom of the tube.
8. Incubate the complete samples for 2 h at room temperature.
9. Add 8 μL of quenching solution (5% hydroxylamine) to each sample and incubate for 15 min at room temperature to quench the reaction.
10. Combine all ten samples of the compound-treated lysates (120 μL each) yielding a total volume of 1200 μL.
11. Combine all ten samples of the vehicle-treated lysates (120 μL each) yielding a total volume of 1200 μL.
12. Evaporate all liquid using a vacuum concentrator at 30 °C to generate a dry, white pellet (This will take approx. 4 h.).
13. Store the samples at −20 °C until fractionation or proceed with Subheading 3.6.3, **step 2**, directly.

3.6.3 High pH Reversed-Phase Fractionation of TMT-Labeled Peptides

High pH reversed-phase fractionation of TMT-labeled peptides reduces sample complexity for subsequent LC-MS/MS analysis. This significantly improves identification rates and accuracy of quantification.

1. Thaw samples at room temperature shortly before analysis.
2. Resuspend the samples in 120 μL of eluent A and subject them to 2 min sonication in an ultrasonic bath and subsequent vortexing for 1 min.

3. Centrifuge the samples at 15,000 × *g* for 3 min and transfer 110 μL of the supernatant to autosampler vials.
4. Inject 50 μL of the sample onto an Xbridge Peptide BEH C18 column.
5. Apply the following linear gradient at a flow rate of 50 μL/min: 0–10 min at 95% eluent A, and 5% eluent B, 10–15 min to 75% eluent A, and 25% eluent B, 15–75 min to 35% eluent A, and 65% eluent B, 75–85 min to 100% eluent B, 85–100 min at 100% eluent B.
6. Record separation at 214 nm.
7. From minute 15 to minute 100 collect fractions in a meandering way over ten collection tubes, changing tube position every 30 s.
8. Evaporate all samples to dryness using a vacuum concentrator at 30 °C (approx. 4 h).
9. Store samples at −20 °C until LC-MS/MS analysis or proceed with Subheading 3.6.4, **step 2**, directly.

3.6.4 LC-MS/MS of Fractionated TMT-Labeled Peptides

High-resolution mass spectrometry has to be exploited to enable identification and quantification of TMT-labeled peptides (*see* **Note 6**).

It might be useful to consult an expert in mass spectrometry or a representative of a mass spectrometry facility prior to analysis and data evaluation if the reader is not familiar with these methods or does not have the expertise or instrumentation to complete the protocol.

1. Thaw samples at room temperature shortly before analysis.
2. Dissolve the samples in 20 μL of eluent E.
3. Sonicate them in an ultrasonic bath at room temperature for 15 min to dissolve the peptides as complete as possible.
4. Centrifuge the samples at 15,000 × *g* for 1 min at room temperature.
5. Transfer the supernatant to autosampler vials.
6. Use the following typical conditions for reversed-phase nano-LC separations: Injection of 3 μL of sample onto a precolumn cartridge using eluent E with a flow rate of 30 μL/min, desalting for 5 min with eluent flow to waste. Backflush the sample from the precolumn to the nano-HPLC column using eluents C and D. Perform the following linear gradient at flow rate of 300 nL/min: 0–5 min at 95% eluent C, and 5% eluent D, 5–125 min to 60% eluent C, and 40% eluent D, 125–130 min to 40% eluent C, and 60% eluent D, 130–135 min to 5% eluent C, and 95% eluent D, 135–140 min at 5% eluent C, and 95% eluent D; column temperature 40 °C.

7. Use the following typical conditions for MS/MS analysis using a Q-Exactive HF mass spectrometer: 2.1 kV spray voltage; mass range of *m/z* 300–1650; resolution of 60,000 for full scan MS performing one microscan using an automated gain control (AGC) target of 3e6 and a maximum injection time (IT) of 20 ms followed by up to 15 high energy collision dissociation (HCD) MS/MS of the most intense at least doubly charged ions at fixed first mass of *m/z* 100 (e.g., resolution 30,000, AGC target 1e5, IT 25 ms, isolation window 1.4 *m/z*, normalized collision energy (NCE) 35.0, intensity threshold 4.0e3, dynamic exclusion 20.0 s) recording profile spectra for full scan and fragmentation spectra as well.

3.6.5 Database Search

Analyze the MS data using the freely available protein identification and quantification software "MaxQuant" [14].

1. Perform a search against a database of the used taxonomy.
2. Use the following typical parameter settings for a search: all ten TMT labels as isobaric labels for lysine and the N-terminus as reporter ion MS2 with a reporter mass tolerance of 0.003 Da, oxidation of methionine and N-terminal acetylation of proteins as variable modifications, carbamidomethylation as fixed modification, max. number of modifications per peptide 5, max. charge 7, min. peptide length 7, specific tryptic digest with max. 2 missed cleavages, false discovery rate of 1%.
3. Use the resulting list of proteins, "proteinGroups.txt" for further processing.

3.6.6 Excel Macro "TPP_Makro_1.0"

Analyze the data file "proteinGroup.txt" resulting from the database search using Microsoft Excel and the embedded macro "TPP_Makro_1.0.xlsm" (*see* **Note 11**) to determine melting curves and differences in melting behaviors of proteins between compound- and vehicle-treated samples (*see* **Notes 12** and **13**).

1. Open the macro by double-clicking the file name.
2. Load the file "proteinGroups.txt" into the Excel sheet "proteinGroups" and insert a new second row. Insert the temperatures corresponding to the labels into the second row under the headers "Reporter intensity corrected ..." for each replicate and start the macro by clicking the "Start" button on the "Instructions" sheet (*see* **Note 13**). Enter the number of samples used per replicate and press "Start."
3. Hide unnecessary sheets by clicking on the "Finish" button on the "Instructions" sheet after the macro has finished its calculations.

3.6.7 Validation of Melting Curves

Validate the curves of the proteins individually. Possible hits show either (1) a shift of the melting point, i.e., the inflection point of the melting curve of the compound-treated sample compared to the curve of the vehicle control, or (2) a difference in the fraction of unmelted protein at the highest temperatures, i.e., the height of the bottom asymptote of the curve resulting from the compound-treated sample differs from the height of the bottom asymptote of the curve resulting from the vehicle one. To reduce the number of proteins of interest, which have to be checked manually and to enhance the comparability of different experiments, all proteins are sorted into four quality classes (*see* **Notes 12** and **14**). Check at least the curves of the proteins of classes two and three, manually. Proteins in these classes have the highest probability of a difference in the melting behavior of the desired protein in the compound- compared to the vehicle-treated sample, but manual evaluation is absolutely necessary to avoid false positive results (*see* **step 2** below in this section).

1. Sort the column "QualityClass" on the sheet "Calculation" from biggest to smallest and click the "Refresh" button on the sheet "Charts." Each chart shows up to three replicates of the uppermost visible protein in the sheet "Calculation." The data points are the reporter intensity normalized to the first temperature. The curves show a fit to the trend of the data points.
2. Validate proteins manually. Sometimes the fitted curves do not match the data points or there are significant differences between the replicates (*see* **Note 15**). Both are indicators for false positives. Sort the proteins to the "Yes," "No," and "Maybe" sheets by clicking the corresponding buttons on the "Charts" sheet (*see* **Notes 14** and **16**).

3.7 Immunoblot Detection

This section will focus on the detection of specific proteins via antibodies, followed by the quantification of the band intensities. The generated values will be used afterward to generate a melting curve. The immunoblot-based analysis can for example be used after mass spectrometry as a follow-up experiment with an antibody directed against the identified (off)target protein to confirm these results.

1. Thaw samples (soluble protein fraction from Subheading 3.5, **step 2**) at room temperature shortly before analysis.
2. Add 6.25 μL 5× SDS sample buffer to 25 μL of each sample. If not recommended otherwise, the samples should be heated at 95 °C for 6–7 min while shaking at 350 rpm using a thermoblock (*see* **Note 17**). Let the samples cool down at least 5 min before you proceed with gel loading.

3. Assemble the clean gel chamber and check for leakage. Add about 5 mL running gel in each chamber up to about 80%. Cover the running gel with about 1 mL isopropanol. Let the running gel polymerize for about 30 min. Remove isopropanol thoroughly and fill about 1.5 mL stacking gel on top of each running gel. Directly, insert comb into the gel and incubate for about 45 min. Proceed to Subheading 3.7, **step 4**. For storage, rap the gels including combs into water soaked paper towels, put them into a plastic bag, and store them at 4 °C.
4. Assemble two gels (one gel for C1–C10, one gel for V1–V10) in one electrophoresis chamber and fill the room between the gels with running buffer. Fill the running buffer also in the surrounding plastic pan.
5. Fill 31.25 μL of each compound-treated lysate sample in the wells of one denaturing SDS gel. Fill 31.25 μL of each vehicle-treated lysate sample in the wells of the other denaturating SDS gel.
6. Fill 5 μL marker in one well of each gel and all empty wells with 31.25 μL 1× SDS sample buffer.
7. Let the gel run at 80 V until the sample runs through the stacking gel and the blue running front reaches the running gel.
8. Increase the voltage to 120 V until the blue front reaches the end of the running gel.
9. Activate the PVDF membrane with about 10 mL methanol for 30 s and store it afterward in prechilled transfer buffer.
10. Soak the blotting paper in prechilled transfer buffer for at least 15 min.
11. Transfer the proteins on a PVDF membrane (e.g., by means of semidry blot touch the membrane with forceps, only): Assemble the semidry blot machine by mounting two layers blotting paper, the PVDF membrane, one gel and two layers of blotting paper on the anode. Do the same for the second gel next to the first setup. Run the transfer for both gels at the same time (if possible) at 24 V for 40 min up to 1.5 h.
12. Check the transfer efficiency by staining the PVDF membrane with 5 mL Ponceau S solution for about 3 min. Use a plastic weighing dish. Wash thoroughly about three times with water. Store the blot in water.
13. Block the blot with the desired blocking solution for 1 h at room temperature while shaking on a roller shaker.
14. Incubate the blot in 5–10 mL primary antibody solution per membrane at 4 °C in a cooling chamber overnight while shaking on a roller shaker.

15. Remove the primary antibody solution (it can be reused up to three times when stored at −20 °C) and wash three times for 10 min with 10 mL TBS buffer, each.
16. Incubate each blot with 5 mL secondary antibody solution for 1 h at room temperature while shaking. Work in the dark for this and the following steps when using the fluorescent detection method.
17. Remove the secondary antibody solution and wash thrice 10 min with 10 mL TBS buffer.
18. Store the blot in TBS buffer until imaging.
19. Image the blot using a suitable detection system.
20. Use a suitable software for densitometric analysis (e.g., Image Studio, LI-COR) to quantify the band intensities on the blot.
21. Plot the intensity against the temperature values to generate a melting curve (e.g., GraphPad Prism).
22. Use a Boltzman fit to calculate the melting temperature (i.e., inflection point).

4 Notes

1. Other suspension cell lines or adherent cells can be used to generate the desired lysate. For adherent cells: grow cells in recommended medium (e.g., see ATCC) until subconfluence (90%) is reached. Wash adherent cells with PBS to eliminate residual media components. Detach cells using either 1.5 mL trypsin solution or 1.5 mL cell dissociation buffer per T75 cell culture flask. Proceed with the resulting cell suspension with Subheading 3.1, **step 2**.
2. NP40 is a detergent that solubilizes membrane proteins without affecting the solubility of cytosolic proteins. We therefore recommend to add 0.4% (v/v) NP40 (alternative) at all times to the lysis buffer to include all potential protein (off)targets, i.e., also membrane proteins. In certain cases, the exclusion of NP40 (alternative) may be necessary, e.g., when the protein of interest interacts with lipophilic residues (e.g., fatty acids).
3. Thick-walled polycarbonate tubes that are suitable for ultracentrifugation are necessary. We recommend Beckman Coulter polycarbonate tubes for the ultracentrifuge rotor TLA-120.1 from Beckman Coulter. These tubes are reusable: use lint-free tissues and 70% (v/v) ethanol and wipe out the tubes at least twice. Let the tubes dry.

4. Choose the temperature gradient according to the melting temperature of your hypothesized target protein (e.g., consider "https://www.proteomicsdb.org/" for human proteins). In case of an unbiased approach, use the recommended temperature range mentioned above. Keep in mind that different species can exhibit different body temperatures. Therefore, the melting temperature of insect proteins can be considerably lower than the melting temperature of mammalian proteins. Heat the lid of the PCR tubes to 100 °C to avoid extensive evaporation/condensation of the probes. It might be necessary to use two mastercyclers (and two programs) in parallel to cover the temperature range properly.
5. TEAB buffer is used instead of ammonium bicarbonate buffer to avoid a nucleophilic attack of the free ammonium nitrogen to the TMT-labeling reagent.
6. As some of the proximate TMT reporter ions only differ by 0.006 Da only Orbitrap instruments capabilities allow a smooth analysis of TMT10plex isobaric mass tags due to their high resolution [16].
7. We recommend to keep the samples on ice the whole time to avoid proteolytic cleavage by cellular proteases. The utilization of protease inhibitor cocktails led in our hands to the same or worse results (e.g., lower amount of detected and quantified proteins).
8. In case the absorbance of the Bradford solution is too high, dilute the sample prior to your next measurement. If the absorbance is too low (lower than the lowest point from the BSA standard curve), increase the ratio of sample to Bradford solution (e.g., use 5 μL sample and 995 μL Bradford solution). Remember to include the dilution factors into your calculations.
9. The final volume of DMSO should not exceed 1% (v/v). The concentration of the compound needed to see a stabilizing effect on the target protein must exceed the dissociation constant. Due to the usage of highly concentrated protein lysates the compound concentrations used in the CETSA approach are usually higher than those applied in functional cell assays [15].
10. To achieve a labeling efficiency of >98%, it is sufficient to use half of the recommended volume (41 μL instead of 82 μL) of the labeling reagent [17].
11. The TPP_Makro_1.0 uses the Solver function of Microsoft Excel, which is not activated by default. If not activated yet, you have to activate the solver function before the first use of the TPP_Makro_1.0. A description for the activation of solver function is available on https://msdn.microsoft.com/en-us/vba/excel-vba/articles/using-the-solver-vba-functions .

12. How the macro works:

Denaturation changes at different temperatures are tracked by the reporter ion intensity and observed in relation to the lowest temperature; therefore the lowest temperature is set to 1. The relative fold changes are calculated as a function of temperature. The following equation is used to iteratively fit a sigmoidal curve through the measuring points:

$$y = \text{bottom plateau} + \frac{(\text{top plateau} - \text{bottom plateau})}{1 + e^{-\left(\frac{a}{\text{Temp}}\right)-b}}$$

Top plateau is fixed to one, Temp is the temperature, bottom plateau is a protein-specific constant that defines the maximal denaturation and *a* and *b* are constants which describe the curve progression.

The next step is to determine the melting point of the proteins by calculating the inflection point of the curve through the maximum of the first derivation. Finally, the proteins are matched to four different quality classes and the area between the curves is calculated. In future versions the area will be an additional validation factor. Proteins which are sorted into class three fulfill the strictest criteria: (1) Each protein must be identified by at least two unique peptides in each replicate, and (2) the normalized intensity of the penultimate vehicle label must be below 0.35 in all replicates, and (3) either the melting point must show a shift of at least 1.0 °C in the same direction in all three replicates or the normalized intensity of the of the penultimate label must be greater than 0.05 in all three replicates.

Class two proteins fulfill medium strict criteria: (1) The proteins must be identified by at least two "razor+unique" peptides in two out of three replicates, and (2) both curves (resulting from compound- and vehicle-treated sample) must be present in two out of three replicates, and (3) either a melting point shift of at least 0.5 °C into the same direction in two out of three replicates or a difference of the normalized intensity of the penultimate label greater than 0.03 in two out of three replicate has to be observed.

To be sorted into the class one, a protein must (1) show both curves (resulting from compound- and vehicle-treated sample) in two out of three replicates and (2) either have a melting point shift of at least 0.5 °C into the same direction in two out of three replicates or have a difference of the normalized intensity of the penultimate label greater than 0.03 in two out of three replicates.

Proteins are listed as class 0, if none of the above mentioned combination of criteria is fulfilled, i.e., there is no hint of a difference in the melting behavior of compound-treated sample and solvent-treated control.

The quality of the protein identification is shown by the color of the chart frames. A green frame indicates that the protein is identified by at least three razor + unique peptides and a reporter ion intensity of more than 1×10^5 for the first three temperatures. An orange frame is given to proteins identified by at least two razor + unique peptides and a reporter ion intensity of more than 1×10^4 for the first three temperatures. All other proteins get a red frame.

13. Microsoft Excel macros are only processed on a single CPU thread; therefore, the calculation time of a full triplicate can be up to 72 h.
14. You can change the quality class parameters by clicking on the "Filter settings" button on the "Charts" sheet. After changing filter parameters or the sorting of the proteins in "Calculation" you must always click on the "Refresh" button on the "Charts" sheet.
15. Possible reasons for false positive results can be outliers that shift the curve into a wrong direction, indicating a thermal shift that is not present. In particular, errors of the first label, to which all other labels are normalized, can create nonsensical results. Even if a fitted curve matches its data points nicely, it is necessary to check if the curves of all replicates are similar to each other. Curve sets that are not alike, are not valid.
16. To check again already validated proteins you can move the whole row of the protein to the sheet "Calculation" into row three or below. Click the "Refresh" button before you proceed with the evaluation of the graphs.
17. For some proteins, it is not recommended to heat the SDS protein sample before loading on the denaturating gel (e.g., membrane proteins). This should be checked beforehand.

Acknowledgment

We thank Dr. Marc Schürmann for the establishment of the TPP assay in our department. Without his large effort, it would not have been possible to perform the described assay in-house.

References

1. Ziegler S, Pries V, Hedberg C, Waldmann H (2013) Target identification for small bioactive molecules: finding the needle in the haystack. Angew Chem Int Ed Engl 52(10):2744–2792. https://doi.org/10.1002/anie.201208749
2. Schurmann M, Janning P, Ziegler S, Waldmann H (2016) Small-molecule target engagement in cells. Cell Chem Biol 23(4):435–441. https://doi.org/10.1016/j.chembiol.2016.03.008

3. Kapoor S, Waldmann H, Ziegler S (2016) Novel approaches to map small molecule-target interactions. Bioorg Med Chem 24 (15):3232–3245. https://doi.org/10.1016/j.bmc.2016.05.020
4. Schenone M, Dancik V, Wagner BK, Clemons PA (2013) Target identification and mechanism of action in chemical biology and drug discovery. Nat Chem Biol 9(4):232–240. https://doi.org/10.1038/nchembio.1199
5. Lee H, Lee JW (2016) Target identification for biologically active small molecules using chemical biology approaches. Arch Pharm Res 39 (9):1193–1201. https://doi.org/10.1007/s12272-016-0791-z
6. Kanoh N, Honda K, Simizu S, Muroi M, Osada H (2005) Photo-cross-linked small-molecule affinity matrix for facilitating forward and reverse chemical genetics. Angew Chem Int Ed Engl 44(23):3559–3562. https://doi.org/10.1002/anie.200462370
7. Lomenick B, Hao R, Jonai N, Chin RM, Aghajan M, Warburton S, Wang JN, Wu RP, Gomez F, Loo JA, Wohlschlegel JA, Vondriska TM, Pelletier J, Herschman HR, Clardy J, Clarke CF, Huang J (2009) Target identification using drug affinity responsive target stability (DARTS). Proc Natl Acad Sci U S A 106 (51):21984–21989. https://doi.org/10.1073/pnas.0910040106
8. Savitski MM, Reinhard FB, Franken H, Werner T, Savitski MF, Eberhard D, Martinez Molina D, Jafari R, Dovega RB, Klaeger S, Kuster B, Nordlund P, Bantscheff M, Drewes G (2014) Tracking cancer drugs in living cells by thermal profiling of the proteome. Science 346(6205):1255784. https://doi.org/10.1126/science.1255784
9. Martinez MD, Jafari R, Ignatushchenko M, Seki T, Larsson EA, Dan C, Sreekumar L, Cao Y, Nordlund P (2013) Monitoring drug target engagement in cells and tissues using the cellular thermal shift assay. Science 341 (6141):84–87. https://doi.org/10.1126/science.1233606 [doi] 341/6141/84 [pii]
10. Vedadi M, Niesen FH, Allali-Hassani A, Fedorov OY, Finerty PJ Jr, Wasney GA, Yeung R, Arrowsmith C, Ball LJ, Berglund H, Hui R, Marsden BD, Nordlund P, Sundstrom M, Weigelt J, Edwards AM (2006) Chemical screening methods to identify ligands that promote protein stability, protein crystallization, and structure determination. Proc Natl Acad Sci U S A 103(43):15835–15840
11. Franken H, Mathieson T, Childs D, Sweetman GMA, Werner T, Togel I, Doce C, Gade S, Bantscheff M, Drewes G, Reinhard FBM, Huber W, Savitski MM (2015) Thermal proteome profiling for unbiased identification of direct and indirect drug targets using multiplexed quantitative mass spectrometry. Nat Protoc 10(10):1567–1593. https://doi.org/10.1038/nprot.2015.101
12. Reinhard FB, Eberhard D, Werner T, Franken H, Childs D, Doce C, Savitski MF, Huber W, Bantscheff M, Savitski MM, Drewes G (2015) Thermal proteome profiling monitors ligand interactions with cellular membrane proteins. Nat Methods 12(12):1129–1131. https://doi.org/10.1038/nmeth.3652
13. Huber KV, Olek KM, Muller AC, Tan CS, Bennett KL, Colinge J, Superti-Furga G (2015) Proteome-wide drug and metabolite interaction mapping by thermal-stability profiling. Nat Methods 12(11):1055–1057. https://doi.org/10.1038/nmeth.3590
14. Cox J, Mann M (2008) MaxQuant enables high peptide identification rates, individualized p.p.b.-range mass accuracies and proteome-wide protein quantification. Nat Biotechnol 26(12):1367–1372. https://doi.org/10.1038/nbt.1511
15. Jafari R, Almqvist H, Axelsson H, Ignatushchenko M, Lundback T, Nordlund P, Martinez Molina D (2014) The cellular thermal shift assay for evaluating drug target interactions in cells. Nat Protoc 9(9):2100–2122. https://doi.org/10.1038/nprot.2014.138
16. Mateus A, Maatta TA, Savitski MM (2016) Thermal proteome profiling: unbiased assessment of protein state through heat-induced stability changes. Proteome Sci 15:13. https://doi.org/10.1186/s12953-017-0122-4
17. Instructions–TMT10plex mass tag labeling kits and reagents. Pierce biotechnology. https://tools.thermofisher.com/content/sfs/manuals/MAN0016969_2162457_TMT10plex_UG.pdf

Chapter 5

Visualizing Bioactive Small Molecules by Alkyne Tagging and Slit-Scanning Raman Microscopy

Jun Ando, Kosuke Dodo, Katsumasa Fujita, and Mikiko Sodeoka

Abstract

An understanding of the intracellular distribution of bioactive small molecules provides insight into their target organelles and biomolecules, and throws light on their molecular mechanisms of action and specificity. Many studies in this area have employed fluorescence imaging, using molecules of interest labeled with fluorescent dyes. However, modification with a bulky fluorophore may significantly alter the properties, including bioactivity, of small molecules. To address this issue, we have recently developed Raman scattering microscopy of alkyne tags to visualize small molecules in biological systems. The tiny alkyne tag should have a minimal effect on the properties of the tagged molecule, and the unique alkyne vibrational properties allow observation of the Raman signals of the tagged molecules in a wavelength region where there is little interference from Raman signals of endogenous biomolecules. Here we summarize the methodology for Raman imaging of alkyne-tagged bioactive small molecules in living biological systems including the development of slit-scanning Raman microscope, which is useful for fast imaging of alkyne-tagged molecules.

Key words Alkyne tag, Raman scattering, Small molecules, Slit-scanning Raman microscopy

1 Introduction

Raman scattering spectroscopy has been used to identify molecular species in a sample by acquiring the information of molecular vibration through scattered light [1]. Since Raman spectroscopy is applicable in an aqueous environment, it can be used to investigate many kinds of biomolecules, such as proteins, lipids, sugars, and nucleic acids, and their interactions with bioactive small molecules [1–4]. Although Raman spectroscopy has had a disadvantage as regards detection sensitivity, due to recent technical advances in Raman spectroscopy, such as development of high-power lasers, high-sensitivity detectors, and optical filters, molecular imaging of biological samples such as cells and tissues by Raman scattering microscopy has become possible [5]. In particular, the development of parallel spectral acquisition systems, such as slit-scanning

Slava Ziegler and Herbert Waldmann (eds.), *Systems Chemical Biology: Methods and Protocols*, Methods in Molecular Biology, vol. 1888, https://doi.org/10.1007/978-1-4939-8891-4_5,

Raman microscopy, has improved the imaging speed by more than two orders of magnitude compared with that of conventional confocal (point-scanning) Raman microscopy [6, 7]. It is now possible to observe various molecules in living biological samples at diffraction-limited spatial resolution [8, 9].

In principle, Raman scattering microscopy can visualize the distribution of bioactive small molecules in cells and tissues, by observing the specific scattering signals [10]. However, Raman peaks of small molecules and endogenous biomolecules appear in a similar spectral region at around 500–1800 cm^{-1}, which is called the Raman fingerprint region, and in practice, the extensive overlaps of signals from multiple molecules makes it difficult to extract the target signal. To overcome this problem, in 2011 we proposed and demonstrated an alkyne-tag method to observe small molecules in biological systems [11]. The alkyne carbon–carbon triple bond shows unique Raman peaks at around 2050–2300 cm^{-1}, which lies within the Raman-silent region of biomolecules. Raman microscopy can therefore selectively observe alkyne-tagged small molecules in complex biological samples [11]. An advantage of Raman microscopy is that the whole spectrum can be acquired for each pixel, and simultaneous imaging of the alkyne-tagged small molecule and biomolecules such as lipid, protein, and cytochromes is possible. The distribution of intrinsic biomolecules works as a reference for that of small molecules, and provides colocalization information with lipid droplets and mitochondria in cells [12]. Furthermore, simultaneous imaging of two or more different alkyne-tagged small molecules is also possible. Molecules that have been observed include an alkyne-tagged analog of thymidine (cell proliferation probe 5-ethynyl-2′-deoxyuridine; EdU) [11], coenzyme Q [12], sphingomyelin [13], cholesterol [14], amino acid [15], glycan [16], and an alkyne-containing anticancer drug [17]. In addition, alkyne-tagged organelle markers have been developed to enable specific observation of mitochondria in live cells by Raman microscopy [18].

In the following sections, we introduce the methodology to visualize bioactive small molecules in biological samples by slit-scanning Raman microscopy using an alkyne tag, including the choice of a suitable tag moiety, procedures of sample preparation, and setting up the slit-scanning Raman microscope.

2 Materials

2.1 Development of Alkyne-Tagged Molecules

1. Candidates for Raman tags

   Several chemical structures, such as alkyne, nitrile, azide, and deuterated methyl/methylene groups, exhibit Raman peaks in the Raman-silent region of biomolecules (1800–2600 cm^{-1}), but it is important to choose a tag that

	Tag structure	Peak position	Peak intensity relative to nitrile
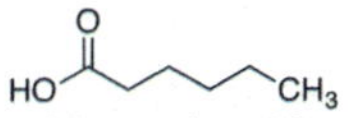 (Hexanoic acid)	—	—	—
	Alkyne	2110	2.34
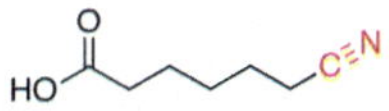	Nitrile	2242	1.00
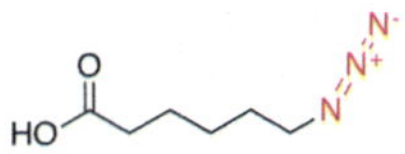	Azide	2096	0.38
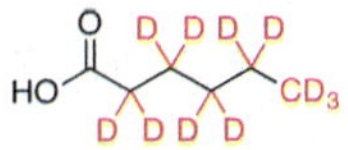	CD_2 / CD_3	2124	1.07

Fig. 1 Peak positions and intensities of hexanoic acid modified with various Raman tags: alkyne, nitrile, azide, and deuterated methyl/methylene group (adapted with permission from ref. 12. Copyright 2012 American Chemical Society)

provides a strong Raman signal. We compared the Raman peak intensities of hexanoic acids modified with various candidates [12]. As shown in Fig. 1, the peak intensity of alkyne is more than twice that of nitrile. The peak intensity of azide is less than half that of nitrile. Since the signal derived from one C-D bond is weak, we prepared a hexanoic acid derivative with eleven deuterium atoms (four deuterated methylene groups and one deuterated methyl group). Several peaks derived from the deuterated methyl/methylene groups are observed in the Raman spectrum of this molecule, and the maximum peak height is similar to that of nitrile. We concluded that alkyne is the best tag among these candidates in terms of signal intensity.

2. Relationship between alkyne structure and Raman peak intensity/position

   The intensity and position of Raman peaks of alkyne tags depend upon the tag structure, and this fact can be utilized to maximize the tag detection capability of Raman microscopy. Suitable choices of Raman tags can also enable multitag imaging for simultaneous observation of two or more different molecules. Figure 2 summarizes the alkyne peak intensity and Raman shift values of 77 compounds [12]. The vertical axis shows the relative Raman intensity of each compound versus EdU (the peak intensity of EdU is taken as 1.0). Figure 2a shows the plot of for terminal alkyne. The Raman peaks of the

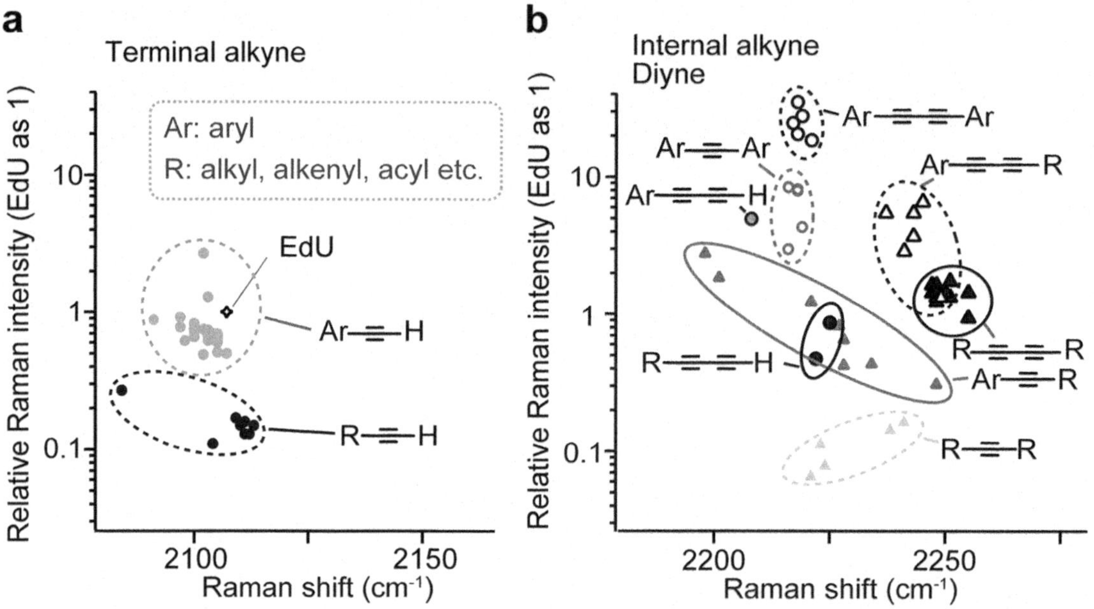

Fig. 2 Relationship between alkyne-tag structure and peak positions/intensities of compounds containing (**a**) terminal alkyne and (**b**) internal alkyne/diyne. The vertical axis shows Raman intensity relative to EdU, and the horizontal axis shows Raman shift (adapted with permission from ref. 12. Copyright 2012 American Chemical Society)

terminal alkynes are located at around 2100 cm^{-1}, and conjugation to an aromatic ring increases the peak intensity severalfold compared to unconjugated alkyne. Figure 2b shows the plot for internal alkyne and diynes. The peak intensity of diyne is several times higher than that of terminal alkyne, and is further increased when the diyne is conjugated to an aromatic ring, although the structure of the tag then becomes relatively large. Peak positions of both internal alkyne and diyne are located around 2200–2250 cm^{-1}.

3. Guideline for the selection of an alkyne tag for small molecules

The selection of a suitable alkyne tag is critical for the success of Raman imaging. Strong Raman intensity is required for imaging, but the biological and physical properties of the original small molecule should be retained in the alkyne-tagged molecule. The Raman scattering properties of alkyne tags shown in Fig. 2 provide guidance for selecting alkyne tags for small molecules. In the case of small molecules that contain an aromatic ring, introduction of an ethynyl group at the ring is favorable. If the target molecule does not contain an aromatic ring, diyne is a good choice [19]. It is also important to avoid introducing the tag at a position that is critical for interaction of the small molecule with its target protein, since this may cause loss or decrease of the biological activity, or alter the intracellular localization of the molecule. Therefore, the biological activity of an alkyne-tagged molecule should always be checked before imaging. Among the compounds verified in Fig. 2, diyne with two aromatic rings shows the highest intensity. However, it may not be suitable as a tag for a small molecule because its large size and structural rigidity are likely to strongly influence the properties of the small molecule. Instead, it could be useful in organelle-specific Raman markers [18]. For multi-color imaging, combinations of alkynes with distinct Raman shifts should be selected. Terminal alkyne (~2100 cm^{-1}) and internal alkyne or diyne (2200–2250 cm^{-1}) would be a suitable choice of tags for two different molecules. To obtain Raman images of more than three molecules, even more careful selection of the alkyne tag is required. In this connection, it is noteworthy that fine-tuning of the Raman shift of the alkyne tag is possible by replacing alkyne carbon with its isotope (^{13}C) [20].

2.2 Cell Culture

1. Cell culture medium: Dulbecco's modified Eagle's medium (DMEM) with 1000 mg glucose/L, phenol red and $NaHCO_3$. 500 mL of DMEM is mixed with 50 mL of fetal bovine serum (FBS) and 5 mL of penicillin–streptomycin–glutamine (penicillin 10,000 unit/mL, streptomycin 10,000 μg/mL, and L-glutamine 29.2 mg/mL).
2. Substrate: 25 mm diameter quartz substrate. The thickness of the substrate is 0.17 mm. The substrate is fixed in place with a stainless steel coverslip holder with an O-ring seal.
3. Medium for Raman imaging: Tyrode's solution composed of 150 mM NaCl, 10 mM glucose, 10 mM HEPES, 4.0 mM KCl, 1.0 mM $MgCl_2$, 1.0 mM $CaCl_2$, and 4.0 mM NaOH.
4. 35 mm diameter sterilized plastic dish.
5. CO_2 incubator with a humidified atmosphere of 5% CO_2 in air, maintained at 37 °C.
6. HeLa cells.
7. 20 mM stock solution of 5-ethynyl-2′-deoxyuridine (EdU): dissolved in dimethyl sulfoxide (DMSO).

2.3 Instrumentation and Standard Samples for Slit-Scanning Raman Microscopy

In slit-scanning Raman microscopy, a line-shaped focus is formed by a cylindrical lens (CL) and is projected onto the sample by an objective lens. The Raman scattering signal along the line-shaped focus is collected by the same objective lens, and led to the entrance slit of a spectrophotometer. Several hundred Raman spectra can be obtained in parallel by using a spectrophotometer equipped with a two-dimensional cooled CCD camera. Key instruments and samples for the construction of a slit-scanning Raman microscope are summarized below. Raman imaging of alkyne-tagged molecules in cells is also possible with commercial slit-scanning Raman microscopes, such as Raman-11 from Nanophoton Corporation.

1. Excitation light source: Nd:YVO_4 laser at wavelength of 532 nm. Line width is narrower than 30 GHz. Laser power is more than 3 W.
2. Objective lens: Water immersion objective lens. Numerical aperture (NA) is more than 1.2, and the magnification is 60.
3. Long-pass edge filter: The transmission band is 539–700 nm with transmittance of more than 97%. Optical density at 532 nm is 6. Diameter of the filter is 25 mm. Two sets of filters will be used.
4. Cylindrical lens: Plano-convex cylindrical lens. Focal length of the lens is 100 mm. The shape of the lens is round, and the diameter of the lens is 25 mm.
5. Detector for Raman scattering: Two-dimensional cooled CCD camera. The temperature of the detector is −70 °C. 1340 × 400 imaging array with 20 × 20 μm pixels. Imaging area is 26.8 × 8.0 mm. Read noise is lower than 3 e^- rms with a readout rate of 100 kHz.
6. Spectrometer: Aberration (astigmatism)-compensated imaging spectrometer with a 600 g mm^{-1} grating. The F number is 4.4, and focal length is 300 mm.
7. Scanning mirror: Galvanometer mirror. Clear aperture of the mirror is 15 mm.
8. Microscope body: Inverted optical microscope with eyepiece and halogen lamp.
9. Table: Vibration-damping optical table with table size of around 1200 × 2000 mm.
10. Microbeads: Polystyrene (or polymethyl methacrylate) beads of 500 nm diameter.
11. Surface-modified cover glass: 25 mm diameter glass substrate. Thickness of the substrate is 0.17 mm. Surface of the glass is modified by silanization using 3-aminopropyltrimethoxysilane. Alternatively, use a commercially available silanized glass substrate.
12. Rhodamine 6G solution: 100 μM rhodamine 6G dissolved in ethanol.
13. Glass-bottomed dish: 35 mm diameter glass bottom dish with clear aperture of 14 mm.
14. Software for image acquisition: In-house software that synchronizes the exposure of the cooled CCD camera and galvanometer mirror scanning. Parameters including exposure time,

imaging width and scan pitch are set in the software. Details of the software are described in Ref. 6 (*see* **Note 1**).

15. Objective micrometer: Crosslines with square grids with 10 μm pitch, 200 μm length.
16. Software for image display: Image J, an image processing software that was designed with an open architecture [21]. No license is required.
17. Pinhole for alignment: Single pinhole with diameter of 75 μm. Pinhole material is stainless steel.
18. Digital-to-analog (DA) converting board: Bipolar, ±5 V with resolution of 16 bit.

3 Methods

3.1 Setup of a Slit-Scanning Raman Microscope

1. [Setting up the optical system]: Align the optical components and instruments, as shown in Fig. 3. Typical choices of the focal length of L1~L6 and CL1~CL2 are as follows: L(spherical lens) 1: f = 50 mm, L2: f = 200 mm, L3: f = 200 mm, L4: f = 200 mm, L5: f = 160 mm, L6: f = 200 mm, CL(cylindrical lens)1: f = 50 mm, CL2: f = 150 mm. Both CL1 and CL2 are tilted at 90° to the third cylindrical lens. It is preferable to use an achromatic doublet lens for L3~L6. For the alignment of the optics, place pinhole with 75 μm diameter on the microscope stage. Illuminate it with the halogen lamp of the microscope. Center the position of the pinhole with the microscope

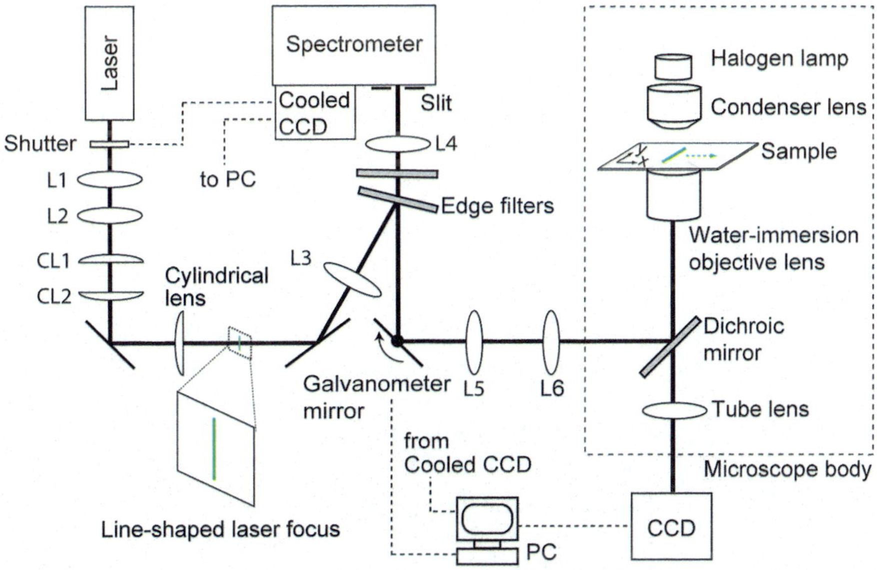

Fig. 3 Schematic illustration of the experimental setup of the slit-scanning Raman microscope

stage by watching the field of view through the ocular lens. Watch the optical path of pinhole light, which exits from the back port of the microscope. Align the optics so that the pinhole light travels at the center of each lens. Align optical path of the laser to overlap the light path of the pinhole light.

2. Turn on the laser, spectrophotometer, cooled CCD camera for spectrometer, CCD camera for microscope body, halogen lamp, shutter and PC. Open the software for cooled CCD camera, CCD camera, and spectrophotometer. Set the target temperature of the cooled CCD camera at −70 °C, and start cooling.
3. [Preparation of the sample for alignment] Mix 500 nm diameter polystyrene beads with the rhodamine 6G solution, drop the mixture on the modified glass substrate, and wait for an hour.
4. Wash the glass substrate with deionized distilled water, and fix it in the stainless steel chamber.
5. Fill the chamber with water, and place it on the sample stage.
6. [Adjusting the detection optics] Adjust the height of the microscope objective lens to focus on the bright-field image of the beads. Use the halogen lamp and CCD camera on the microscope body to obtain a bright-field image.
7. Rotate the grating of the spectrometer to set the central wavelength to 0 nm, and open the slit of the spectrometer to more than a few millimeters.
8. Set the exposure time of the cooled CCD camera for the spectrometer to 0.5 s. Start acquiring the bright-field image with the cooled CCD camera. Adjust the light intensity of the halogen lamp as necessary, and acquire the bright-field image of the beads continuously.
9. Adjust the position of the spectrophotometer to obtain a sharp focus of the cooled CCD camera on the beads.
10. Narrow the slit width of the spectrometer to around 20 μm, and confirm the center position of the slit on the cooled CCD camera. Open the slit of the spectrophotometer until part of the bright-field image is recognized by the camera.
11. Move a particular bead in the bright-field image to the center of the field of view of the microscope. Adjust the position of the spectrometer to locate the bright-field image of the targeted bead on the center position of the slit. Stop exposing the cooled CCD camera.
12. [Adjusting the illumination laser] Open the shutter of the laser, set the laser power to ~10 mW, and observe the line-shaped laser beam focused on the sample with the CCD camera on the

microscope body. The line-shaped laser beam is reflected by the interface between the upper surface of glass and water and focused on the CCD camera.

13. Check that the focus positions of the line-shaped laser beam and bright-field image of the beads are same.
14. Turn off the halogen lamp, and start exposing the cooled CCD camera of the spectrophotometer. Acquire the fluorescence image of the sample continuously. Adjust the position of the cylindrical lens to maximize the fluorescence intensity of the image. Check if the line-shaped fluorescence image is parallel to the slit of spectrometer. If not, rotate the cylindrical lens to make them parallel.
15. Narrow the slit width of the spectrometer to 50 μm. Check if the line-shaped fluorescence image is located within the clear aperture of the slit. Close the shutter of the laser, and stop image acquisition of the cooled CCD camera.
16. Replace the beads sample on the microscope stage with an objective micrometer. Turn on the halogen lamp, and observe the bright-field image of the micrometer grid with the CCD camera on the microscope body. Open the shutter of the laser.
17. Measure the line length of the illumination. Adjust the line length to around 80 μm, using the iris placed at the focus position of the cylindrical lens.
18. Adjust the laser power at ~3.5 mW/μm^2.
19. [Setting the galvanometer mirror] Scan the galvanometer mirror to confirm the actual scanning pitch and scanning width. Set the scanning pitch as ~330 nm. Turn off the halogen lamp. Remove the objective micrometer.
20. [Wavenumber calibration] Place a glass-bottomed dish filled with ethanol on the microscope stage. Adjust the height of the objective lens to be more than a few micrometers above the interface of the glass substrate and ethanol.
21. Rotate the grating of spectrometer to set the center wavelength to around 620 nm.
22. Set the exposure time of the cooled CCD camera on the spectrometer to 5 s. Start exposing the cooled CCD camera on the spectrometer to Raman scattering light from the sample. Shutter opening of the laser is synchronized with the exposure of the cooled CCD camera.
23. Average the scattering intensity of the cooled CCD image along the direction parallel to the slit, to obtain the Raman spectrum of ethanol.
24. Calibrate the horizontal axis of the cooled CCD camera as Raman shift (cm^{-1}). Use the pixel numbers of three

characteristic peaks of the obtained ethanol spectrum and calibrate the wavenumber based on the known spectrum of ethanol (it is preferable to choose 434, 1454, and 2930 cm^{-1}).

3.2 Raman Imaging of Alkyne-Tagged Small Molecules

3.2.1 Imaging of Alkyne-Tagged Thymidine Analog (EdU) in Living HeLa Cells (See **Note 2**)

1. [Sample preparation] Set a 25 mm diameter quartz substrate at the bottom of a 35 mm diameter plastic dish. Add 2 mL of cell culture medium. Seed HeLa cells on the quartz substrate (*see* **Note 3**).
2. Culture HeLa cells in the incubator at 37 °C in a humidified 5% CO_2 environment for ~2 days.
3. Mix the EdU stock solution with cell culture medium to obtain an EdU solution with a concentration of 20 μM.
4. Remove cell culture medium from the plastic dish and replace with 2 mL of the 20 μM EdU solution.
5. Culture HeLa cells with EdU in the incubator at 37 °C in a humidified 5% CO_2 environment for 12 h to 1 day.
6. Remove the EdU-containing cell culture medium from the plastic dish, and wash cells with 2 mL of Tyrode's solution several times (*see* **Note 4**).
7. Take the quartz substrate from the plastic dish, and fix it in the stainless steel chamber. Fill the chamber with Tyrode's solution. Wipe the lower surface of the quartz substrate with wet paper.
8. [Setting the sample] Place the stainless steel chamber on the sample stage of the microscope.
9. Adjust the position of the microscope objective lens by observing the bright-field image of the HeLa cells (*see* **Note 5**). Use the halogen lamp and the CCD camera on the microscope body to obtain the bright-field image.
10. [Raman imaging] Set the exposure time of the cooled CCD camera to 10–20 s. Set the scanning width to 30–60 μm, and the scanning pitch as ~330 nm. Laser intensity at the sample is set as ~3.5 $mW/\mu m^2$ (*see* **Note 6**).
11. Turn off the halogen lamp and start the measurement.

3.2.2 Imaging of Alkyne-Tagged Coenzyme Q Analog (AltQ4) in Living HeLa Cells

1. Set a quartz substrate with a diameter of 25 mm at the bottom of a 35 mm diameter plastic dish. Add 2 mL of cell culture medium. Seed HeLa cells on the quartz substrate.
2. Culture HeLa cells in the incubator at 37 °C in a humidified 5% CO_2 environment for ~2 days.
3. Prepare 6 μM AltQ4 dissolved in Tyrode's solution. Details of the procedure for chemical synthesis of AltQ4 are given in Ref. 12 (*see* **Note** 7 and Scheme 1).

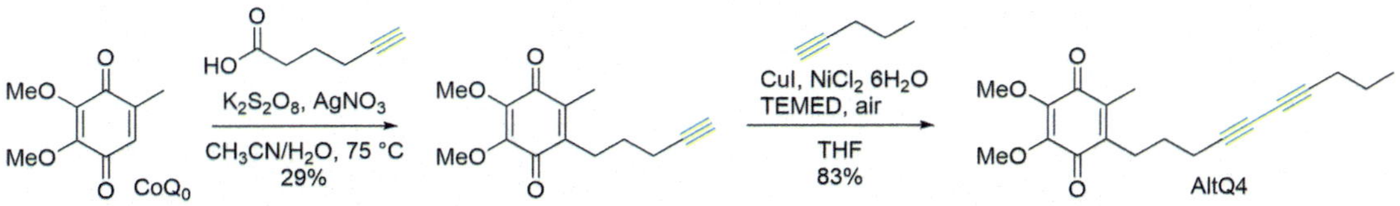

Scheme 1 Synthesis of AltQ4 from CoQ$_0$

4. Remove cell culture medium from the plastic dish, and wash the dish with 2 mL of Tyrode's solution several times.
5. Take the quartz substrate from the plastic dish, and fix it in the stainless steel chamber. Fill the chamber with 6 μM AltQ4-dissolved Tyrode's solution. Wipe the lower surface of the quartz substrate with wet paper.
6. Wait for 50 min, and place the stainless steel chamber on the sample stage of the microscope.
7. Adjust the position of the microscope objective lens by observing the bright-field image of the HeLa cells. Use the halogen lamp and the CCD camera on the microscope body to obtain the bright-field image.
8. Set the exposure time of the cooled CCD camera to 5 s. Set the scanning width to 30–60 μm, and the scanning pitch as ~330 nm. Laser intensity at the sample is set to ~3.0 mW/μm^2.
9. Turn off the halogen lamp, and start the measurement.

3.3 Data Processing of Obtained Raman Imaging Data and Image Reconstruction

1. Remove bright pixels due to cosmic ray exposure in each Raman image by using a median filter.
2. For Raman imaging of alkyne, choose the appropriate wavenumber region: 1800–2300 cm^{-1} is preferable to exclude spectral information of endogenous molecules.
3. Process the Raman data with singular value decomposition (SVD) [22]. Choose an appropriate number of loading vectors that contribute substantially to the image contrast, and reconstruct the Raman data.
4. Process the Raman data by a modified polyfit background removal technique at each pixel [23] to subtract the fluorescence signal from the Raman spectra.
5. Reconstruct the Raman image of alkyne-tagged molecules by using the alkyne's peak intensity distribution. A typical peak position of EdU is 2123 cm^{-1} (Fig. 4), and that of AltQ4 is 2258 cm^{-1} (Fig. 5).
6. Save the image in an Image J-readable format, such as txt or tiff.
7. Perform the same procedures (**steps 1–6**) at the fingerprint region (500–1800 cm^{-1}) and the high wavenumber region

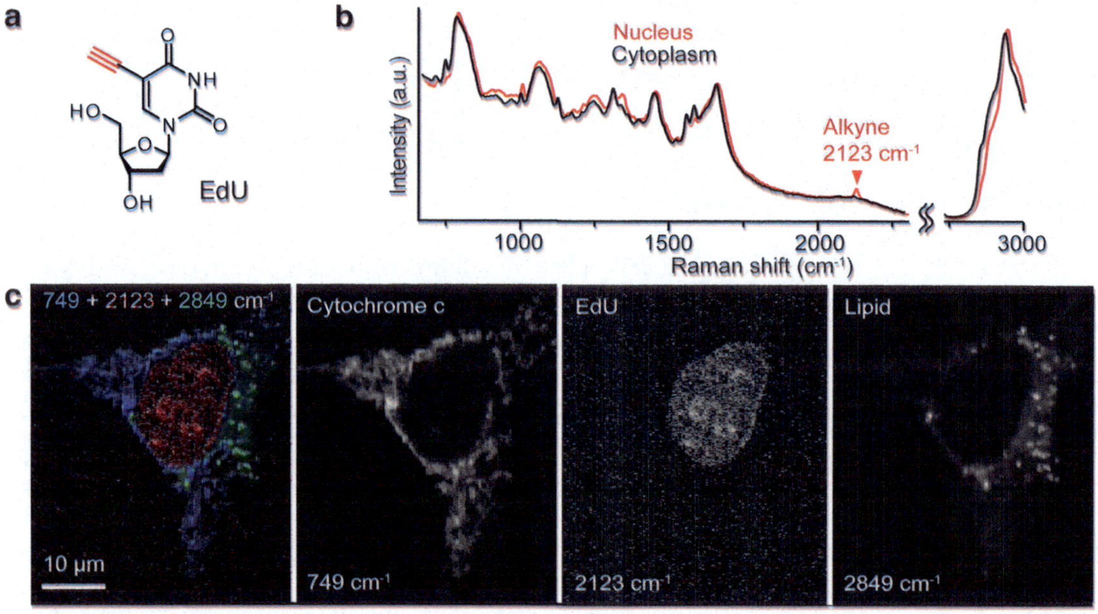

Fig. 4 (**a**) Chemical structure of EdU (**b**) Raman spectra and (**c**) Raman images of a living HeLa cell treated with 20 μM EdU for 6 h. Laser intensity was 3.5 mW/μm^2. Exposure time was 20 s/line. Number of lines was 127. Total imaging time was 49 min (adapted with permission from ref. 11. Copyright 2011 American Chemical Society)

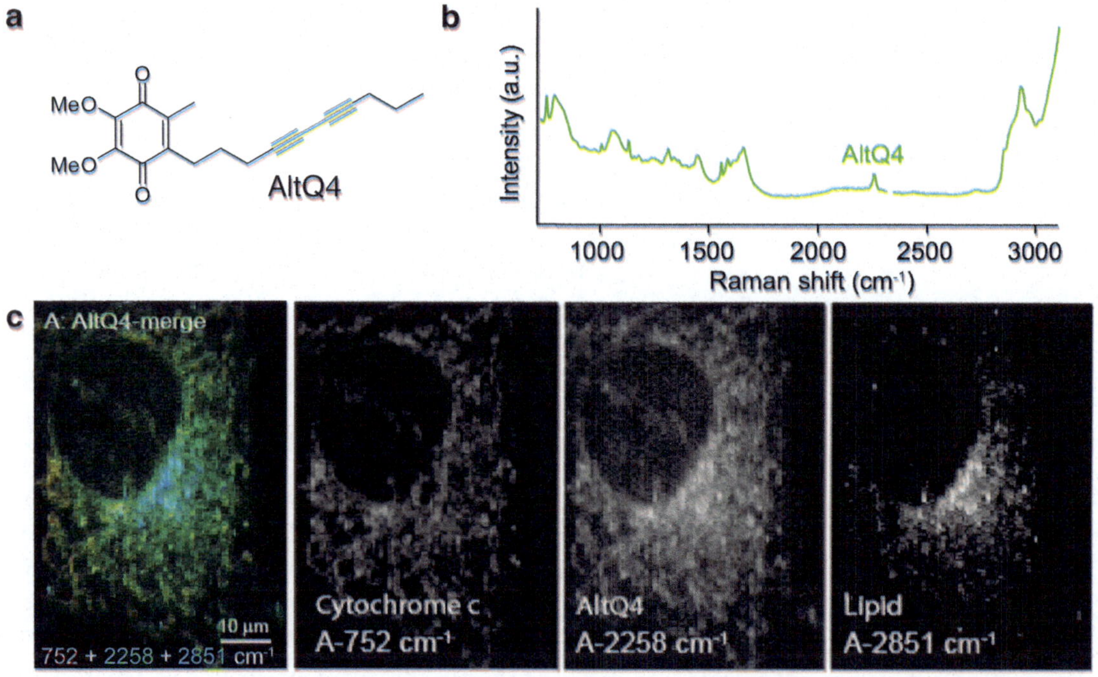

Fig. 5 (**a**) Chemical structure of AltQ4 (**b**) Raman spectrum at cytoplasm and (**c**) Raman images of a living HeLa cell treated with 6 μM AltQ4 for 50 min. Laser intensity was 3.0 mW/μm^2. Exposure time was 5 s/line. Number of lines was 95. Total imaging time was 13 min (adapted with permission from ref. 12. Copyright 2012 American Chemical Society)

(2700–3100 cm^{-1}). Distribution of cytochrome *c* is obtained at ~750 cm^{-1}, that of protein beta sheet is obtained at ~1684 cm^{-1} and that of lipid is obtained at ~2850 cm^{-1}.

8. Open Image J.
9. Open Raman images of alkyne, cytochrome *c*, and lipid.
10. Adjust the aspect ratio of the image.
11. Adjust the image contrast for each image.
12. Merge the three channels (for EdU, red: alkyne, green: lipid, blue: cytochrome *c*).
13. Stack the three channels into one.
14. Save each image as TIFF and BMP file.

4 Notes

1. To obtain Raman images, the CCD camera exposure and laser illumination should be synchronized. Output trigger from the CCD camera is connected to the controller of a mechanical shutter for synchronization. The camera exposure should also be synchronized with the scanning of the line-shaped laser at the sample. Laser scanning is performed with a galvanometer mirror. In our case, image acquisition is controlled by a home-made software. It generates an output voltage through a DA board, which is connected to the galvanometer mirror. The output voltage is changed to relocate the angle of the galvanometer mirror, resulting in the relocation of the line-shaped laser at the sample. The camera exposure is then started to acquire a spectroscopic image at each line, and the images are saved. Parameters including exposure time, imaging width and scan pitch are set in the software before starting Raman imaging. Based on the parameters, the output voltage from DA board is determined.
2. Raman signal of alkyne-tagged small molecules in cells is often weak compared with that of intrinsic biomolecules such as lipids and proteins. To confirm if the system is sensitive enough for detection of alkyne signal, EdU is a good sample. This is because EdU is localized at the nucleus in cells and does not diffuse during imaging since it is incorporated into DNA. For sample preparation, culture HeLa cells with 20 μM EdU for 1 day. Since doubling time of HeLa cells is ~20 h [24], almost all HeLa cells should contain EdU in their nuclei. After the sample preparation, direct the line-shaped laser to the cell nucleus. The location of the nucleus can be confirmed by a phase contrast image of the cell (or bright-field image, although the contrast is low). Expose the laser light for 1 min

at ~3.5 mW/μm^2 and acquire a Raman spectral image along the line. At the position of the nucleus, a Raman peak of EdU at ~2123 cm^{-1} should be seen. Once the signal is obtained, shorten the exposure time as much as possible. At the optimized exposure time, perform Raman imaging of the cells. If the measurement works properly, the Raman scattering at 2123 cm^{-1} localizes in the nucleus.

3. The cell-adhering substrate is another major background source for Raman imaging. Selection of the material for substrate is important to reduce the background signal. For Raman microscope based on an inverted optical microscope body, the substrate should be transparent to both excitation and scattering light. A typical wavelength of Raman excitation light is 532 nm, and that of scattering light is around 550–630 nm. Glass, quartz, and calcium fluoride (CaF_2) are possible candidates for the substrate. However, the background signal of a glass substrate is large, compared with quartz and CaF_2. Therefore, quartz or CaF_2 should be used for Raman imaging. CaF_2 is much more fragile than quartz. It is not easy to fabricate a CaF_2 substrate at the thickness of ~0.2 mm, which is required for most of the aberration-compensated objective lenses. Quartz substrate is therefore a first choice for the material of cell-adhering substrate.
4. The Raman scattering signal of small molecules in cells is quite weak. It is easily buried in the background signal, especially the fluorescence signal from the medium. The medium for live cells during Raman imaging should be carefully chosen. Cell culture medium usually contains molecules that exhibit fluorescence signal, such as phenol red. In addition, various biomolecules from serum can cause background Raman signal. Therefore, Tyrode's solution, which does not contain fluorescent molecules or serum, is used as the medium to minimize the background signal. It is important to completely remove remaining cell culture medium during washing processes (*see* Subheadings 3.2.1 and 3.2.2). This medium can keep HeLa cells viable for a few hours. It is recommended to finish Raman imaging within 2 h of replacement of the medium with Tyrode's solution.
5. Weak fluorescence background signal can be seen from water for a water immersion objective lens when the same water is used for a long time. It is recommended to use fresh distilled deionized water for each day of the experiment. Dust remaining on the surface of the water immersion objective lens can also be a source for fluorescence background signal. The objective lens should be thoroughly cleaned before and after the experiment with lens cleaning paper and ethanol.

6. To obtain alkyne's Raman signal from small molecules in cells, a typical exposure time per line is more than 10 s. It often takes several tenths of minutes for acquiring one Raman image of alkyne-tagged small molecules in cells. It is important to maintain the focus position of the objective lens during long imaging time. There are commercially available systems that automatically maintain the focus position, such as Perfect Focus System developed by Nikon. The system uses LED light that is focused on the surface of the glass substrate through the objective lens. Reflected light is used as a feedback system to maintain the distance between the glass substrate and the objective lens. It is valuable to use these systems to maintain the focus position during imaging. Wavelength of the LED, and the reflection wavelength region of the dichroic filter for LED, should be checked to avoid overlap of the wavelengths of LED and Raman scattering.
7. Synthesis of alkyne-tagged molecule: Various reactions are available to introduce alkyne tags into molecules, as described below. The Sonogashira reaction [25] is one of the most powerful reactions to introduce an alkyne tag into an aromatic ring. Indeed, EdU was synthesized by means of Sonogashira coupling with 5-iodo-2′-deoxyuridine [26]. To construct a diyne tag, acetylenic coupling reaction [27] is useful to introduce the additional alkyne into an alkyne tag. The Bestmann–Ohira reagent [28] can transform aldehyde into terminal alkyne on an aromatic ring or another alkyne.

Acknowledgment

This work was partially supported by JST-ERATO (Sodeoka Live Cell Chemistry Project), AMED-CREST (No. JP17gm0710004), RIKEN and JSPS KAKENHI Grant Number 26600117 (J.A.).

References

1. Parker FS (ed) (1983) Applications of Infrared, Raman, and Resonance Raman spectroscopy in biochemistry. Springer, Heidelberg
2. Carey P (1982) Biochemical applications of Raman and Resonance Raman spectroscopes. Academic Press, New York
3. Karthigeyan D, Siddhanta S, Kishore AH, Perumal SSRR, Ågren H, Sudevan S, Bhat AV, Balasubramanyam K, Subbegowda RK, Kundu TK, Narayana C (2014) SERS and MD simulation studies of a kinase inhibitor demonstrate the emergence of a potential drug discovery tool. Proc Natl Acad Sci U S A 111:10416–10421
4. Ando J, Asanuma M, Dodo K, Yamakoshi H, Kawata S, Fujita K, Sodeoka M (2016) Alkyne-tag SERS screening and identification of small-molecule-binding sites in protein. J Am Chem Soc 138:13901–13910
5. Huang YS, Karashima T, Yamamoto M, Hamaguchi H (2003) Molecular-level pursuit of yeast mitosis by time- and space-resolved Raman spectroscopy. J Raman Spectrosc 34:1–3
6. Palonpon AF, Ando J, Yamakoshi H, Dodo K, Sodeoka M, Kawata S, Fujita K (2013) Raman and SERS microscopy for molecular imaging of live cells. Nat Protoc 8:677–692

7. Ando J, Palonpon AF, Sodeoka M, Fujita K (2016) High-speed Raman imaging of cellular processes. Curr Opin Chem Biol 33:16–24
8. Hamada K, Fujita K, Smith NI, Kobayashi M, Inouye Y, Kawata S (2008) Raman microscopy for dynamic molecular imaging of living cells. J Biomed Opt 13:044027
9. Okada M, Smith NI, Palonpon AF, Endo H, Kawata S, Sodeoka M, Fujita K (2012) Label-free Raman observation of cytochrome *c* dynamic during apoptosis. Proc Natl Acad Sci U S A 109:28–32
10. Ling J, Weitman SD, Miller MA, Moore RV, Bovik AC (2002) Direct Raman imaging techniques for study of the subcellular distribution of a drug. Appl Opt 41:6006–6017
11. Yamakoshi H, Dodo K, Okada M, Ando J, Palonpon A, Fujita K, Kawata S, Sodeoka M (2011) Imaging of EdU, an alkyne-tagged cell proliferation probe, by Raman microscopy. J Am Chem Soc 133:6102–6105
12. Yamakoshi H, Dodo K, Palonpon A, Ando J, Fujita K, Kawata S, Sodeoka M (2012) Alkyne-tag Raman imaging for visualization of mobile small molecules in live cells. J Am Chem Soc 134:20681–20689
13. Ando J, Kinoshita M, Cui J, Yamakoshi H, Dodo K, Fujita K, Murata M, Sodeoka M (2015) Sphingomyelin distribution in lipid rafts of artificial monolayer membranes visualized by Raman microscopy. Proc Natl Acad Sci U S A 112:4558–4563
14. Lee J, Zhang W, Zhang D, Yang Y, Liu B, Barker EL, Buhman KK, Slipchenko LV, Dai M, Cheng JX (2015) Assessing cholesterol storage in live cells and C. elegans by stimulated Raman scattering imaging of phenyl-diyne cholesterol. Sci Rep 5:7930
15. Wei L, Hu F, Shen Y, Chen Z, Yu Y, Lin CC, Wang MC, Min W (2014) Live-cell imaging of alkyne-tagged small biomolecules by stimulated Raman scattering. Nat Methods 11:410–412
16. Hong S, Chen T, Zhu Y, Li A, Huang Y, Chen X (2014) Live-cell stimulated Raman scattering imaging of alkyne-tagged biomolecules. Angew Chem Int Ed 53:5827–5831
17. El-Mashtoly SF, Petersen D, Yosef HK, Mosig A, Reinacher-Schick A, Kötting C, Gerwert K (2014) Label-free imaging of drug distribution and metabolism in colon cancer cells by Raman microscopy. Analyst 139:1155–1161
18. Yamakoshi H, Palonpon A, Dodo K, Ando J, Kawata S, Fujita K, Sodeoka M (2015) A sensitive and specific Raman probe based on bisarylbutadiyne for live cell imaging of mitochondria. Bioorg Med Chem Lett 25:664–667
19. Ueda M, Egoshi S, Dodo K, Ishimaru Y, Yamakoshi H, Nakano T, Takaoka Y, Tsukiji S, Sodeoka M (2017) Noncanonical function of a small-molecular virulence factor coronatine against plant immunity: an in vivo Raman imaging approach. ACS Cent Sci 3:462–472
20. Chen Z, Paley DW, Wei L, Weisman AL, Friesner RA, Nuckolls C, Min W (2014) Multicolor live-cell chemical imaging by isotopically edited alkyne vibrational palette. J Am Chem Soc 136:8027–8033
21. Schneider CA, Rasband WS, Eliceiri KW (2012) NIH Image to ImageJ: 25 years of image analysis. Nat Methods 9:671–675
22. Manen HJ, Kraan YM, Roos D, Otto C (2004) Intracellular chemical imaging of heme containing enzymes involved in innate immunity using resonance Raman microscopy. J Phys Chem B 108:18762–18771
23. Lieber CA, Mahadevan-Jansen A (2003) Automated method for subtraction of fluorescence from biological Raman spectra. Appl Spectrosc 57:1363–1367
24. Mu ZM, Le XF, Vallian S, Glassman AB, Chang KS (1997) Stable overexpression of PML alters regulation of cell cycle progression in HeLa cells. Carcinogenesis 18:2063–2069
25. Doucet H, Hierso J-C (2007) Palladium-Based Catalytic Systems for the Synthesis of Conjugated Enynes by Sonogashira Reactions and Related Alkynylations. Angew Chem Int Ed 46:834–871
26. Shealy YF, O'Dell CA, Arnett G, Shannon WM (1986) Synthesis and antiviral activity of the carbocyclic analogues of 5-Ethyl-2-′-deoxyuridine and of 5-Ethynyl-2-′-deoxyuridine. J Med Chem 29:79–84
27. Siemsen P, Livingston RC, Diederich F (2000) Acetylenic coupling: a powerful tool in molecular construction. Angew Chem Int Ed 39:2632–2657
28. Pietruszka J, Witt A (2006) Synthesis of the Bestmann-Ohira reagent. Synthesis 24:4266–4268

Chapter 6

The Cell Painting Assay as a Screening Tool for the Discovery of Bioactivities in New Chemical Matter

Axel Pahl and Sonja Sievers

Abstract

Multiparametric phenotypic screening based on cellular morphology interrogates many biological pathways simultaneously and is therefore a valuable screening tool for the discovery of new biological activities. The cell painting assay stains various cellular features using six different dyes in one well. By automated image analysis, hundreds of parameters are calculated from the images which deliver a phenotypic profile of the cell. It has been shown that compounds with similar modes of action deliver similar phenotypic profiles. Using a reference set of compounds with known modes of action, it is possible to assign probable modes of action to new compounds and to discover compounds with potentially new modes of action.

Here we describe the cell painting assay as a screening tool using a hit identification workflow which has been implemented using open-source software.

Key words Cell painting assay, Phenotypic profile, Screening, Hit identification, Morphological profiling

1 Introduction

Traditionally, high-throughput screening (HTS) has been divided into target-based and phenotypic approaches. Target-based screens are often carried out in in vitro set ups and therefore suffer from a lack of physiological relevance, while offering excellent opportunities for hit optimization and hit-to-lead development especially when structural information on the target is available. Phenotypic screens are in most cases carried out in a target-agnostic manner and offer a higher degree of physiological relevance, especially when pluripotent stem cell models are used [1]. On the flipside, hit optimization may be difficult and target identification via for example proteomic methods is a huge effort requiring many resources [2]. While phenotypic screens offer the chance to analyze the cell as a whole, investigators still often choose to limit the readout to a single or only a few parameters, often picked because of special significance for a pathway of interest [3]. As a consequence, most of

Slava Ziegler and Herbert Waldmann (eds.), *Systems Chemical Biology: Methods and Protocols*, Methods in Molecular Biology, vol. 1888, https://doi.org/10.1007/978-1-4939-8891-4_6,

the information which lies in the phenotype is disregarded in conventional phenotypic screening.

Multiparameter phenotypic screens based on cellular morphology hold promise in enabling unbiased screening for the discovery of (new) bioactivities and have the potential to replace a whole fleet of single parameter assays. Thus multiparametric phenotypic screening might be more efficient than conventional screening [4].

The comprehensive description of hundreds of cellular features generates a phenotypic profile of a chemical compound. By comparing phenotypic profiles between agents with related modes of action, it could be shown that similar perturbations lead to similar changes in cellular morphology. For example, structurally unrelated tubulin inhibitors or HDAC inhibitors, respectively, were found to induce particular phenotypes and cluster together after hierarchical clustering analysis [5, 6]. Multiparametric phenotypic screening also has the potential to distinguish between new and already known bioactivities when a set of annotated reference compounds is screened alongside the new compound collection. For screening compounds with a phenotypic profile similar to a reference profile, multiparametric screening can deliver information about possible targets or mechanisms of action. Any compound with a phenotypic profile that significantly differs from the DMSO control but shows little or no similarity to the reference profiles of known compounds potentially modulates a new kind of bioactivity. It must be noted that in every case the screening results depend heavily on the reference compound set used. Focused libraries with drugs or drug candidates like the LOPAC library, the Prestwick Chemical library or the US-Drug collection are mostly used for this purpose [7]. However, it should be pointed out that drugs often do not only have one cellular target but that therapeutic efficacy may be linked to the modulation of several targets [8, 9]. Compound annotations seldom contain this polypharmacology information. Thus, it may be very well the case that an observed phenotype might be derived from the modulation of a target other than the annotated one [10].

The cell painting assay (CPA) has been described as a morphological profiling tool for the generation of multiparametric profiles of cells [5]. While not being the first example of a morphological multiparametric profiling assay [11], CPA has the potential to be applied to large compound sets [12] because of its relative ease of use and fairly low costs associated with the used dyes as opposed to antibody-based staining in other approaches [6]. It is also highly versatile in the adaptation to different cell lines. CPA is highly multiplexed, it highlights various cellular features by staining eight cellular compartments (nucleus, nucleoli, mitochondria, ER, Golgi, plasma membrane, actin cytoskeleton, cytosolic RNA) in five different fluorescent channels in one well [13]. By subsequent image analysis hundreds of parameters are deduced from every channel which are combined into the morphological profile of the cell.

We combined the cell painting assay with a hit identification workflow to efficiently identify bioactive compounds in large compound sets.

2 Materials

2.1 Cell Staining

1. U2OS cells.
2. DMEM containing 10% fetal bovine serum, 1% L-glutamine, 1% sodium pyruvate, and 1% nonessential amino acids.
3. Black clear-bottomed 384 well plates.
4. Automated Dispenser: Multidrop Combi (Thermo).
5. Acoustic dispensing machine: Echo 520 dispenser (Labcyte Inc.).
6. Automated cell washer: Elx405 (Biotek).
7. Dulbecco's Phosphate-buffered saline (PBS): 137 mM NaCl, 2.7 mM KCl, 10 mM Na_2HPO_4*2 H_2O, 2 mM KH_2PO_4 in ultrapure water H_2O, adjust pH to 7.2 to 7.4 and autoclave.
8. MitoTracker staining solution: dilute MitoTracker stock solution (1 mM) to 100 nM with DMEM containing 10% fetal bovine serum freshly every screening day.
9. Fixing solution: dilute formaldehyde (37% stock solution) to 18.5% with PBS freshly very screening day.
10. Permeabilization solution: 0.1% (v/v) Triton X-100 in PBS.
11. Staining solution: 5 μg/ml Hoechst 33342, 1.5 μg/ml WGA, 25 μg/ml concanavalin A, 5 μl/ml phalloidin solution, 1.5 μM SYTO 14, and 1% BSA in PBS (*see* **Notes 1** and **2**). Prepare freshly every screening day.
12. Self-adhesive aluminum foil.
13. Reference set of compounds: e.g., LOPAC (Library of Pharmacologically Active Compounds) (Sigma), Prestwick Chemical Library (Prestwick Chemical), US-Drug collection (MicroSource Discovery Systems), Selleckchem kinase inhibitor library (Selleckchem), the published kinase inhibitor set (PKIS [14]).
14. Automated microscope: ImageXpress Micro XL (Molecular Devices) with filter sets for DAPI, FITC, SybrGold, TxRed, and Cy5 (*see* Table 1).
15. CellProfiler (http://cellprofiler.org/ and https://github.com/CellProfiler/CellProfiler last accessed 18 July 2017).

Table 1
Overview about the different dyes and filter sets used for the cell painting assay

Dye	Cellular component(s)	Filter set	Ex/nm	Em/nm
Hoechst 33342	Nucleus	DAPI	350–400	410–480
Concanavalin A/Alexa Fluor 488 conjugate	Endoplasmic reticulum	FITC	470–500	510–540
SYTO 14 green fluorescent nucleic acid stain	Nucleoli, cytoplasmic RNA	Spectrum Gold	520–545	560–585
Phalloidin/Alexa Fluor 568	F-actin cytoskeleton	TxRed	535–585	600–650
Wheat-germ agglutinin/Alexa Fluor 555	Golgi, plasma membrane	TxRed	535–585	600–650
MitoTracker Deep Red	Mitochondria	Cy5	605–650	670–715

2.2 Software

1. Python 2.7 (https://www.python.org/; for CellProfiler on the Cluster in an Anaconda (https://www.continuum.io/downloads) environment),
2. Python 3.5 (for the processing workflow).
3. Nim (https://nim-lang.org/; for small scripts on the cluster).
4. Jupyter (http://jupyter.org/; as execution environment for the data processing workflow).
5. Pandas (http://pandas.pydata.org/; Python Data Analysis Library).
6. RDKit (http://rdkit.org/; Cheminformatics toolkit for working with chemical structures).

All last accessed 18 July 2017.

3 Methods

3.1 Cell Culture and Compound Treatment

1. Grow U2OS cells in supplemented DMEM. Split twice a week, using 600 cells per cm^2 for reseeding of the cells.
2. Seed U2OS cells at 1600 cells/ well in 25 μl medium in 384 well plates using a multichannel pipette or automated dispensing and incubate for 4 h at 37 °C and 5% CO_2 (*see* **Notes 3–5**).
3. Add compounds to a final screening concentration of 10 μM, for example using acoustic dispensing (*see* **Note 6**).
4. Incubate cells with compounds for approx. 20 h at 37 °C with 5% CO_2.

3.2 Live Cell Staining

Mitochondrial staining has to be performed in live cells. MitoTracker dyes diffuse passively across membranes and accumulate in active mitochondria. Most of the MitoTracker dyes are also well retained in the mitochondria after fixation.

1. For the live cell staining, aspirate medium from the plates leaving ca. 10 μl of residual volume using an automated cell washer.
2. Add 25 μl of MitoTracker staining solution using a multichannel pipette or automated dispensing.
3. Incubate plates for 30 min in the dark at 37 °C.

3.3 Fixation and Permeabilization

1. Fix the cells by addition of 7 μl of fixing solution to give a final concentration of 3.1% (v/v) formaldehyde using a multichannel pipette or automated dispensing and incubate at room temperature for 20 min (*see* **Note 7**).
2. Wash the plates three times with 70 μl PBS with final aspiration using an automated washer.
3. Permeabilize the cells by addition of 25 μl of permeabilization solution using a multichannel pipette or automated dispensing and incubate in the dark at room temperature for 10–20 min.
4. Wash the plates three times with 70 μl PBS with final aspiration using an automated washer.

3.4 Staining

1. Stain the cells by addition of 25 μl of staining solution to each well using a multichannel pipette or automated dispensing and incubate in the dark at room temperature for 30 min.
2. Wash the plates three times with 70 μl PBS with final aspiration using an automated washer.
3. Seal the plate with self-adhesive aluminum foil.

3.5 Imaging

1. Image the plate at 20× magnification using an automated microscope with filter sets for the 5 wavelengths as shown in Table 1. Take nine pictures per well to cover a large number of cells (>1000 cells in total) (*see* **Notes 8** and **9**).

3.6 Image Analysis

CellProfiler is an open source image analysis software which is fully compatible with high-throughput screening [15]. CellProfiler offers a modular approach to single cell image analysis even to non-image analysis experts. Different modules for image processing, object detection and measurements are combined into a pipeline. Measured parameters include size, shape, intensity, and texture features of the defined objects. For a small number of plates CellProfiler can easily be run on a desktop PC. However, for HTS mode CellProfiler needs to be installed and run on a cluster which might require the help of IT experts.

1. Analyze the images with CellProfiler software using the analysis pipeline in https://github.com/mpimp-comas/cellpainting. This pipeline was based on Bray et al. [13] with slight modifications. First, images are loaded into the pipeline and an illumination correction is performed. Then, object detection is carried out: primary objects (nuclei) are detected in the Hoechst channel, secondary objects (cytoplasm) are identified from the TxRed channel using a watershed algorithm and tertiary objects (cells) are defined as the sum of primary and secondary objects. CellProfiler then measures more than 2300 parameters for each individual cell. During Cell Profiler analysis, parameters are aggregated by the median values of the microscope sites, resulting in 9 values per microtiter well for every parameter.
2. For high-throughput analysis, install CellProfiler on a cluster. In our case, the processing of each image set from one plate is distributed over 96 parallel jobs. The processing of one plate takes approx. 5 h. At the end of a distributed processing job on the cluster, the result files from the individual processing jobs are concatenated to one result file per plate of 3456 lines (384 wells × 9 sites) and 1937 columns (~90 MB). The CellProfiler pipeline is exported und reused in every run on the cluster (*see* **Notes 10** and **11**).

3.7 Data Mining

To further develop CPA into a screening assay, we established a data mining procedure to transform phenotypic profiles into a direct measure of compound activity ("CPA activity score") without losing similarity information gained by reference compounds (Pahl et al., manuscript in preparation). For the CPA activity score, first a phenotypic profile is established using median per well values of the 307 parameters that were found to be nonredundant (for selection of parameters *see* Subheading 2). For profile comparison, we assign equal weight to all parameters and therefore, the absolute change in the parameters is not taken into account. Rather, for every parameter it is checked if it is identical to the DMSO control (activity code = 1) or if it has a significantly (>7.5 × Median Absolute Deviation, MAD) lower or higher numeric value than the DMSO control (activity code = 0 or activity code = 2, respectively). All activity codes are concatenated resulting in a 307 digit-activity string. The CPA activity score is then calculated from the sum of parameters with an activity code different from 1 divided by the total number of parameters and is expressed in percent. Thus, the CPA activity score indicates in how many parameters the phenotypic profile of a compound differs from the DMSO profile.

The data mining functionalities (Parts 1–4) are implemented as a Python module and the workflows are run in Jupyter Notebooks. All scripts have been uploaded to GitHub (*see* **Note 10**).

1. Data preparation

   First, the data is prepared for further analysis (*see* **Note 12**). In this script, the values for every parameter of each well are grouped on the median, skipped wells from the Echo compound transfer are removed, the well type (Control/Compound) is assigned by their position on the plate (DMSO controls are in columns 11 and 12) and results for toxic compounds are flagged (a compound is considered toxic when it has less than 50% of the median cell count of the controls) (*see* **Note 13**).

   Finally, data from all plates of a screening batch is concatenated into one result.

2. Finding relevant parameters

   In this step, the relevant, uncorrelated parameters that are used for calculation of the activity score are determined from the compound set, from which toxic compounds have been removed (*see* **Note 14**).

   The measured parameters are first filtered for their relevance, that is, parameters are removed that among the controls have a standard deviation of less than 0.05% or more than 2.5%, effectively removing parameters with constant values or very small variability as well as parameters with high variance already in the controls which would make them unsuitable for characterizing test compounds.

   For our data set this results in 773 parameters, which are then submitted to a correlation filter which filters out parameters that are highly correlated to each other. A correlation matrix is generated; the parameters which are correlated to the highest number of other parameters are analyzed and only the parameter with the lowest standard deviation is kept and the other parameters are removed. From the remaining parameters, a new correlation matrix is generated and the process is repeated until no correlated parameters remain in the data set, leading to a set of 307 uncorrelated parameters at a cut-off of 0.9 for the correlation factor.

3. Calculation of CPA activity score for the reference set

   In this part the phenotypic profiles are calculated for the reference compounds from the set of relevant parameters. The phenotypic profile is a fingerprint-like property that is calculated from each of the relevant parameters as follows: if the value of a compound for a specific parameter is higher (or lower) than the median of the controls for that parameter +/−7.5 times the MAD, then the fingerprint at that position gets assigned a 2 (or 0). Otherwise a 1 is assigned for inactivity. The resulting fingerprint is a string with the length of the number of relevant parameters that at each position has either a 0 (lower than control), a 1 (inactive) or a 2 (higher than control). Then, the activity score is calculated which consists of the absolute number of active parameters in the activity profile divided by the total number of parameters, and is expressed in percent.

This categorization of parameters makes it very easy to compare compounds by their activity profiles (*see* **Note 15**). The profile similarity score is calculated by the Tanimoto method [16] as follows: two activity profiles are compared position by position. For all significant parameters (with value 0 or 2) that have the same value at the same position, a counter is incremented. The similarity is then the final value of the counter divided by the total number of significant positions from both profiles, resulting in values from 0 (different in every position) to 1 (completely identical).

The script for the calculation of the CPA activity score of the reference compounds first processes the raw data as described in **step 1**. In addition, the plate layout is joined (linking plate position and Compound_ID), the structures are joined as Smiles, low-purity or toxic reference compounds are removed, activity profiles and activity scores are calculated for the determined relevant parameter set, annotations (trivial names, known activities) are joined to the reference compound set, the data set is saved to a TSV file (tab-separated).

4. Comparison of the screening compounds to the reference compounds

   A screening plate containing test compounds is processed as in Subheading 3 whereby removal of impure and toxic compounds is omitted. For active compounds (minimum activity score of 5%), profile similarity scores are calculated.
   Finally, a report is generated for each test compound. If a test compound shows similarity to a reference compound (minimum similarity score of 25%), a list of reference compounds and their annotations is also reported, otherwise it is noted that no similar references could be found.

4 Notes

1. As a cheaper alternative, Concanavalin A FITC conjugate can also be used. We used it at the same concentration as the Alexa 488 conjugate and achieved similar staining results.
2. As a cheaper alternative, MFP-DY-594-Phalloidin conjugate (Dyomics) can also be used. We used it at the same

concentration as the Alexa 594 conjugate and achieved similar staining results.

3. For most gentle cell seeding, standard tube cassettes should be used with dispensing speed set to "medium."
4. Even cell seeding can be achieved by first dispensing 5 μl medium only, followed by a second dispensing step of 20 μl cell suspension. After that, the plate should be left at room temperature for 10–20 min before putting it into the incubator [17].
5. CellCarrier Ultra plates outperformed all other plates tested because their bottom is highly even and no drift in focusing was observed across the images.
6. When screening focused libraries of for example kinase inhibitors or certain classes of natural compounds, which tend to be toxic at high concentrations, the screening concentration should be reduced to 1–2 μM. At a screening concentration of 10 μM for known drugs and 2 μM for kinase inhibitors, we found 39% of our reference compound set to be active (activity score >10%).
7. Formaldehyde should be freshly diluted to 18.5% from the 37% stock every screening day.
8. Exposure times may vary for the different dyes. Make sure to not saturate your images. We usually choose the brightest objects to have an intensity of 10,000 of the 64,000 grey shades.
9. If possible, use camera binning set to 2. This will quarter image size and reduce storage space as well as data transfer and processing times.
10. The setup of CellProfiler on a cluster may need the help of IT experts. It is recommended to install the CellProfiler Python program in an Anaconda virtual environment directly from the CellProfiler GitHub sources. Scripts and helper tools to run parallel jobs for the Sun Grid Engine (SGE) can be found on the COMAS GitHub page (https://github.com/mpimp-comas/cellpainting) and need to be adapted to the individual setup.

    After the parallel processing of the images by CellProfiler and the concatenation of the results on the cluster, the result data is transferred to a normal desktop computer on which all downstream processing and reporting is performed.
11. CellProfiler offers multiple ways to handle configurations and to output results. For parallel processing on a cluster, we found the best solution to be working with *.cppipe configuration files and to generate the result data in plain CSV (or TSV) files. The CellProfiler pipeline (*.cppipe) is generated by

configuring and exporting a workflow in the graphical interface of CellProfiler which is accessed on the cluster via VNC client.

12. To construct a phenotypic profile for a compound from the single cell data of the primary image analysis, the data needs to be normalized (to account for plate and batch differences) and aggregated: different statistical methods can be used to combine data on the image level, the well level (several sites are imaged per well) and the compound level (if replicates are measured). Comparison of different statistical methods showed that using a simple mean or median value per well already delivers robust results, even for heterogeneous phenotypes [18, 19]. However, improvements in the accuracy of classification of a set of test compounds could be achieved using the median combined with median absolute deviation (MAD) or percentile values [18, 19].
13. We chose the toxicity cutoff at <50% of the median cell count of the DMSO control. Admittedly, a potent antimitotic compound could also lead to such a reduction in cell count over 20 h compound incubation time. However, using a screening test set of 1400 proprietary compounds only 4% of compounds were flagged to be toxic and in our reference set of 4220 compounds only 3.3% of compounds were flagged to be toxic.
14. During image analysis, several hundred parameters are calculated for every cell delivering the basis for a phenotypic profile. Many approaches for the generation of phenotypic profiles (also known as fingerprints) have been described. Thereby, some approaches work with a high number of individual parameters; while others use dimension reduction or feature selection methods to come up with a screening result that can be biologically interpreted and is more accessible to the human mind [20, 21]. Reducing the number of highly correlated parameters can be beneficial in subsequent analysis, for example, in reducing processing times or because of higher importance of the chosen parameters which in turn can result in a more stable phenotypic classification [19]. However, dimension reduction or feature selection also bears the risk of losing information which could be valuable for phenotype discrimination. This is especially true for unbiased screening with an undefined phenotype; thus it is expected that for such approaches a high number of parameters should be included into the phenotypic profile to cover a broad range of phenotypes that might be induced by the screening compounds. In every case caution should be applied to preserve as much of the variance as possible.
15. Phenotypic profiles of the test compounds need to be compared to those of DMSO controls and reference compounds.

Because of the multiparametric nature of the phenotypic profiles, multivariate statistical measures are often applied to detect relevant differences and similarities between phenotypic profiles. For small molecule screening purposes, the Mahalanobis distance has often been used [22]. A study comparing different distance- and correlation-based methods found that correlation-based methods work best for different purposes while distance-based methods are stronger in detecting all phenotypes different from the neutral controls [23]. Multivariate similarity measures are also a prerequisite for any kind of clustering approaches which group similar phenotypic profiles together. However, clustering workflows usually do not implement an activity threshold which is needed for screening purposes to differentiate active from inactive compounds.

To overcome this problem, an arbitrary cutoff has been applied to the Mahalanobis distance to define a hit [22]. In another effort to create an activity score for phenotypic profiles, a "multidimensional perturbation value" (mp-value) has been designed to define a hit [24, 25]. Mp-values are significance measurements that are calculated from Mahalanobis distances by permutation analysis. Like for nominal significance tests, a cutoff of 0.05 can be applied as a cutoff for active compounds.

References

1. Vincent F, Loria P, Pregel M, Stanton R, Kitching L, Nocka K, Doyonnas R, Steppan C, Gilbert A, Schroeter T, Peakman MC (2015) Developing predictive assays: the phenotypic screening "rule of 3". Sci Transl Med 7(293):293ps215. https://doi.org/10.1126/scitranslmed.aab1201
2. Vendrell-Navarro G, Brockmeyer A, Waldmann H, Janning P, Ziegler S (2015) Identification of the targets of biologically active small molecules using quantitative proteomics. Methods Mol Biol 1263:263–286. https://doi.org/10.1007/978-1-4939-2269-7_21
3. Singh S, Carpenter AE, Genovesio A (2014) Increasing the content of high-content screening: an overview. J Biomol Screen 19(5):640–650. https://doi.org/10.1177/1087057114528537
4. Pennisi E (2016) IMAGING. 'Cell painting' highlights responses to drugs and toxins. Science 352(6288):877–878. https://doi.org/10.1126/science.352.6288.877
5. Gustafsdottir SM, Ljosa V, Sokolnicki KL, Anthony Wilson J, Walpita D, Kemp MM, Petri Seiler K, Carrel HA, Golub TR, Schreiber SL, Clemons PA, Carpenter AE, Shamji AF (2013) Multiplex cytological profiling assay to measure diverse cellular states. PLoS One 8(12):e80999. https://doi.org/10.1371/journal.pone.0080999
6. Reisen F, Sauty de Chalon A, Pfeifer M, Zhang X, Gabriel D, Selzer P (2015) Linking phenotypes and modes of action through high-content screen fingerprints. Assay Drug Dev Technol 13(7):415–427. https://doi.org/10.1089/adt.2015.656
7. Jones LH, Bunnage ME (2017) Applications of chemogenomic library screening in drug discovery. Nat Rev Drug Discov 16(4):285–296. https://doi.org/10.1038/nrd.2016.244
8. Boran AD, Iyengar R (2010) Systems approaches to polypharmacology and drug discovery. Curr Opin Drug Discov Devel 13(3):297–309
9. Hu Y, Bajorath J (2013) High-resolution view of compound promiscuity. F1000 Res 2:144. https://doi.org/10.12688/f1000research.2-144.v2
10. Ursu A, Illich DJ, Takemoto Y, Porfetye AT, Zhang M, Brockmeyer A, Janning P, Watanabe N, Osada H, Vetter IR, Ziegler S, Scholer HR, Waldmann H (2016) Epiblastin A

induces reprogramming of epiblast stem cells into embryonic stem cells by inhibition of casein kinase 1. Cell Chem Biol 23 (4):494–507. https://doi.org/10.1016/j.chembiol.2016.02.015

11. Perlman ZE, Slack MD, Feng Y, Mitchison TJ, Wu LF, Altschuler SJ (2004) Multidimensional drug profiling by automated microscopy. Science 306(5699):1194–1198. https://doi.org/10.1126/science.1100709
12. Bray MA, Gustafsdottir SM, Ljosa V, Singh S, Sokolnicki KL, Bittker JA, Bodycombe NE, Dancik V, Hasaka TP, Hon CS, Kemp MM, Li K, Walpita D, Wawer MJ, Golub TR, Schreiber SL, Clemons PA, Shamji AF, Carpenter AE (2017) A dataset of images and morphological profiles of 30,000 small-molecule treatments using the Cell Painting assay. Gigascience 6 (12):1–5. https://doi.org/10.1093/gigascience/giw014
13. Bray MA, Singh S, Han H, Davis CT, Borgeson B, Hartland C, Kost-Alimova M, Gustafsdottir SM, Gibson CC, Carpenter AE (2016) Cell Painting, a high-content image-based assay for morphological profiling using multiplexed fluorescent dyes. Nat Protoc 11 (9):1757–1774. https://doi.org/10.1038/nprot.2016.105
14. Knapp S, Arruda P, Blagg J, Burley S, Drewry DH, Edwards A, Fabbro D, Gillespie P, Gray NS, Kuster B, Lackey KE, Mazzafera P, Tomkinson NC, Willson TM, Workman P, Zuercher WJ (2013) A public-private partnership to unlock the untargeted kinome. Nat Chem Biol 9(1):3–6. https://doi.org/10.1038/nchembio.1113
15. Carpenter AE, Jones TR, Lamprecht MR, Clarke C, Kang IH, Friman O, Guertin DA, Chang JH, Lindquist RA, Moffat J, Golland P, Sabatini DM (2006) CellProfiler: image analysis software for identifying and quantifying cell phenotypes. Genome Biol 7(10):R100. https://doi.org/10.1186/gb-2006-7-10-r100
16. Rogers DJ, Tanimoto TT (1960) A Computer program for classifying plants. Science 132 (3434):1115–1118. https://doi.org/10.1126/science.132.3434.1115
17. Lundholt BK, Scudder KM, Pagliaro L (2003) A simple technique for reducing edge effect in cell-based assays. J Biomol Screen 8 (5):566–570. https://doi.org/10.1177/1087057103256465
18. Ljosa V, Caie PD, Ter Horst R, Sokolnicki KL, Jenkins EL, Daya S, Roberts ME, Jones TR, Singh S, Genovesio A, Clemons PA, Carragher NO, Carpenter AE (2013) Comparison of methods for image-based profiling of cellular morphological responses to small-molecule treatment. J Biomol Screen 18 (10):1321–1329. https://doi.org/10.1177/1087057113503553
19. Kummel A, Selzer P, Beibel M, Gubler H, Parker CN, Gabriel D (2011) Comparison of multivariate data analysis strategies for high-content screening. J Biomol Screen 16 (3):338–347. https://doi.org/10.1177/1087057110395390
20. Twarog NR, Low JA, Currier DG, Miller G, Chen T, Shelat AA (2016) Robust classification of small-molecule mechanism of action using a minimalist high-content microscopy screen and multidimensional phenotypic trajectory analysis. PLoS One 11(2):e0149439. https://doi.org/10.1371/journal.pone.0149439
21. Young DW, Bender A, Hoyt J, McWhinnie E, Chirn GW, Tao CY, Tallarico JA, Labow M, Jenkins JL, Mitchison TJ, Feng Y (2008) Integrating high-content screening and ligand-target prediction to identify mechanism of action. Nat Chem Biol 4(1):59–68. https://doi.org/10.1038/nchembio.2007.53
22. Gerry CJ, Hua BK, Wawer MJ, Knowles JP, Nelson SD Jr, Verho O, Dandapani S, Wagner BK, Clemons PA, Booker-Milburn KI, Boskovic ZV, Schreiber SL (2016) Real-time biological annotation of synthetic compounds. J Am Chem Soc 138(28):8920–8927. https://doi.org/10.1021/jacs.6b04614
23. Reisen F, Zhang X, Gabriel D, Selzer P (2013) Benchmarking of multivariate similarity measures for high-content screening fingerprints in phenotypic drug discovery. J Biomol Screen 18(10):1284–1297. https://doi.org/10.1177/1087057113501390
24. Hutz JE, Nelson T, Wu H, McAllister G, Moutsatsos I, Jaeger SA, Bandyopadhyay S, Nigsch F, Cornett B, Jenkins JL, Selinger DW (2013) The multidimensional perturbation value: a single metric to measure similarity and activity of treatments in high-throughput multidimensional screens. J Biomol Screen 18 (4):367–377. https://doi.org/10.1177/1087057112469257
25. Wawer MJ, Li K, Gustafsdottir SM, Ljosa V, Bodycombe NE, Marton MA, Sokolnicki KL, Bray MA, Kemp MM, Winchester E, Taylor B, Grant GB, Hon CS, Duvall JR, Wilson JA, Bittker JA, Dancik V, Narayan R, Subramanian A, Winckler W, Golub TR, Carpenter AE, Shamji AF, Schreiber SL, Clemons PA (2014) Toward performance-diverse small-molecule libraries for cell-based phenotypic screening using multiplexed high-dimensional profiling. Proc Natl Acad Sci U S A 111 (30):10911–10916. https://doi.org/10.1073/pnas.1410933111

Chapter 7

Proteomic Profiling for Target Identification of Biologically Active Small Molecules Using 2D DIGE

Makoto Muroi and Hiroyuki Osada

Abstract

Recent improvements in technologies such as omics analysis have enabled us to acquire a large amount of data regarding the biological changes in cells treated with bioactive small molecules. Using such data, a variety of profiling methods have been established for target identification of such bioactive compounds. In this chapter, we describe a proteomic profiling system, ChemProteoBase, based on proteome analysis using two-dimensional difference gel electrophoresis. This system compares the similarities in protein expression of 296 spots detected in the gel among the test compounds.

Key words Proteomics, Two-dimensional florescent electrophoresis (2D DIGE), Profiling, Target identification of compounds, Multivariate analysis

1 Introduction

Target identification of compounds is one of the most important steps in chemical biology research. However, it is a laborious and time-consuming process because we have to identify the target of a compound from a wide variety of candidates among cellular molecules. There are two fundamentally different approaches for target identification of new compounds. One directly detects the interaction between a compound and its target and involves processes such as analysis of the binding protein using chemical beads. The other is the indirect approach, which identifies the compound's target based on the phenotype induced by the compound. Recent improvements in technologies such as omics analysis has enabled us to acquire a large amount of data regarding biological changes induced by active compounds; using these data, a variety of profiling methods have been established [1]. The differential sensitivity of cancer cells to drugs is used for target profiling methods such as NCI-60 [2] and JFCR39 [3] cell line panels. Cytotoxicity profiling along with transcriptome, proteome, and metabolome

Slava Ziegler and Herbert Waldmann (eds.), *Systems Chemical Biology: Methods and Protocols*, Methods in Molecular Biology, vol. 1888, https://doi.org/10.1007/978-1-4939-8891-4_7, © Springer Science+Business Media, LLC, part of Springer Nature 2019

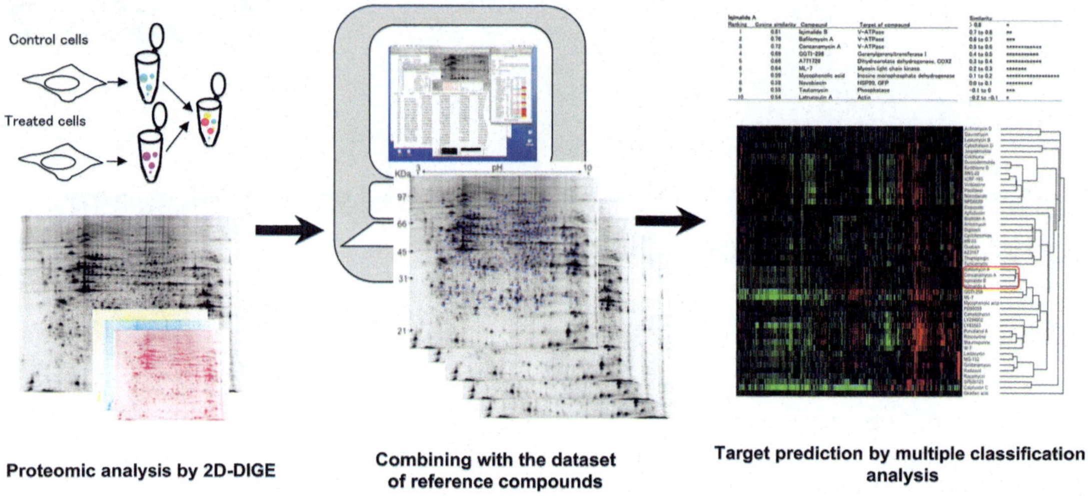

Fig. 1 Outline of compound target prediction using the proteomic profiling system ChemProteoBase

analyses and cell morphological analysis have been utilized for the profiling to identify a compound's target.

However, these profiling methods are an indirect approach; therefore, the results obtained using these profiling methods have to be validated using different methods such as affinity beads experiments or enzyme assays. The development of a direct approach to detect the binding between a small molecule and a protein is very important, but some problems related to the methodology such as nonspecific binding or binding not related to the activity remain a challenge. Indirect approaches are based on the biological change as a result of the activity of a compound. Moreover, the target of a compound may not be limited to a protein. Therefore, indirect approaches are useful to complement direct approaches. In this chapter, we describe a proteomic profiling system, ChemProteoBase (Fig. 1), which is an indirect approach for target identification of bioactive compounds [4, 5].

ChemProteoBase is based on proteome analysis using two-dimensional difference gel electrophoresis (2D DIGE). The method for proteomic analysis is classified into two approaches: a gel-dependent approach and a gel-independent mass-based approach. Recently, gel-independent mass-based approaches, which generally analyze digested peptides using LC-MS/MS, have been greatly improved and include isobaric tags for relative and absolute quantification (iTRAQ) and stable isotope labeling by amino acids in cell culture (SILAC); these approaches have been widely used for quantitative proteomics. On the other hand, gel-dependent proteomic analysis methods including two-dimensional gel electrophoresis (2-DE) and differential 1D gel analysis using affinity beads have been used conventionally, but they remain a powerful tool for the quantitative analysis of

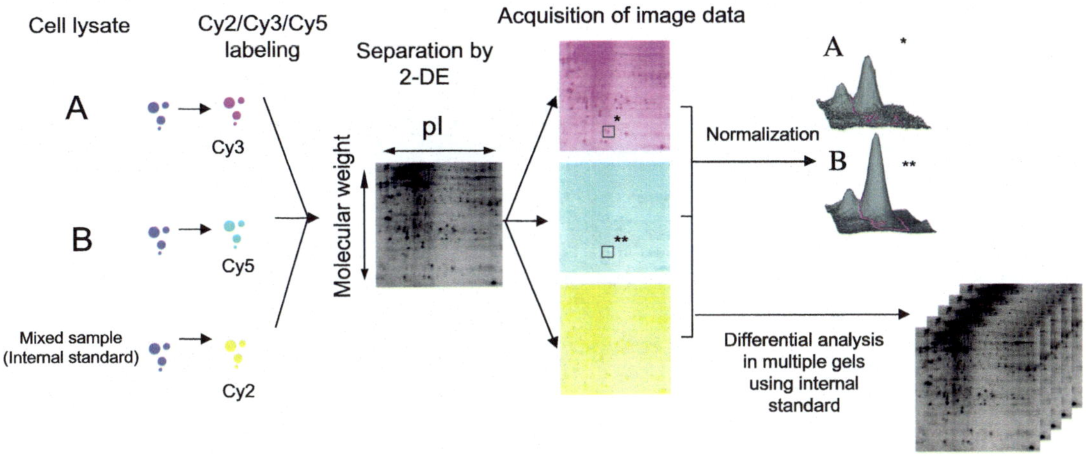

Fig. 2 Two-dimensional fluorescent electrophoresis (2D DIGE)

proteins with high-resolution separation in complex biological samples. A limited overlap of the results obtained using iTRAQ and 2-DE has been reported in previous studies, which indicates that both these approaches are complementary [6, 7]. Advances in 2-DE (i.e., 2D DIGE) have allowed the measurement of the abundance of each protein spot reproducibly [8]. In 2D DIGE, protein lysates are labeled with different fluorescent dyes such as Cy3 and Cy5 and because mixed samples are analyzed in a gel, the ratio of spots between samples can be accurately measured. Further, using the same lysate labeled with Cy2 as an internal standard, the abundance of each protein spot between different gels can be accurately measured [9] (Fig. 2). With 2D DIGE, reproducible abundant proteomic expression data after compound treatment can be collected in a dataset, and patterns can be compared. ChemProteoBase compares the similarities in the protein expression of 296 spots detected in a gel among test compounds (Figs. 1 and 3). Many well-known inhibitors are classified by cluster analysis according to their mechanism of action, which can be estimated if the test compound induces a proteomic profile similar to that induced by the reference compound in the database. To compare the data between separate experiments in a database, the data for the test compound are used as the log-fold ratio of the mean values in triplicate, which is normalized by using a DMSO-control in each experiment.

We have been screening new compounds with novel biological activity using a cell-based screening system. ChemProteoBase has been utilized for the target identification of the new compounds such as BNS-22 (topoisomerase II inhibitor) [10], pyrrolizilactone (proteasome inhibitor) [11], and GN39482 (tubulin inhibitor) [12]. Collismycin A, a microbial product, has antiproliferative activity against cancer cells. The use of ChemProteoBase revealed

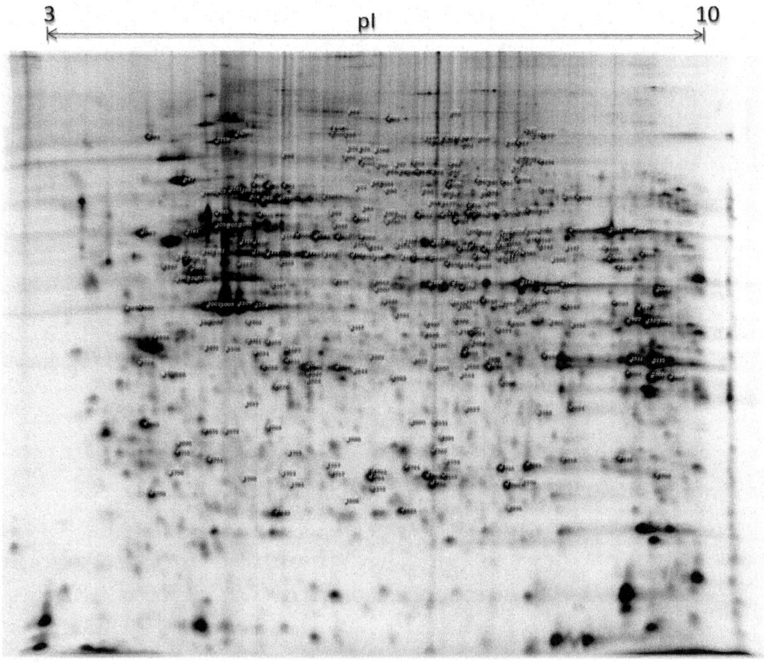

Fig. 3 The master gel image of ChemProteoBase including 296 spots. The 296 spots are reproducibly detected in every experiment and the master gel including the spots is used for matching the spots of all other gels in order to analyze the 296 spots

that collismycin A acts as a specific iron chelator, which leads to the inhibition of cancer cell growth [13]. Our findings showed that ChemProteoBase can be applied for nonprotein targets as well as protein targets. Furthermore, we identified tubulin as the off-target of TH287, which is known to be an inhibitor of MTH1 [14]. Together, our findings indicate that proteomic profiling such as that using ChemProteoBase is useful for target identification.

2 Materials

Prepare all solutions using ultrapure water (resistivity at 25 °C is more than 18 MΩ cm) and analytical grade reagents.

2.1 Determination of the Concentration of the Test Compounds

1. HeLa cells (*see* **Note 1**).
2. HeLa cell culture medium: High-glucose (4.5 g/L) Dulbecco's Modified Eagle Medium (DMEM) supplemented with 10% (v/v) fetal bovine serum (FBS), 4 mM L-glutamine, and 1 mM pyruvate. Add 50 mL of FBS, 2.5 mL of penicillin/streptomycin (10,000 U/mL penicillin and 10,000 μg/mL streptomycin) to 500 mL of high-glucose DMEM. Mix and store at 4 °C for up to 1 month.

3. Phosphate-buffered saline (PBS): 10 mM Na_2HPO_4, 137 mM NaCl, 27 mM KCl, 8.1 mM Na_2HPO_4, and 14.7 mM KH_2PO_4, pH 7.4. Add 8.0 g of NaCl, 2.0 g of KCl, 29.0 g of $Na_2HPO_4 \cdot 12H_2O$, and 2.0 g of KH_2PO_4 to ultrapure water. Adjust the pH to 7.4 using 10 N NaOH and sterilize in an autoclave.
4. Trypsin solution: 0.25% trypsin in PBS.
5. Cell culture dishes (100 mm diameter).
6. Cell culture microplates (96-well, flat bottom).
7. Commercial WST-8 reagent for cell counting (mixed solution of WST and 1-Methoxy PMS).
8. A microplate reader capable of measuring absorbance at 450 nm.
9. Test compound dissolved in DMSO.

2.2 Preparation of Cell Lysates

1. Cell culture dishes (60 mm diameters).
2. Cell scrapers.
3. Lysis buffer: 30 mM Tris, 8 M urea, 4% CHAPS, pH 8.5. Add 0.363 g of tris(hydroxymethyl)aminomethane (Tris), 48.0 g of urea, and 4.0 g of CHAPS to about 100 mL of ultrapure water. Adjust the pH to 8.5 using 1 N HCl and make up the volume to 100 mL using ultrapure water. Store at −20 °C in 1 mL aliquots for up to 3 months.
4. Sonicator with a microtip probe for cell disruption.
5. Benzonase® Nuclease (>25 units/μL).
6. Commercial Bradford protein determination reagent.

2.3 Labeling the Cell Lysate with CyDye DIGE Fluors and Minimal Labeling Dyes

1. *N*, *N*-Dimethylformamide, anhydrous (anhydrous DMF).
2. Stock solution of CyDye DIGE Fluor: Add 4 μL of anhydrous DMF to 5 nmol of CyDye DIGE Fluor, minimal labeling kit (GE Healthcare) (Cy2, Cy3, or Cy5) and mix using a vortex mixer. Aliquot 1 μL of the solution into microtubes, replace the air in the tubes with nitrogen, wrap with Parafilm, and store in a plastic bag at −80 °C.
3. 10 mM L-lysine: Add 1.5 mg lysine to 1 mL of ultrapure water and store at −20 °C.

2.4 Isoelectric Focusing

1. Immobiline DryStrip pH 3–10 NL, 24 cm (GE Healthcare).
2. DeStreak Reagent (GE Healthcare).
3. IPG Buffer (GE Healthcare).
4. Immobiline DryStrip Cover Fluid (GE Healthcare).
5. Rehydration buffer: 8 M urea, 4% CHAPS. Add 48.0 g of urea and 4.0 g of CHAPS to about 100 mL of ultrapure water. Add

a small amount of bromophenol blue for coloring the buffer. Make up the volume to 100 mL using ultrapure water. Store at −20 °C in 1 mL aliquots for up to 6 months.

6. Ettan IPGphor 3 IEF System (GE Healthcare).

2.5 SDS-PAGE and Gel Image Acquisition

1. Acrylamide solution: 30% acrylamide, 0.8% *N,N*,-methylene-bisacrylamide. Add 300 g of acrylamide and 8 g of *N,N*,-methylenebisacrylamide to about 800 mL of ultrapure water and make up the volume to 1 L using ultrapure water after completely solubilizing the mixture using a stirrer. Filter the solution using a filter with a 0.22 μm pore size and store at 4 °C protected from light.
2. 1.5 M Tris buffer (pH 8.8): Add 182 g of Tris to about 950 mL of ultrapure water. Adjust the pH to 8.8 using 6 N HCl. Filter using a filter with a 0.22 μm pore size and store at 4 °C.
3. 10% SDS solution: Add 10 g of sodium dodecyl sulfate (SDS) to ultrapure water and make up the volume to 100 mL. Store at room temperature for up to 6 months.
4. 10% APS solution: Add 0.5 g of ammonium peroxodisulfate (APS) to ultrapure water and make up the volume to 5 mL. Prepare at the time of use.
5. 10% TEMED solution: Add 100 μL of tetramethylethylenediamine (TEMED) to 900 μL of ultrapure water. Prepare at the time of use.
6. Water-saturated butanol.
7. 10× Running buffer: 250 mM Tris, 1.92 M glycine, 1% SDS. Add 605 g of Tris, 2882 g of glycine, and 200 g of SDS to ultrapure water and make up the volume to 20 L.
8. Agarose overlay solution: 0.7% agarose. Add 0.7 g of agarose and a small amount of bromophenol blue to 100 mL of 1× Running buffer.
9. Equilibration buffer: 50 mM Tris, 6 M urea, 30% (v/v) glycerol, and 2% (w/v) SDS. Add 6.7 mL of 1.5 M Tris buffer (pH 8.8), 72.07 g of urea, 60 mL of glycerol, and 4 g of SDS to ultrapure water and make up the volume to 200 mL. Add a small amount of bromophenol blue for coloring the buffer. Store at −20 °C.
10. 10 mg/mL dithiothreitol (DTT).
11. 25 mg/mL iodoacetic acid (IAA).
12. Ettan Dalt *twelve* or Ettan Dalt *six* (GE Healthcare).
13. Imaging system for 2D DIGE.

2.6 Data Processing

1. Quantification software for 2D DIGE analysis, such as SameSpots (Totallab).

2. A master gel image of HeLa cells including the selected spots that are measured (*see* **Note 2**).
3. A computer installed with a spreadsheet program such as Microsoft Excel and a programing language for the calculation, such as R or php. If necessary, database software can be used to manage a dataset of standard compounds and samples.
4. Software for cluster analysis, such as Cluster 3.0 [15] and Java tree view [16].

3 Methods

3.1 Determination of the Concentration of the Test Compounds

Because the analysis by 2D DIGE is time consuming and expensive, the numbers of experiments are limited. Comparison of the test compound and the compounds in the dataset should be performed under the same experimental condition as far as possible. Therefore, the concentrations of the compounds are unified, at which HeLa cell growth is almost completely inhibited at 48 h incubation by measuring the number of living cells by WST-8 assay [5]. WST-8, a water-soluble tetrazolium salt, is reduced by dehydrogenase in cells to give a yellow color formazan dye, which is proportional to the number of living cells.

1. Grow HeLa cells to subconfluence (80–90%) in 100-mm cell culture dishes using HeLa cell culture medium.
2. Wash the cells with 5 mL of PBS twice, detach the cells using 2 mL of trypsin solution, and add 8 mL of fresh culture medium.
3. Centrifuge at 300 × *g* for 3 min, resuspend with 10 mL of fresh culture medium, and count cell number.
4. Seed 1.5×10^4 HeLa cells in 150 μL of medium per well in cell culture 96-well microplates.
5. Incubate the cells at 37 °C and 5% CO_2 for 24 h.
6. Add 75 μL of the medium with a test sample (*see* **Note 3**) to the first lane of the wells in the microplates and perform threefold serial dilution in ten steps.
7. Incubate the cells at 37 °C and 5% CO_2 for 48 h.
8. Add 15 μL of WST-8 reagent to the cells.
9. Incubate the cells at 37 °C and 5% CO_2 for 1 h.
10. Measure the absorbance at 450 nm using a microplate reader and calculate the 100% inhibitory concentration of HeLa cell growth.

3.2 Preparation of Cell Lysates

1. Grow HeLa cells to subconfluence (80–90%) in 100-mm cell culture dishes using HeLa cell culture medium.

2. Wash the cells with 5 mL of PBS twice, detach the cells using 2 mL of trypsin solution, and add 8 mL of fresh culture medium.
3. Centrifuge at 300 × *g* for 3 min, resuspend with 10 mL of fresh culture medium, and count cell number.
4. Seed 30 × 10^4 HeLa cells in 3 mL of medium per dish in 60-mm cell culture dishes (*see* **Note 4**).
5. Incubate the cells at 37 °C and 5% CO_2 for 24 h.
6. Add 3 μL of DMSO or 3 μL of a test compound dissolved in DMSO at the 1000-fold concentration determined in Subheading 3.1.
7. Incubate the cells at 37 °C and 5% CO_2 for 18 h.
8. Discard the medium and wash with 1 mL of ice-cold PBS.
9. Add 500 μL of ice-cold PBS, detach the cells using a cell scraper, and place the cell suspension to a microtube.
10. Centrifuge at 1500 × *g* for 3 min at 4 °C and discard the supernatant.
11. Add 1 mL of ice-cold PBS to the cells, centrifuge at 1500 × *g* for 3 min at 4 °C, and remove as much of the supernatant as possible.
12. Add 50 μL of lysis buffer and disrupt the cells by sonication.
13. Add 2.5 μL of Bensonase® and incubate for 1 h on ice.
14. Centrifuge at 14,000 × *g* for 3 min at 4 °C.
15. Collect the supernatant and determine the protein concentration using the Bradford protein determination reagent.
16. Store the lysates at −80 °C.

3.3 Labeling the Cell Lysate with CyDye DIGE Fluors, Minimal Labeling Dyes

1. Add 50 μg of each cell lysate to one microtube to make an internal standard and mix.
2. Add 5 μL of anhydrous DMF to the stock solution of CyDye DIGE Fluor and mix.
3. Add 50 μg of the lysate to a microtube and add 1 μL of stock solution of Cy3 or Cy5 of CyDye DIGE Fluor (*see* **Note 5**). Take 50 μg of internal standard and add 1 μL of stock solution of Cy2 of CyDye DIGE Fluor.
4. Incubate for 30 min on ice protected from light.
5. Add 1 μL of 1 mM L-lysine, mix, and incubate for 10 min on ice protected from light.

3.4 Isoelectric Focusing

1. Mix the lysates labeled with Cy2, Cy3, or Cy5 of CyDye DIGE Fluor (*see* Subheading 3.3).
2. Add 5.4 μL of DeStreak Reagent and 2.25 μL of IPG Buffer to the mixture and make up the volume to 450 μL using rehydration buffer.

3. Place the mixed lysate to the strip folder of the Ettan IPGphor 3 IEF System, put Immobiline DryStrip on the mixed lysate, and overlay with cover fluid.
4. Run the protocol given in the table below.

Step	Voltage change pattern	Voltage (V)	Time or volt-hours
Rehydration		0	12 h
S1	Linear gradient	0 – 100	2 h
S2	Increase and hold	500	50 min
S3	Increase and hold	1000	1 h and 50 min
S4	Increase and hold	8000	50,000 Vh

5. After sample focusing, store the strips at −80 °C for up to a week or run immediately on the secondary dimension gel apparatus.

3.5 SDS-PAGE and Gel Image Acquisition

1. To make 11% acrylamide gel, add 183 mL of acrylamide solution, 125 mL of 1.5 M Tris-buffer (pH 8.8), 5 mL of 10% SDS, and 5 mL of 10% APS to 180.8 mL of ultrapure water and mix. Add 5 mL of TEMED to the mixed solution, pour it into a gel caster, and overlay 1–2 mL of water-saturated butanol on to the gels. Incubate at room temperature until the gels are completely polymerized (*see* **Note 6**).
2. Place the strips in individual 100-mm dishes and incubate with 10 mL of 10 mg/mL DTT in equilibration buffer for 15 min. Replace the buffer with 10 mL of 25 mg/mL of IAA in equilibration buffer and incubate for further 15 min.
3. Place the strips on the individual gels and overlay with melted agarose overlay solution.
4. Fill the tank of Ettan Dalt *twelve* or Ettan Dalt *six* with 1× running buffer. Place the gels and fill with 2× running buffer on the top of the buffer.
5. Run overnight at 2 W per gel at 20 °C.
6. Scan the gels immediately after they have run.

3.6 Data Processing

1. Using a quantification software for 2D DIGE, align all the acquired images with the master image, and obtain the normalized volume of each selected spot according to the user's manual of the software.
2. Calculate the average of the normalized volume of each selected spot and calculate the log-fold ratio of each spot of sample against the corresponding spot of the control.

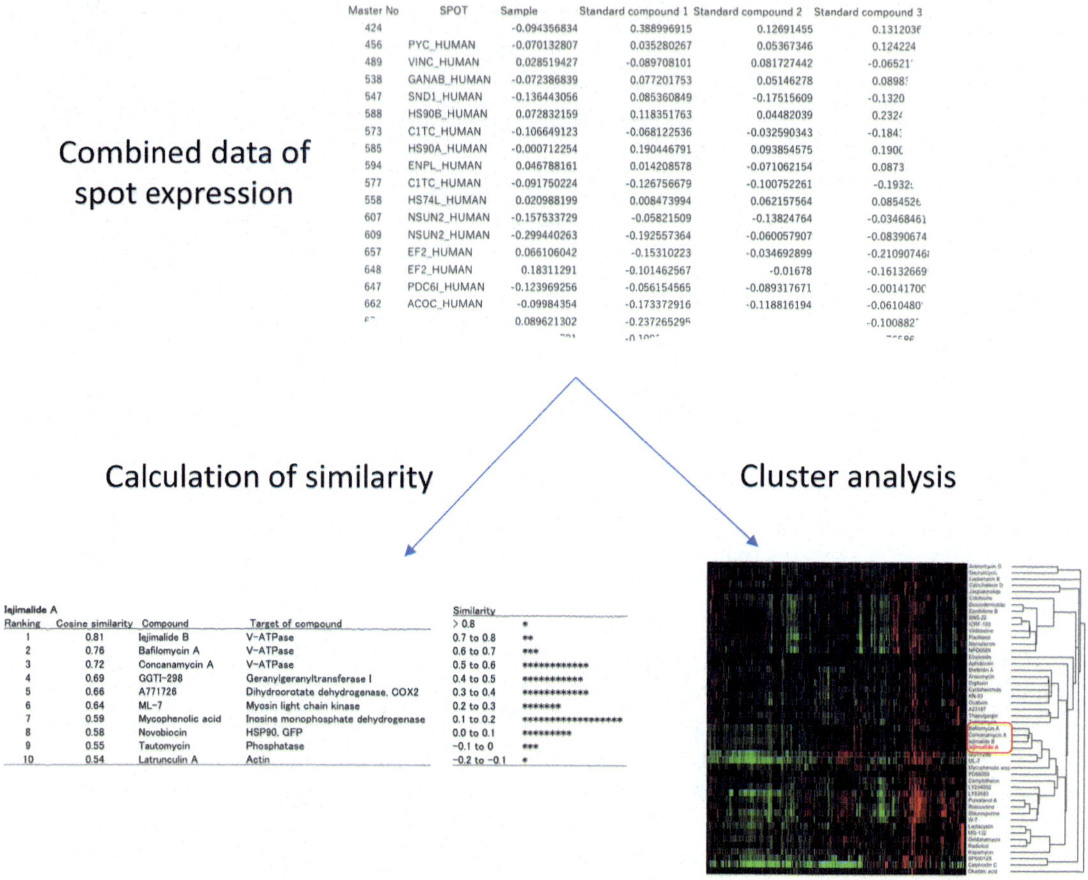

Master No	SPOT	Sample	Standard compound 1	Standard compound 2	Standard compound 3
424		-0.094356834	0.388996915	0.12691455	0.131203
456	PYC_HUMAN	-0.070132807	0.035280267	0.05367346	0.124224
489	VINC_HUMAN	0.028519427	-0.089708101	0.081727442	-0.06521
538	GANAB_HUMAN	-0.072386839	0.077201753	0.05146278	0.0898
547	SND1_HUMAN	-0.136443056	0.085360849	-0.17515609	-0.1320
588	HS90B_HUMAN	0.072832159	0.118351763	0.04482039	0.232
573	C1TC_HUMAN	-0.106649123	-0.068122536	-0.032590343	-0.184
585	HS90A_HUMAN	-0.000712254	0.190446791	0.093854575	0.190
594	ENPL_HUMAN	0.046788161	0.014208578	-0.071062154	0.0873
577	C1TC_HUMAN	-0.091750224	-0.126756679	-0.100752261	-0.1932
558	HS74L_HUMAN	0.020988199	0.008473994	0.062157564	0.085452
607	NSUN2_HUMAN	-0.157533729	-0.05821509	-0.13824764	-0.0346846
609	NSUN2_HUMAN	-0.299440263	-0.192557364	-0.060057907	-0.08390674
657	EF2_HUMAN	0.066106042	-0.15310223	-0.034692899	-0.2109074
648	EF2_HUMAN	0.18311291	-0.101462567	-0.01678	-0.16132669
647	PDC6I_HUMAN	-0.123969256	-0.056154565	-0.089317671	-0.0014170
662	ACOC_HUMAN	-0.09984354	-0.173372916	-0.118816194	-0.0610480
		0.089621302	-0.237265295		-0.100882

Iejimalide A

Ranking	Cosine similarity	Compound	Target of compound
1	0.81	Iejimalide B	V-ATPase
2	0.76	Bafilomycin A	V-ATPase
3	0.72	Concanamycin A	V-ATPase
4	0.69	GGTI-298	Geranylgeranyltransferase I
5	0.66	A771726	Dihydroorotate dehydrogenase, COX2
6	0.64	ML-7	Myosin light chain kinase
7	0.59	Mycophenolic acid	Inosine monophosphate dehydrogenase
8	0.58	Novobiocin	HSP90, GFP
9	0.55	Tautomycin	Phosphatase
10	0.54	Latrunculin A	Actin

Similarity	
> 0.8	*
0.7 to 0.8	**
0.6 to 0.7	***
0.5 to 0.6	************
0.4 to 0.5	***********
0.3 to 0.4	************
0.2 to 0.3	*******
0.1 to 0.2	*****************
0.0 to 0.1	*********
−0.1 to 0	***
−0.2 to −0.1	*

Fig. 4 Combined data of the protein expression of the 296 spots is used for the calculation of similarity and for cluster analysis to predict the target of a compound. Using the expression data for the 296 spots, the values of cosine similarities are calculated between a sample and each standard compound and sorted by the order of the value. The standard compound with the same target as that of a sample is highly ranked. The matrix of expression data is used for hierarchical cluster analysis. If a sample has the same target, a cluster will be formed with the standard compound and the sample

$$\text{Log fold ratio} = \log\left(\frac{\text{Average of nomalized spot volume of sample}}{\text{Average of nomalized spot volume of control}}\right)$$

3. Calculate cosine similarities between the sample and each standard compound to compare using the formula, and sort standard compounds based on the similarity to identify the compound most similar to the sample.

$$\text{Similarity} = \frac{\sum_{i=1}^{n} (\text{Log ratio of sample})_i (\text{Log ratio of a standard compound})_i}{\sqrt{\sum_{i=1}^{n} (\text{Log ratio of sample})_i^2}\sqrt{\sum_{i=1}^{n} (\text{Log ratio of a standard compound})_i^2}}$$

4. Combine the result with a dataset of standard compounds to compare in a spreadsheet (Fig. 4) (*see* **Note 7**).
5. Perform hierarchical cluster analysis with parameters such as correlation for distance metric and centroid linkage for clustering method (*see* **Note 8**).
6. Predict sample target from the targets of standard compounds with high similarity and perform a validation study (*see* **Note 9**).

4 Notes

1. If necessary, check the identity of the cells using short tandem repeat (STR) analysis.
2. The master gel image of our system includes the 296 spots that were reproducibly detected in several experiments (Fig. 3). Once a master gel image is determined, it is used as the master image for every experiment.
3. Sample compound is resolved with DMSO for the stock solution at 1000-fold concentration and is kept at −20 °C. The stock solution of sample should be diluted with culture medium in advance, added to the first well of the culture plate, and diluted serially on the culture plate.
4. Controls must be prepared for every experiment. Cells are prepared in triplicate for each sample. We usually prepare six dishes in the case of one sample or 12 dishes in the case of three samples.
5. Avoid fixing one CyDye to lysates from one sample if possible. For example, two lysates of a sample and a lysate of control are labeled with Cy3, and two lysates of a control and a lysate of samples are labeled with Cy5.
6. Gels are usually polymerized for 4 h at room temperature.
7. The data can be added to a dataset on a spreadsheet for cluster analysis. However, a database software with a scripting language such as MySQL (https://www-jp.mysql.com) is useful to manage the dataset smartly.
8. Appropriate parameter for clustering methods and distance metrics may be chosen to obtain a good result which is clearly classified in standard compounds according to their mechanism of action.
9. If a compound has multiple targets or its target does not exist in HeLa cells, we could not predict the target. Therefore, the target of any compound cannot be determined using only this method; a validation study is necessary.

Acknowledgments

We thank Ms. H. Kondo, Ms. K. Noda, Ms. Y. Nakata, Ms. Y. Hirata, and Ms. M. Tanaka for conducting proteomic analysis. This work was supported in part by JSPS KAKENHI Grant Numbers JP16H06276, JP17H06412, JP18H03945, JP17K07783, AMED under Grant Number JP18cm0106112 and the NARO Bio-oriented Technology Research Advancement Institution (Research program on development of innovative technology).

References

1. Futamura Y, Muroi M, Osada H (2013) Target identification of small molecules based on chemical biology approaches. Mol BioSyst 9 (5):897–914. https://doi.org/10.1039/c2mb25468a
2. Shoemaker RH (2006) The NCI60 human tumour cell line anticancer drug screen. Nat Rev Cancer 6(10):813–823. https://doi.org/10.1038/nrc1951
3. Nakatsu N, Nakamura T, Yamazaki K, Sadahiro S, Makuuchi H, Kanno J, Yamori T (2007) Evaluation of action mechanisms of toxic chemicals using JFCR39, a panel of human cancer cell lines. Mol Pharmacol 72 (5):1171–1180. https://doi.org/10.1124/mol.107.038836
4. Muroi M, Futamura Y, Osada H (2016) Integrated profiling methods for identifying the targets of bioactive compounds: MorphoBase and ChemProteoBase. Nat Prod Rep 33 (5):621–625. https://doi.org/10.1039/c5np00106d
5. Muroi M, Kazami S, Noda K, Kondo H, Takayama H, Kawatani M, Usui T, Osada H (2010) Application of proteomic profiling based on 2D-DIGE for classification of compounds according to the mechanism of action. Chem Biol 17(5):460–470. https://doi.org/10.1016/j.chembiol.2010.03.016
6. Ning F, Wu X, Wang W (2016) Exploiting the potential of 2DE in proteomics analyses. Expert Rev Proteomics:1–3. https://doi.org/10.1080/14789450.2016.1230498
7. Benesova M, Hola D, Fischer L, Jedelsky PL, Hnilicka F, Wilhelmova N, Rothova O, Kocova M, Prochazkova D, Honnerova J, Fridrichova L, Hnilickova H (2012) The physiology and proteomics of drought tolerance in maize: early stomatal closure as a cause of lower tolerance to short-term dehydration? PLoS One 7(6):e38017. https://doi.org/10.1371/journal.pone.0038017
8. Kondo T, Hirohashi S (2006) Application of highly sensitive fluorescent dyes (CyDye DIGE Fluor saturation dyes) to laser microdissection and two-dimensional difference gel electrophoresis (2D-DIGE) for cancer proteomics. Nat Protoc 1(6):2940–2956. https://doi.org/10.1038/nprot.2006.421
9. Scherp P, Ku G, Coleman L, Kheterpal I (2011) Gel-based and gel-free proteomic technologies. Methods Mol Biol 702:163–190. https://doi.org/10.1007/978-1-61737-960-4_13
10. Kawatani M, Takayama H, Muroi M, Kimura S, Maekawa T, Osada H (2011) Identification of a small-molecule inhibitor of DNA topoisomerase II by proteomic profiling. Chem Biol 18 (6):743–751. https://doi.org/10.1016/j.chembiol.2011.03.012
11. Futamura Y, Kawatani M, Muroi M, Aono H, Nogawa T, Osada H (2013) Identification of a molecular target of a novel fungal metabolite, pyrrolizilactone, by phenotypic profiling systems. Chembiochem 14(18):2456–2463. https://doi.org/10.1002/cbic.201300499
12. Minegishi H, Futamura Y, Fukashiro S, Muroi M, Kawatani M, Osada H, Nakamura H (2015) Methyl 3-((6-methoxy-1,4-dihydroindeno[1,2-c]pyrazol-3-yl)amino)benzoate (GN39482) as a tubulin polymerization inhibitor identified by MorphoBase and ChemProteoBase profiling methods. J Med Chem 58 (10):4230–4241. https://doi.org/10.1021/acs.jmedchem.5b00035
13. Kawatani M, Muroi M, Wada A, Inoue G, Futamura Y, Aono H, Shimizu K, Shimizu T, Igarashi Y, Takahashi-Ando N, Osada H (2016) Proteomic profiling reveals that collismycin A is an iron chelator. Sci Rep 6:38385. https://doi.org/10.1038/srep38385
14. Kawamura T, Kawatani M, Muroi M, Kondoh Y, Futamura Y, Aono H, Tanaka M, Honda K, Osada H (2016) Proteomic profiling of small-molecule inhibitors reveals dispensability of MTH1 for cancer cell survival. Sci Rep 6:26521. https://doi.org/10.1038/srep26521

15. de Hoon MJ, Imoto S, Nolan J, Miyano S (2004) Open source clustering software. Bioinformatics 20(9):1453–1454. https://doi.org/10.1093/bioinformatics/bth078

16. Saldanha AJ (2004) Java Treeview–extensible visualization of microarray data. Bioinformatics 20(17):3246–3248. https://doi.org/10.1093/bioinformatics/bth349

Chapter 8

Examining Cellular Responses to Kinase Drug Inhibition Through Phosphoproteome Mapping of Substrates

Daniel Bucio-Noble, Crystal Semaan, and Mark P. Molloy

Abstract

This chapter provides a detailed description for mass spectrometry-based phosphoproteomics analysis. We describe sample preparation, phosphopeptide enrichment, mass spectrometry acquisition, label-free data analysis and statistical analysis. This technique can be employed to characterize cell signaling networks, and is particularly useful to monitor cellular responses to kinase drug inhibitors.

Key words Phosphoproteomics, Phosphopeptide enrichment, Mass spectrometry, Label-free

1 Introduction

The study of protein phosphorylation is of importance in understanding many vital aspects of cell biology such as cell growth, differentiation and cell death. Chemical modification of proteins through reversible phosphorylation plays a major role in modulating protein functions including activation state, cellular localization, degradation, and interactions with proteins and other biomolecules [1]. Phosphorylation of serine, threonine and tyrosine amino acids is considered as one of the most common post-translational modifications (PTM) that involves sequence-specific kinases and phosphatases. This modification has been evidenced as central to cancer biology due to its roles in cell proliferation, oncogenic kinase signaling [2], and transcriptional regulation [3]. The fact that phosphorylation sites are often altered in cancers by gene mutations, or structurally mimicked by mutation to provide oncogenes with structural similarity to activated kinases [4], has made this area highly relevant to cancer researchers.

A study investigating somatic cancer mutation datasets found that there was significant enrichment for mutations which resulted in gain/loss of phosphorylation when compared to other mutations (e.g., random amino acid substitutions) [5]. Mutations in

Slava Ziegler and Herbert Waldmann (eds.), *Systems Chemical Biology: Methods and Protocols*, Methods in Molecular Biology, vol. 1888, https://doi.org/10.1007/978-1-4939-8891-4_8, © Springer Science+Business Media, LLC, part of Springer Nature 2019

kinase genes in particular represented the highest number of mutations that disrupted phosphorylation, suggesting that phosphorylation target site mutations are associated with aberrant phosphorylation [5]. Phosphorylation has also been shown to be pivotal in treatment, as it is a pharmacologically targetable mechanism [6, 7]. For instance, Chapman et al. found that BRAF kinase inhibitor vemurafenib (PLX4032) showed promise for patients with metastatic melanoma carrying the BRAF V600E mutation [6]. β-catenin, a protein associated with several cancers is phosphorylated on its tyrosine residues, which is thought to be associated with tumor invasiveness [8]. Furthermore, a link has been shown between β-catenin phosphorylation status and intestinal tumorigenesis in mice [8].

Mass spectrometry (MS) has emerged as a powerful tool in proteomics over the past decade, useful for both peptide identification and quantitation [9–11]. Quantitative MS allows for comparisons between biological samples to determine molecular changes at the expressed protein level. Addition of a phosphate group to the protein sequence is an event that often occurs at low stoichiometry and is also a reversible and dynamic mechanism. In the background of nonphosphorylated peptides, phosphopeptides are poorly detected in positive ionization mode electrospray MS. These elements complicate identification of phosphopeptides by MS.

In order to address these challenges, phosphopeptide enrichment prior to MS-based analysis is necessary. Phosphopeptide enrichment using metal oxides beads, such as titanium dioxide (TiO_2), enables the efficient capture of these modified peptides from the high background of nonphosphorylated peptides [12]. There are different MS-based strategies for the quantitation of phosphorylated peptides. Labeling techniques such as stable isotope labeling by amino acids in cell culture (SILAC) have proven successful for quantitation of phosphopeptides. SILAC is a reliable strategy for simultaneous identification and quantitation of two or more samples of complex protein mixtures, such as cell lysates [13]. On the other hand, SILAC can only assess a limited number of biological conditions in a single experiment and its usage is restricted to live samples that can be metabolically labeled [14]. In addition, the need of a minimum passage number for complete isotope incorporation limits SILAC use in primary cell lines [15].

Label-free quantitation is a simple and cost effective strategy compatible with a wide range of sample types. It is also applicable to a larger number of samples compared to the limited number of samples afforded by labeling techniques [16]. Label-free quantitation of the phosphoproteome has provided discoveries in the elucidation of cellular mechanisms. This approach showed the potential role of casein kinase 2 (CK2) in the modulation of the AKT signaling pathway in BRAFV600E mutated thyroid cancer and melanoma cells [17]. Further, phosphoproteomics analysis can be employed in the characterization of cell signaling in response to treatment with kinase inhibitors in the context of cancer research and other pathologies [18]. Deeper understanding of the phosphoproteome will provide more information in regards to the integrated system of cell signaling, particularly in pharmacological intervention. In this protocol we present a straight forward methodology that we use for phosphopeptide enrichment, data acquisition and informatic analysis. Special consideration is taken particularly in the technical aspects and the bioinformatic tools used in this protocol.

2 Materials

2.1 Protein Lysis and Sample Preparation

1. 500 μg of protein from cell lysate (*see* **Note 1**).
2. Phosphate-buffered saline (PBS). 0.14 M NaCl, 10 mM Na_2HPO_4, 2.7 mM KCl, 1.8 mM KH_2PO_4, pH 7.4 dissolved in ultrapure water and sterilized.
3. Lysis buffer: 1% (w/v) sodium deoxycholate, 0.1 M triethylammonium bicarbonate dissolved in water.
4. Benzonase (>250 units/μL).
5. Bicinchoninic acid (BCA) protein concentration assay kit.
6. 500 mM dithiothreitol (DTT) dissolved in lysis buffer.
7. 250 mM iodoacetamide dissolved in lysis buffer.
8. 10 μg of mass spectrometry grade trypsin dissolved in 100 μL of lysis buffer.
9. Ultra-sonic probe.

2.2 Phosphopeptide Enrichment

1. Formic acid.
2. Loading buffer: 80% (v/v) acetonitrile, 5% (v/v) trifluoroacetic acid (TFA), 76 mg/mL glycolic acid dissolved in water.
3. Washing buffer 1: 80% (v/v) acetonitrile, 1% (v/v) TFA dissolved in water.
4. Washing buffer 2: 10% (v/v) acetonitrile, 0.1% (v/v) TFA dissolved in water.
5. Elution buffer: 1% (v/v) ammonia dissolved in water.
6. 2% (v/v) acetonitrile, 0.1% (v/v) TFA dissolved in water.
7. 70% (v/v) acetonitrile, 0.1% (v/v) TFA dissolved in water.
8. Titansphere TiO_2 beads 5 μm (GL Sciences, Japan).

2.3 C18 Cleaning

1. Methanol.
2. 3 M Empore™ C18 disk.
3. Protein low binding microcentrifuge tubes.
4. Vacuum centrifuge.

2.4 Mass Spectrometry and Informatic Analysis

1. Halo® C18 column, 2.7 μm, 160 Å pore size.
2. Q-Exactive mass spectrometer (Thermo Fisher Scientific).
3. Buffer B: 99.9% (v/v) acetonitrile, 0.1% (v/v) formic acid.
4. MaxQuant software V1.5.
5. Perseus software V1.5.
6. NetworKIN website (networkin.info).

3 Methods

Our group commonly conducts phosphopeptide enrichment from cancer cell lines treated with kinase inhibitor drugs [17] and from tissues of human origin. Apart from the selection of a suitable lysis buffer depending on the nature of the sample (*see* **Note 2**), this methodology has been standardized for human cancer cell lines. However, optimization of this protocol might be necessary when other types of samples are analyzed. A schematic representation of the main steps of this protocol is shown in Fig. 1. Principal component analysis, hierarchical clustering and volcano plot presented here are taken from a phosphopeptide enrichment performed on SW480 human colon carcinoma cells treated with the inflammatory inducer lipopolysaccharides (LPS) derived from *Escherichia coli* and the phytochemical resveratrol (RSV).

3.1 Protein Lysis and Sample Preparation

1. Mammalian cell lines are grown at 37 °C and 5% CO_2. To investigate the effects of kinase inhibition on the cellular phosphoproteome, cells are incubated with a specific small molecule

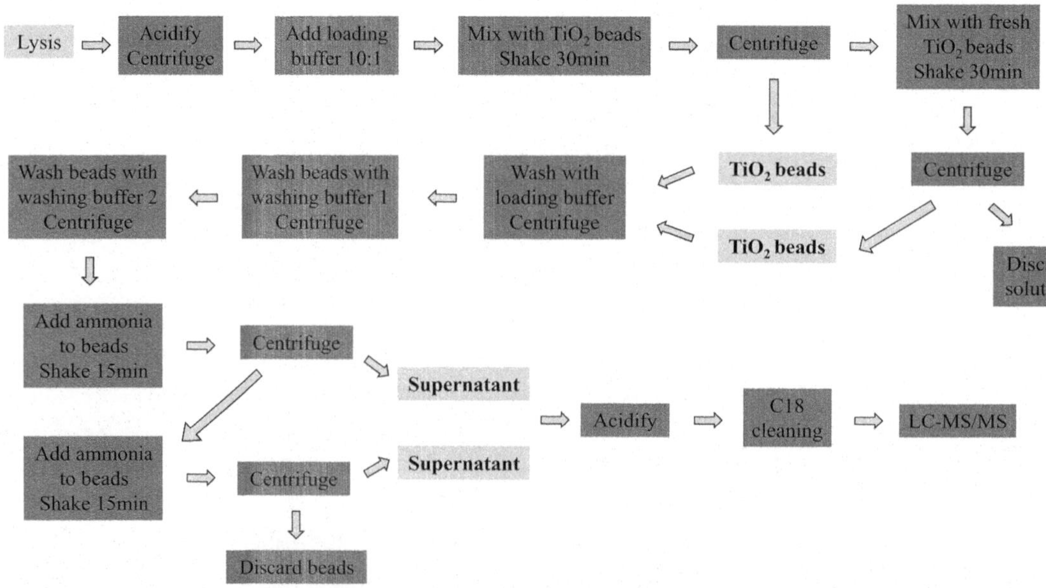

Fig. 1 Schematic of phosphopeptide enrichment protocol

kinase inhibitor for up to 30 min for comparison against vehicle treated control cells (*see* **Note 3**).

2. After the incubation period with the kinase inhibitor is complete, cells are collected by scrapping in ice-cold sterile PBS and centrifuged at 5000 × *g* at 4 °C for 5 min. The supernatant is removed and the cell pellet can be stored at −80 °C or immediately lysed.
3. Resuspend cell pellet or tissue in lysis buffer (*see* **Note 2**).
4. Heat lysate at 95 °C for 1 min (*see* **Note 4**), then cool on ice.
5. Probe-sonicate to ensure complete cell disruption, then cool on ice.
6. Add 2 μL of benzonase to degrade DNA.
7. Centrifuge at 10,000 × *g* at 4 °C for 10 min and collect supernatant.
8. Determine protein concentration using the BCA assay.
9. Aliquot 500 μg of protein and reduce proteins by adding DTT to a final concentration of 10 mM followed by incubation at 60 °C for 30 min.
10. Alkylate proteins by adding iodoacetamide at a final concentration of 25 mM followed by incubation at room temperature for 20 min in the dark.
11. Digest proteins with trypsin in a 1:50 enzyme: protein ratio (i.e., 10 μg of trypsin). Incubate at 37 °C overnight.

3.2 Phosphopeptide Enrichment

1. Acidify samples with formic acid to a final concentration of 2% (v/v) (*see* **Note 5**). Centrifuge at 10,000 × *g* for 10 min and collect supernatant.
2. Add loading buffer to the sample in a 10:1 ratio (*see* **Notes 6** and 7).
3. Wash 3 mg of TiO_2 beads with 100 μL of loading buffer. Mix for a few seconds and centrifuge (1000 × *g* for 10 s) to pellet down beads. Discard supernatant.
4. Add sample to the beads and incubate at room temperature with constant shaking for 30 min. Centrifuge (1000 × *g* for 10 s) to pellet down beads.
5. Wash a new batch of 3 mg of TiO_2 beads as in **step 3**.
6. Add supernatant from **step 4** to the new beads and incubate at room temperature with constant shaking for 30 min. Centrifuge (1000 × *g* for 10 s) to pellet down beads. Discard supernatant.
7. Combine the TiO_2 beads from the two incubations in a new low binding microcentrifuge tube and wash with 100 μL of

loading buffer. Mix for a few seconds and centrifuge (1000 × *g* for 10 s) to pellet down beads. Discard supernatant.

8. Wash beads with 100 μL washing buffer 1. Mix for a few seconds and centrifuge (1000 × *g* for 10 s) to pellet down beads. Discard supernatant.
9. Wash beads with 100 μL washing buffer 2. Mix for a few seconds and centrifuge (1000 × *g* for 10 s) to pellet down beads. Discard supernatant.
10. Remove any remnant of solvent by placing sample in the vacuum centrifuge for 5 min.
11. Elute phosphopeptides by adding 100 μL of elution buffer (*see* **Note 8**). Incubate for 15 min under constant shaking at room temperature. Centrifuge (1000 × *g* for 10 s) to pellet down beads and collect supernatant in a new low binding microcentrifuge tube.
12. Add for a second time 100 μL of elution buffer. Incubate for 15 min under constant shaking at room temperature. Centrifuge (1000 × *g* for 10 s) to pellet down beads and pool the two supernatants. Acidify with 16 μL of formic acid.

3.3 C18 Cleaning

1. Cut 1 cm above the end of a 200 μL tip. With a needle tip cut two stack layers of C18 material and insert them into the tip. Place the stacked tip inside a 1.5 mL tube.
2. Equilibrate C18 tips by adding 100 μL of methanol into the tip and centrifuge at 1000 × *g* for 2 min or until the complete volume passes through the C18 material. Discard flow-through.
3. Wash C18 tips with 100 μL of 2% (v/v) acetonitrile, 0.1% (v/v) TFA, centrifuge (1000 × *g* for 2 min) and discard flow-through.
4. Add phosphopeptide solution (*see* Subheading 3.2, **step 12**), centrifuge (1000 × *g* for 2 min) and discard supernatant.
5. Wash with 100 μL of 2% (v/v) acetonitrile, 0.1% (v/v) TFA, centrifuge (1000 × *g* for 2 min) and discard flow-through.
6. Elute with 70% (v/v) acetonitrile, 0.1% (v/v) TFA, centrifuge (1000 × *g* for 2 min) and collect flow-through in a new low-binding tube.
7. Remove solvent by complete drying in a vacuum centrifuge.

3.4 Mass Spectrometry (LC-MS/MS)

1. Reconstitute dried phosphopeptides in 20 μL of 0.1% (v/v) formic acid.
2. Load phosphopeptides onto a C18 reversed phase chromatography column to conduct nanoLC-MS.

3. Elute phosphopeptides with an A-B buffer (A–water, B–99.9% (v/v) acetonitrile, 0.1% (v/v) formic acid) gradient of 5–35% in a 100 min run directly into a *Q*-Exactive mass spectrometer using positive mode electrospray ionization (*see* **Note 9**).
4. Phosphopeptide analysis occurs via data-dependent acquisition mode consisting of a full MS scan at a resolution of 70,000 over the 350–1800 *m/z* range. The 10 most abundant ions per scan are selected for MS/MS at a resolution setting of 17,500. Peptide fragmentation is conducted by high collision dissociation using normalized collision energy of 35 eV. The automatic gain control (AGC) is set at 1×10^6 ions for a complete full MS scan.

3.5 Data Processing and Analysis

1. Files (*.raw*) are processed using MaxQuant software [16] with a 1% peptide and protein False Discovery Rate (FDR).
2. Database search uses the following parameters: 2 missed cleavages, peptide mass tolerance of 4.5 ppm, carbamidomethylation (C) as fixed modification and oxidation (M), acetylation (Protein *N*-term), and phosphorylation (STY) as variable modifications.
3. Label-free quantitation mode and match between runs functions are activated.

3.6 Statistical Analysis

1. Statistical analysis using Perseus software [19] is performed on the "phospho (STY) sites" files (*.txt*).
2. Intensity values are moved to the 'main columns' box.
3. Reverse and contaminant peptides are removed with the "filter rows based on categorical column" function. Phosphopeptides with a localization probability lower than 75% are also removed with the "filter rows based on numerical/main column" function.
4. Intensities are log2(×) transformed.
5. Using the "filter rows based on valid values" proteins with high number of missing values are removed. Missing values are replaced by imputation (*see* **Note 10**).
6. Data is normalized by median subtraction.
7. Principal component analysis (Fig. 2) and volcano plots (Fig. 3) can be generated by using the corresponding functions.
8. "Multiple-sample test" using the ANOVA test and the selection of an appropriate FDR cutoff permits the identification of phosphosites whose expression is significantly altered among all the conditions. "Filter rows based on categorical column" enables the filtration of significant phosphosites in the matrix from nonsignificant phosphosites. After data filtration, "*Z*-score" normalizes the phosphosites expression values

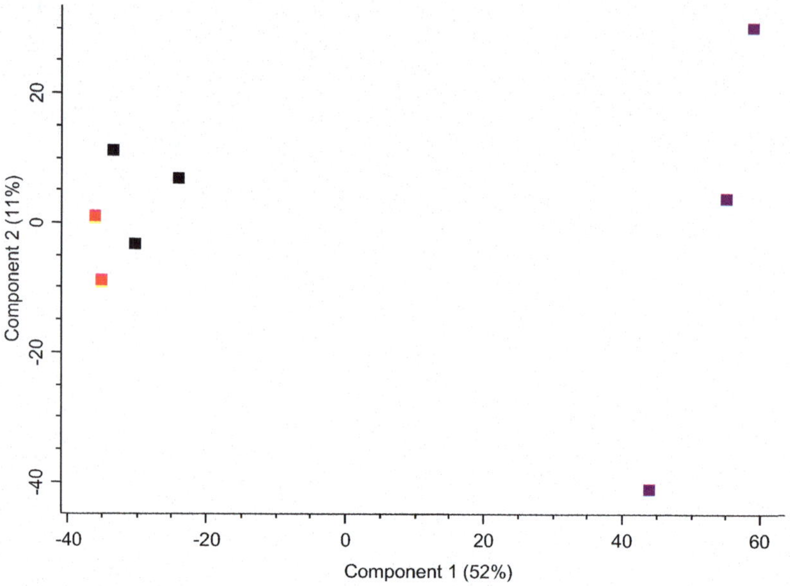

Fig. 2 Principal component analysis representing overall phosphopeptide expression in response to lipopolysaccharides (red), and lipopolysaccharides + resveratrol (purple) SW480-treated cells; untreated cells in black

according to mean and standard deviation. These significant changes can then be visualized with the "hierarchical clustering" function (Fig. 4).

9. Significant regulated phosphosites between two conditions are determined by "two-sample test" and the selection of an appropriate FDR cutoff. Fold change (located in the "Student's *t*-test Difference" column) thresholds can help in the identification of statistical significant changes.

10. Gene enrichment analysis on significant regulated sites can be performed using the Fisher exact test. Data is first filtered using two criteria: *Q*-value and fold change with the objective of selecting statistically significantly upregulated and downregulated values. Fisher exact test is applied using an appropriate FDR cutoff. At this point, Perseus calculates an enrichment factor for the overrepresented kinase motifs.

11. If desired, kinase sequence motifs can be generated for differentially detected phosphopeptides by selecting the function "Sequence logos" and activating "motifs" on the column drop-down list.

12. Alternatively, other bioinformatic tools such as NetworKIN determine the occurrence of kinase motifs in relation to significantly regulated phosphosites. By selecting the phosphorylated amino acids, NetworKIN performs a prediction of the most likely kinases responsible for these modifications.

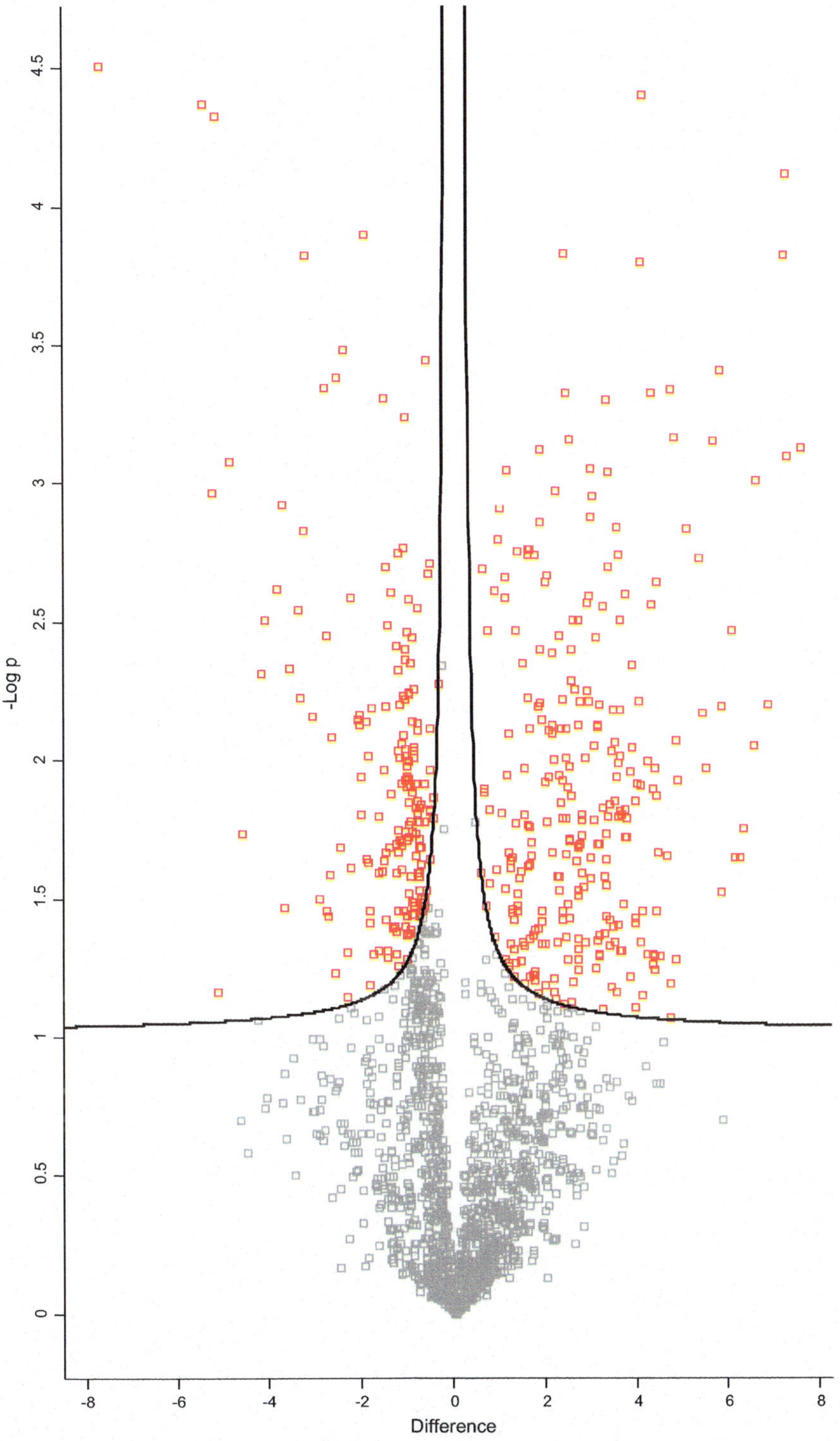

Fig. 3 Volcano plot representing significantly regulated phosphopeptides in response to resveratrol compared to lipopolysaccharides SW480-treated cells (red). Two-sample *t*-test, FDR < 0.05

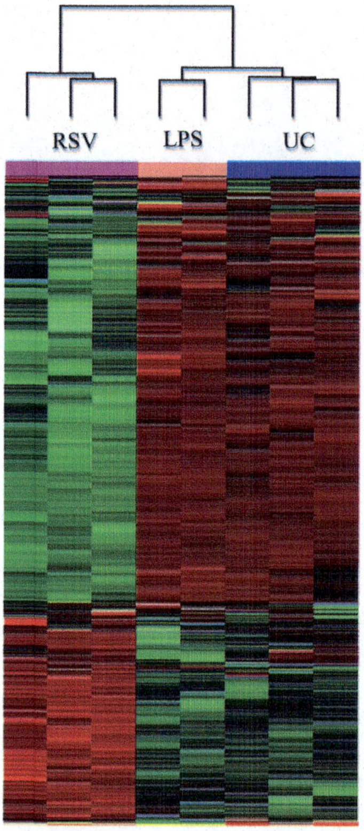

Fig. 4 Hierarchical clustering representing significantly regulated phosphopeptides in response to resveratrol (RSV) and lipopolysaccharides (LPS) SW480-treated cells; untreated cells (UC). ANOVA multiple sample test, FDR < 0.05

13. To validate these predictions, genetic manipulation of predicted kinases using siRNA or other knockout techniques or kinase drug treatment can provide further evidence for the role of impacted kinases as we have shown previously [17].

4 Notes

1. To obtain 500 μg of protein from SW480 colon cancer cells, this can be achieved by seeding 1×10^5 cells in a 100 mm cell culture plate overnight followed by kinase inhibitor treatment and cell harvesting the next day.
2. Using 1% (w/v) sodium deoxycholate, 0.1 M triethylammonium bicarbonate as a lysis buffer provides compatibility with TiO_2 enrichment [20] and trypsin-mediated digestion [21]. Selection of a suitable lysis buffer will depend on the nature of sample.

3. Selecting the timepoint for analysis post drug treatment is important for phosphoproteomics studies. To avoid secondary responses it is typical to sample after 15–30 min of drug treatment. It should be noted that measureable changes in the phosphoproteome can occur in just a few minutes.
4. Excessive heating induces protein precipitation. Addition of protease or phosphatase inhibitors may not be necessary as rapid heating halts enzyme activity. Additionally, it is reported that phosphatases show great affinity for TiO_2 and therefore reduce phosphopeptide enrichment efficiency [20].
5. Addition of formic acid precipitates sodium deoxycholate providing an efficient means to remove the detergent following tryptic digestion.
6. It is recommended to prepare loading buffer and washing buffers fresh in order to avoid any potential contamination.
7. Acidic conditions permit better recovery of phosphopeptides by reducing interference from acidic nonphosphorylated peptides [22].
8. Preparation of fresh ammonia solution is recommended given that high pH is critical for complete elution of phosphopeptides.
9. Phosphopeptides due to their hydrophilic nature tend to elute early in the gradient at low concentrations of acetonitrile.
10. Perseus offers multiple options for the imputation of missing values. Replacement of missing values from normal distribution permits their simulation from a typical abundance region if they had been measured and their random selection from a normal distribution [19].

References

1. Hunter T (2000) Signaling—2000 and beyond. Cell 100(1):113–127
2. Smith CC, Wang Q, Chin CS, Salerno S, Damon LE, Levis MJ, Perl AE, Travers KJ, Wang S, Hunt JP, Zarrinkar PP, Schadt EE, Kasarskis A, Kuriyan J, Shah NP (2012) Validation of ITD mutations in FLT3 as a therapeutic target in human acute myeloid leukaemia. Nature 485(7397):260–263. https://doi.org/10.1038/nature11016
3. Morin PJ, Sparks AB, Korinek V, Barker N, Clevers H, Vogelstein B, Kinzler KW (1997) Activation of beta-catenin-Tcf signaling in colon cancer by mutations in beta-catenin or APC. Science 275(5307):1787–1790
4. Reimand J, Wagih O, Bader GD (2013) The mutational landscape of phosphorylation signaling in cancer. Sci Rep 3:2651. https://doi.org/10.1038/srep02651
5. Radivojac P, Baenziger PH, Kann MG, Mort ME, Hahn MW, Mooney SD (2008) Gain and loss of phosphorylation sites in human cancer. Bioinformatics 24(16):i241–i247. https://doi.org/10.1093/bioinformatics/btn267
6. Chapman PB, Hauschild A, Robert C, Haanen JB, Ascierto P, Larkin J, Dummer R, Garbe C, Testori A, Maio M, Hogg D, Lorigan P, Lebbe C, Jouary T, Schadendorf D, Ribas A, O'Day SJ, Sosman JA, Kirkwood JM, Eggermont AM, Dreno B, Nolop K, Li J, Nelson B, Hou J, Lee RJ, Flaherty KT, McArthur GA, Group B-S (2011) Improved survival with vemurafenib in melanoma with BRAF V600E mutation. N Engl J Med 364(26):2507–2516. https://doi.org/10.1056/NEJMoa1103782

7. Tiacci E, Trifonov V, Schiavoni G, Holmes A, Kern W, Martelli MP, Pucciarini A, Bigerna B, Pacini R, Wells VA, Sportoletti P, Pettirossi V, Mannucci R, Elliott O, Liso A, Ambrosetti A, Pulsoni A, Forconi F, Trentin L, Semenzato G, Inghirami G, Capponi M, Di Raimondo F, Patti C, Arcaini L, Musto P, Pileri S, Haferlach C, Schnittger S, Pizzolo G, Foa R, Farinelli L, Haferlach T, Pasqualucci L, Rabadan R, Falini B (2011) BRAF mutations in hairy-cell leukemia. N Engl J Med 364 (24):2305–2315. https://doi.org/10.1056/NEJMoa1014209
8. van Veelen W, Le NH, Helvensteijn W, Blonden L, Theeuwes M, Bakker ER, Franken PF, van Gurp L, Meijlink F, van der Valk MA, Kuipers EJ, Fodde R, Smits R (2011) beta-catenin tyrosine 654 phosphorylation increases Wnt signalling and intestinal tumorigenesis. Gut 60(9):1204–1212. https://doi.org/10.1136/gut.2010.233460
9. Mann M, Hendrickson RC, Pandey A (2001) Analysis of proteins and proteomes by mass spectrometry. Annu Rev Biochem 70:437–473. https://doi.org/10.1146/annurev.biochem.70.1.437
10. Kuster B, Mortensen P, Andersen JS, Mann M (2001) Mass spectrometry allows direct identification of proteins in large genomes. Proteomics 1(5):641–650. https://doi.org/10.1002/1615-9861(200104)1:5<641::AID-PROT641>3.0.CO;2-R
11. Aebersold R, Mann M (2003) Mass spectrometry-based proteomics. Nature 422 (6928):198–207. https://doi.org/10.1038/nature01511
12. Larsen MR, Thingholm TE, Jensen ON, Roepstorff P, Jørgensen TJ (2005) Highly selective enrichment of phosphorylated peptides from peptide mixtures using titanium dioxide microcolumns. Mol Cell Proteomics 4 (7):873–886
13. Ong SE, Blagoev B, Kratchmarova I, Kristensen DB, Steen H, Pandey A, Mann M (2002) Stable isotope labeling by amino acids in cell culture, SILAC, as a simple and accurate approach to expression proteomics. Mol Cell Proteomics 1(5):376–386
14. Mueller LN, Brusniak M-Y, Mani D, Aebersold R (2008) An assessment of software solutions for the analysis of mass spectrometry based quantitative proteomics data. J Proteome Res 7(01):51–61
15. Eyrich B, Sickmann A, Zahedi RP (2011) Catch me if you can: Mass spectrometry-based phosphoproteomics and quantification strategies. Proteomics 11(4):554–570
16. Cox J, Hein MY, Luber CA, Paron I, Nagaraj N, Mann M (2014) Accurate proteome-wide label-free quantification by delayed normalization and maximal peptide ratio extraction, termed MaxLFQ. Mol Cell Proteomics 13(9):2513–2526
17. Parker R, Clifton-Bligh R, Molloy MP (2014) Phosphoproteomics of MAPK inhibition in BRAF-mutated cells and a role for the lethal synergism of dual BRAF and CK2 inhibition. Mol Cancer Ther 13(7):1894–1906
18. Casado P, Hijazi M, Britton D, Cutillas PR (2016) Impact of phosphoproteomics in the translation of kinase targeted therapies. Proteomics. https://doi.org/10.1002/pmic.201600235
19. Tyanova S, Temu T, Sinitcyn P, Carlson A, Hein MY, Geiger T, Mann M, Cox J (2016) The Perseus computational platform for comprehensive analysis of (prote) omics data. Nat Methods 13(9):731–740
20. Rogers LD, Fang Y, Foster LJ (2010) An integrated global strategy for cell lysis, fractionation, enrichment and mass spectrometric analysis of phosphorylated peptides. Mol BioSyst 6 (5):822–829
21. Zhou J, Zhou T, Cao R, Liu Z, Shen J, Chen P, Wang X, Liang S (2006) Evaluation of the application of sodium deoxycholate to proteomic analysis of rat hippocampal plasma membrane. J Proteome Res 5(10):2547–2553
22. Sugiyama N, Masuda T, Shinoda K, Nakamura A, Tomita M, Ishihama Y (2007) Phosphopeptide enrichment by aliphatic hydroxy acid-modified metal oxide chromatography for nano-LC-MS/MS in proteomics applications. Mol Cell Proteomics 6 (6):1103–1109

Chapter 9

CRISPR/Cas9-Based Chemogenomic Profiling in Mammalian Cells

Dominic Hoepfner, Gregory McAllister, and Gregory R. Hoffman

Abstract

Chemogenomic profiling is a powerful and unbiased approach to elucidate pharmacological targets and the mechanism of bioactive compounds. It is based on identifying cellular hypersensitivity and resistance caused by individual gene modulations with genome-wide coverage. Due to the requirement of bar-coded, genome-wide deletion collections, high-resolution experiments of this nature have historically been limited to fungal systems. Pooled RNAi reagents have enabled similar attempts in mammalian cells but efforts have been hampered by significant off-target effects and experimental noise. The CRISPR/Cas9 system for the first time enables precise DNA editing at defined loci in a genome-wide fashion. Here we present the detailed protocol that leverages the CRISPR/Cas9 system for chemogenomic profiling and target identification of diverse chemical probes.

Key words CRISPR, Cas9, Chemogenomic profiling, Target identification, Gene editing, Genetics, Genomics, Haploinsufficiency, Resistance, NAMPT, Signal peptidase

1 Introduction

The study of cellular responses to a multitude of bioactive chemicals (chemical genetics or chemical biology) in recent years facilitated the elucidation of pathways and targets involved in the pharmacology of potential drug molecules. Low molecular weight probes have revealed novel phenotypes and pathways in systems that have been extensively studied using classical genetics. Thus, it is becoming apparent that integrating classical, functional genetics with chemical genetics is a powerful approach to understand more detailed aspects of any biological process.

Identification of the target modulated by chemical probes can be a daunting task. It took more than 70 years to elucidate the mechanism of action of aspirin after its first mass production by Bayer [1].

Biochemical approaches like pull-down experiments using immobilized compounds have shown to be successful in some

Slava Ziegler and Herbert Waldmann (eds.), *Systems Chemical Biology: Methods and Protocols*, Methods in Molecular Biology, vol. 1888, https://doi.org/10.1007/978-1-4939-8891-4_9, © Springer Science+Business Media, LLC, part of Springer Nature 2019

cases [2] but fail if the target has a low expression level, is a membrane protein, or a nonprotein target like lipids or DNA. Genetic approaches have shown to be more successful. One popular example is the identification of the target of the immunosuppressive and anticancer drug Rapamycin by identification of resistance-conferring residues in the target [3]. But identification of the residues in a genome-wide context is cumbersome and even in the times of powerful full genome sequencing far from being a high-throughput method. With availability of the genome-wide hetero- and homozygous deletion collection of the model eukaryote *Saccharomyces cerevisiae* genetic methodologies with increased throughput have become possible. Drug-induced haploinsufficiency and homozygous profiling (HIP HOP) has first been pioneered by the team of Ron Davis [4] and the developed HIP HOP assay was demonstrated to be able to identify the target and pathway modulated by bioactive chemical probes [5]. The usefulness of the assay to identify modulated targets and pathways has meanwhile been demonstrated for thousands of cases [6–8]. Lack of equivalent deletion collections in mammalian cell lines however has prevented the use of this assay in higher systems. Genome-wide RNAi approaches have enabled similar approaches in higher eukaryotes [9–11] but efforts have been hampered by lack of on target activity and significant off-target effects resulting in difficult to interpret and noisy data.

During the last few years, the CRISPR/Cas9 (clustered regularly interspaced short palindromic repeats-CRISPR associated) system has sparked a revolution in the biological sciences. Among the many impacts of the CRISPR/Cas9 system is its application to targeted loss-of-function forward genetic screens in mammalian cells, a long-standing goal in genetic research. Recently, the successful application of the technology in mammalian cells for chemogenomic profiling purposes has been demonstrated by identification of a novel NAMPT compound and the first eukaryotic signal peptidase inhibitor [12, 13]). Here we describe in detail how virus-based CRISPR/Cas9 reagents allow the in-situ generation of genome-wide deletion collections in cell lines of choice and to recapitulate key features of drug-induced haploinsufficiency and homozygous profiling for target identification of compounds. The protocol provides step-by-step guidance how to generate virus pools, transduce cells, determine quality of the transduced cell pools and how to define cell density and splitting paradigms for the experiment. The chemogenomic profiling assay is based on assessing the relative proliferative fitness of each CRISPR-mutated cell in the presence and absence of compound. Each mutated cell line is unambiguously tagged by the unique, lentiviral integrated, 20 base-pair target locus sequence of the gRNA. At the end of the experiment, the sequence and relative abundance of all gRNA sequences are determined by next generation sequencing and taken as proxy to deduce gene mutations and impact on cell proliferation capacity. The protocol lists steps and bioinformatic resources that can be leveraged to process the sequencing data into descriptive chemogenomic profiles. These allow to identify pathways and genes leading to compound-induced hypersensitivity or resistance and thus delineate the mechanism of action of the bioactive compound.

2 Materials

2.1 Test Compound Potency Determination

1. Subconfluent culture of HCT116 cells (or cell line relevant for the mechanism of action of the test compound) in a T175 flask (or equivalent).
2. Trypsin–EDTA-based cell dissociation solution.
3. Phosphate buffer saline (PBS): 8.0 g/l NaCl, 0.2 g/l KCl, 1.42 g/l Na_2HPO_4, 0.24 KH_2PO_4 adjusted to pH 7.4 using HCl).
4. A high nutrient concentration tissue culture medium (e.g., Dulbecco's Modified Eagle Medium–high glucose), containing 4.5 g/l glucose, supplemented with 10% fetal bovine serum, 50 units/ml of penicillin, and 50 μg/ml of streptomycin.
5. A cell-counting chamber or device.
6. One 384-well glass bottom plate (if a cell counter is used for final readout), and/or one white-bottom 384-well plate (if CellTiter-Glo readout is used).
7. Test compound(s).
8. Cytotoxic, positive control compound (e.g., dequalinium chloride).
9. Dimethylsulfoxide (DMSO) as compound diluent.
10. Robotic pipetting equipment or multi-channel pipettes.
11. CellTiter-Glo reagent for ATP level-dependent viability assay (Promega).
12. 384-well suitable cell-counting device (cell number readout) or luminescence plate reader (CellTiter-Glo readout option).
13. A microplate shaker.
14. Curve-fitting software (e.g., Graphpad, XLfit, Tibco).

2.2 Antibiotic Kill Curve Determination

1. Puromycin stock concentration (10 mg/ml in 20 mM HEPES puffer, pH 7.2–7.5) or other antibiotic corresponding to the resistance marker(s) on the vector(s) used in Subheadings 3.3 and 3.5.
2. HCT116 cells or cell line in question.

3. 6-well tissue culture plates.
4. A high nutrient concentration tissue culture medium (*see* Subheading 2.1, **item 4**).
5. CellTiter-Glo reagent for ATP level-dependent viability assay (Promega).

2.3 Viral Packaging of Lentiviral Cas9 Plasmids

1. HEK293T cells.
2. Cas9 encoding lentiviral vector (*see* **Note 1**).
3. Packaging medium: a high nutrient concentration tissue culture medium (e.g., Dulbecco's Modified Eagle Medium–high glucose), containing 4.5 g/l glucose, supplemented with 10% fetal bovine serum, and 1% nonessential amino acid solution.
4. Transfection medium: buffered, reduced serum tissue culture medium for transfection (e.g., OptiMEM Serum Free Medium, Life Technologies).
5. Packaging mix (4:1 mix of the Gag/Pol and VSV-G) (Cellecta).
6. Collagen I coated 15 cm^2 tissue culture plates.
7. 0.45 μM cellulose acetate filter.
8. TransIT-293 Transfection Reagent (Mirus Bio).

2.4 Generation of Stable Cas9-Expressing Cell Line

1. HCT116 cells or cell line in question.
2. Cas9 encoding vector packaged into lentiviral particles as described in Subheading 3.3.
3. Antibiotic corresponding to the selection marker of your Cas9 vector.
4. Minimal killing concentration of your antibiotic as determined in Subheading 3.2.
5. A high nutrient concentration tissue culture medium (see Subheading 2.1, **item 4**).
6. 10 mg/ml polybrene (hexadimethrine bromide) stock solution to increase transduction efficiency.
7. 6-well tissue culture plates.
8. Cryotubes for storage of cells in liquid nitrogen.

2.5 Viral Packaging of Lentiviral gRNA Plasmids

1. HEK293T cells.
2. A pool of lentiviral vectors encoding the desired gRNAs under control of a Pol III promoter (*see* **Note 2**).
3. Packaging medium (*see* Subheading 2.3, **item 3**).
4. Transfection medium: buffered, reduced serum tissue culture medium for transfection (e.g., OptiMEM Serum Free Medium, Life Technologies).

5. Packaging mix (4:1 mix of the Gag/Pol and VSV-G) (Cellecta).
6. Collagen I coated 15 cm^2 tissue culture plates.
7. 0.45 μM cellulose acetate filter.
8. 5-layer cellstack tissue culture flask (Corning).
9. LentiX qPCR kit (Clontech).
10. TransIT-293 Transfection Reagent (Mirus Bio).

2.6 Determination of Multiplicity of Infection Parameters

1. Flow cytometer (*see* **Note 3**).
2. 6-well tissue culture plates.
3. A high nutrient concentration tissue culture medium (*see* Subheading 2.1, **item 4**).
4. 10 mg/ml polybrene (hexadimethrine bromide) stock solution to increase transduction efficiency.
5. Trypsin–EDTA-based cell dissociation solution.
6. Tubes/plates compatible with used flow cytometer.

2.7 Chemogenomic Profiling

1. Cas9 positive cell line as generated in Subheading 3.4.
2. Lentiviral gRNA pool as generated in Subheading 3.5.
3. Test compound.
4. Dimethyl sulfoxide (DMSO) or diluent of test compound.
5. 5-layer cellstack flasks (Corning).
6. A high nutrient concentration tissue culture medium (*see* Subheading 2.1, **item 4**).
7. 10 mg/ml polybrene (hexadimethrine bromide) stock solution to increase transduction efficiency.
8. Phosphate buffer saline (PBS, *see* Subheading 2.1, **item 3**).
9. 70% ethanol.
10. Trypsin–EDTA-based dissociation solution.

2.8 Genomic DNA Extraction

1. Kit for large scale gDNA extraction from mammalian cells (e.g., QIAmp DNA Blood Maxi Kit, Qiagen).

2.9 Determination of gRNA Abundance by Next Generation Sequencing

1. Quant-iT PicoGreen dsDNA Assay Kit (Thermo Fisher Scientific).
2. Illumina compatible PCR primers
   Primer P1 5′-AATGATACGGCGACCACCGAGATCTACACTCGATTTCTTGGCTTTATATATCTTGTGGAAAGGA-3′ and
   Primer P2 5′-CAAGCAGAAGACGGCATACGAGATXXXXXXXXXXGTGACTGGAGTTCAGACGTGTGCTCTTCCGATC-3′ (*see* **Note 4**).

3. 0.5 mM dNTPs.
4. Titanium Taq DNA polymerase and buffer.
5. 96-well format PCR cleanup kit (e.g., Agencourt AMPure XL beads and Agencourt SPRIPlate Super Magnet Plate, Beckman Coulter).
6. PCR thermocycler.
7. Consumables as outlined in the "Sequencing Library qPCR Quantification Guide" (11322363 C, Illumina).
8. HiSeq 2500 instrument (Illumina) or (*see* **Note 5**).
9. Custom read sequencing primer

   Primer A 5′-TCGATTTCTTGGCTTTATATATCTTGTGG AAAGGACGAAACACCG-3′.
10. Index read standard Illumina indexing primer 5′-GATCGGAAGAGCACACGTCTGAACTCCAGTCAC-3′.

2.10 Data Analysis

Software required for data processing:

1. bcl2fastq v2.18 (https://support.illumina.com/downloads/bcl2fastq-conversion-software-v2-18.html).
2. fastx_toolkit (http://hannonlab.cshl.edu/fastx_toolkit/).
3. Bowtie (http://bowtie-bio.sourceforge.net/index.shtml).
4. R programming language (https://www.r-project.org/) with DESeq2 package (http://bioconductor.org/packages/release/bioc/html/DESeq2.html).

3 Methods

This protocol is suitable for chemogenomic profiling of compounds with cytotoxic or cytostatic effects on a test cell line. As the method is based on gene-dosage dependent hypersensitivity and resistance, the selection of a suitable cell line is important. Ideally, the test cell line should be near diploid as aneuploidies or gene amplification can potentially mask hits. The cell line should demonstrate good growth characteristics in culture with rapid doubling times as the assay is based on competitive proliferation advantage and disadvantage leading to over- and underrepresentation of individual mutants. Good lentiviral transduction rates are a prerequisite. All experiments need to be carried out in a biosafety level 2 environment and by a certified operator in accordance with the local law for work with lentivirus.

3.1 Test Compound Potency Determination

To correctly dose the test substance at a sublethal concentration it is important to first assess the cytotoxic/cytostatic potential on the test cell line. Although conditions may vary from compound to compound, pilot experiments with compounds of different mechanism of actions have identified IC_{50} and IC_{30} concentrations to results in significant separation of mutants selectively affected by the test substance and the rest of the mutants in the pool.

The below depicted protocol is adapted for 384-well pates and suitable to record triplicate datasets of seven different compounds in parallel. As low volumes need to be pipetted, it is recommended to use robotic liquid handlers. If no liquid handling and plate reader equipment for experimentation in multi-well format is available, the same protocol for determination of antibiotic potency as in Subheading 3.2 can be used to determine the potency of the test compound in 6-well plates. It is however recommended to run the experiment in duplicates and extend the tested concentrations to a bigger range (if no information on the compound potency is known). If dose-response analyses have been run in 384-well or 96-well format it is recommended to still verify the $IC_{30/50}$ in 6-well format as compounds tend to exert different inhibition characteristics between a 384-well and a cellstack culture flask.

3.1.1 Day [0]: Cell Seeding

1. Take a subconfluent culture of HCT116 (or cell line in question) in a T175 flask (or equivalent) in high nutrient concentration tissue culture medium, remove medium and wash cells with 10 ml phosphate buffer saline.
2. Detach cells with 2 ml of cell dissociation solution.
3. Add 8 ml tissue culture medium to stop the reaction.
4. Spin down for 5 min at 200 × *g*, remove supernatant and resuspend into 10 ml high nutrient concentration tissue culture medium.
5. Determine cell number using a cell counter.
6. Adjust cell number to 37,500 cells/ml and a volume of >40 ml (enough for two 384-well plates) in high nutrient concentration tissue culture medium.
7. Seed 40 μl/well of the cells (1500 cells) using robotics or multi-channel equipment into one white-bottom 384-well plate and one clear-bottom 384-well.
8. Incubate cells in 37 °C/5% CO_2 incubator for 24 h.

3.1.2 Day [1]: Preparation of Serial Dilution Series and Addition of Compound

1. Using the appropriate equipment available to you, prepare 13-point master serial dilution(s) of your test compound(s) in DMSO. If no compound potency range is already known, start with a 10 mM top concentration and use a dilution factor of 2.5.
2. Prepare a 13-point serial dilution of a compound with known cyototoxic properties (positive control).

3. Remove the cell-containing plates prepared the day before from the incubator.
4. In a row-by-row fashion, dispense 80 nl/well of your master-dose response series (in triplicates, e.g., compound 1: row 1, 2, 3; compound 2: row 4, 5, 6, etc.) into the corresponding wells A–O of the two cell-containing plates. This will result in effective compound concentrations ranging from 20 μM to 0.37 nM.
5. Dispense 80 nl/well of your positive control master-dose response series into the corresponding wells A–O of rows 23 and 24 of the two cell-containing plates (duplicates).
6. Dispense 80 nl/well of DMSO into the P row (no drug control).
7. Mix on a plate shaker.
8. Incubate cells in 37 °C/5% CO_2 incubator for 72 h.

3.1.3 Day [4]: Viability Assessment

1. Remove cells from 37 °C/5% CO_2 incubation and allow cells to equilibrate to room temperature. Observe cells in the clear-bottom plate under the microscope and verify dose-dependent killing. If available, directly quantify cell numbers using a cell counter. If no viability assessment using CellTiter-Glo is performed continue at **step 8**.
2. Dilute the stock solution of CellTiter-Glo provided by the manufacturer 1:1 with H_20, mix and dispense 40 μl/well of the white-bottom plate.
3. Quickly spin the plate to sediment cells.
4. Incubate for 15 min at room temperature.
5. Measure luminescence on a plate reader following the instructions outlined in the CellTiter-Glo manual.
6. Calculate the average of the relative top luminescence intensity by setting the no-drug samples to 100% viability (to obtain "High") and the top concentrations of your cytotoxic, positive control (to obtain "Low").
7. Calculate the % inhibition for each compound concentration ("Sample") using the following equation:

$$\% \text{ Inhibition} = ((\text{High} - \text{Sample}/\text{High} - \text{Low}) \times 100)$$

8. Measure cell numbers on the clear bottom plate on a cell counter. Use the formula depicted above to calculate the % inhibition for each compound concentration.
9. Plot the normalized inhibition values into a curve fitting program (e.g., Excel fit, Graphpad, Tibco), perform a linear regression analysis and extrapolate the IC_{30} and IC_{50} concentrations.

3.2 Antibiotic Kill Curve Determination

The gRNA library vectors typically contain an antibiotic resistance marker (in the case of the used pRSG16-U6-sg-HTS6C-UbiC-TagRFP-2A-Puro it is puromycin). To select for a purely transduced population of cells it is important to determine the minimum amount of antibiotic required to kill off nontransduced cells. This concentration will then be used to select for successfully transduced cells. The displayed protocol will test for three different seeding densities.

3.2.1 Day [0]: Cell Seeding

1. Seed cells into each well of three 6-well plates, one for each of the following seeding densities: 0.1×10^6, 0.2×10^6, 0.4×10^6 cells per well of a standard 6-well plate.
2. Adjust volumes/well to 2 ml of high nutrient concentration tissue culture medium.
3. Incubate cells in 37 °C/5% CO_2 incubator for 48 h.

3.2.2 Day [2]: Medium Change and Addition of Antibiotic

1. Prepare a mastermix at 4 μg/ml of puromycin in high nutrient concentration tissue culture medium and do a series of 1:2 dilutions for the remaining puromycin concentrations (2, 1, 0.5, 0.25, and 0 μg/ml). Puromycin working stock is at 1 mg/ml in H_20.
2. Aspirate medium from each well and replace with 2 ml of the appropriate puromycin mastermix. Dispense puromycin toward the side wall of each well to not detach the cells.
3. Incubate cells in 37 °C/5% CO_2 incubator for 72 h.

3.2.3 Day [5]: Viability Assessment

1. Remove cells from 37 °C/5% CO_2 incubation and allow cells to equilibrate to room temperature. Observe cells under the microscope and verify dose-dependent killing. If available, directly quantify cell numbers using a cell counter and continue at **step 8** (enter cell numbers instead of CellTiter-Glo signal).
2. Aspirate medium from each well.
3. Dispense 1 ml of PBS and 1 ml of CellTiter-Glo reagent directly to the center of each well.
4. Incubate for 1 h at room temperature.
5. Mix each well by repeat pipetting and transfer 100 μl of lysate in triplicate from each well into a 96-well white opaque tissue culture plate (note down the selected pipetting scheme).
6. Measure luminescence on a plate reader following the instructions outlined in the CellTiter-Glo manual.
7. Calculate the relative average luminescence intensity by setting the 0 μg/ml sample to 100% viability (to obtain "High").
8. For each puromycin concentration calculate the % cell death using the following equation:

$$\%\text{Cell Death} = 100 - ((\text{CellTiter} - \text{Glo signal}/\text{High}) \times 100)$$

9. Repeat this for all seeding densities and finally take the minimal inhibitory antibiotic concentration at the seeding density best reflecting the final experiment.
10. Select the optimal seeding density (based on confluence measured visually under the microscope and by looking at the raw CellTiter-Glo data) to determine the minimal concentration of puromycin to be used.

3.3 Viral Packaging of Lentiviral Cas9 Plasmids

A DNA–TransIT ratio of 1:3 will be used and the TransIT-293 transfection master mix is made in a glass bottle, as the transfection lipid will bind to plastic, thereby decreasing transfection efficiency. Warm all reagents to room temperature before use. Penicillin/streptomycin should not be added to the medium, as it will hinder viral production efficiency.

3.3.1 Day [−1]: Seeding of 293T Cells for Virus Production

1. Seed 1.2×10^7 cells in 40 ml of packaging medium into a 15 cm^2 plate.
2. Incubate plate overnight at 37 °C in 5% CO_2.

3.3.2 Day [0]: Transfection

1. Mix 24.3 μl TransIT-293 transfection reagent with 875.7 μl transfection culture medium.
2. Mix solution, and incubate for 5 min at room temperature.
3. Add 4.5 μg of packaging mix and 3.6 μg of Cas9 plasmid to the transfection mix and swirl bottle to mix reagents together.
4. Incubate this reagent mix for 15 min at room temperature.
5. Add the reagent mix dropwise to the cells.
6. Return cells to 37 °C/5% CO_2 incubator.

3.3.3 Day [1]: Medium Change

1. Remove medium from cells and replace with 16 ml of fresh packaging medium.
2. Return cells to 37 °C/5% CO_2 incubator and incubate until day 4.

3.3.4 Day [4]: Viral Supernatant Harvest

1. Harvest viral supernatant from cells.
2. Filter the supernatant through a 0.45 μM cellulose acetate filter to remove any cells that may be in the supernatant.
3. Determine viral titers using the LentiX qPCR kit according to manufacturer's instructions. Titers are typically in the range of 5.0×10^6 TU/ml.
4. Aliquot the newly filtered virus in 1 ml aliquots and store at −80 °C.

3.4 Generation of Stable Cas9 Cell Line

The presented protocol is based on a two-vector system: one lentiviral vector containing the Cas9 protein under control of a suitable promoter (*see* **Note 1**) the other containing the gRNAs (with genome-wide coverage, (*see* **Note 2**). There are several reasons for separate delivery of the gRNA from Cas9: (1) limited packaging capacity of lentivirus (~9.7 kb), (2) robustness of the experiment (allows the verification of proper Cas9 function before conducting the genome-wide experiment), and (3) safety (unlikely that the operator is accidentally infected by both constructs and thus a functional Cas9 complex is assembled). If unknown for the cell line in question, first determine the required minimal killing concentration for the antibiotic resistance marker on your Cas9 construct following the protocol described under Subheading 3.2.

3.4.1 Day [− 1]: Cell Seeding

1. Plate two wells of a 6-well tissue culture multi-well plate with 0.1×10^6 cells in 2 ml/well of high nutrient concentration tissue culture medium (*see* **Note 6**).
2. Grow cells at 37 °C/5% CO_2 for 24 h.

3.4.2 Day [0]: Infection

1. For each well, prepare a master mix as follows: high nutrient concentration tissue culture medium and 5 μg/ml polybrene.
2. Add 2 ml of the master mix to each well. To one of the wells add calculated amount of virus to reach MOI of 1 (*see* Subheading 3.6).
3. Gently swirl the plates and return cells to 37 °C/5% CO_2.

3.4.3 Day [1]: Medium Change and Antibiotic Selection

1. Prepare a master mix of high nutrient concentration tissue culture medium, and the appropriate antibiotic concentration as determined by Subheading 3.2 to kill 100% of nontransduced cells.
2. Aspirate medium and replace with 2 ml of the antibiotic-containing master mix.
3. Gently swirl the plates and return cells to 37 °C/5% CO_2.

3.4.4 Day [4–14]: Medium Change, Subculture, and Antibiotic Selection

1. Replace antibiotic-containing medium every 3–4 days or split cells in the presence of antibiotic-containing medium. Select for at least for 10 days before banking cells.
2. The "no virus" containing well should be used as a guide to determine how long to keep the cells in selection before the cells can be frozen back. Once all the cells in the "no virus" well have been killed off, cells can be banked.
3. Alternatively, dilute/sort for single clones and continue culturing to expand the culture to workable cell numbers for further characterization.

4. Characterize Cas9 expression and function by immunofluorescence, Western blotting and FLAER assay (*see* **Note 7**).
5. Store Cas9 positive cells in cryotubes in liquid nitrogen.

3.5 Viral Packaging of Lentiviral gRNA Plasmids

A DNA–TransIT ratio of 1:3 will be used and the TransIT-293 transfection master mix is made in a glass bottle, as the transfection lipid will bind to plastic, thereby decreasing transfection efficiency. Warm all reagents to room temperature before use. Penicillin-Streptomycin should not be added to the medium, as it will hinder viral production efficiency.

3.5.1 Day [−1]: Seeding of 293T Cells for Virus Production

1. Seed 2.1×10^8 cells in 1 l of packaging medium per each 5 layer cellstack flask.
2. Incubate plate overnight at 37 °C in 5% CO_2.

3.5.2 Day [0]: Transfection

1. For each cell stack mix 510.3 μl TransIT-293 transfection reagent with 18.4 ml transfection medium.
2. Mix solution, and incubate for 5 min at room temperature.
3. Add 94.5 μg of packaging mix and 75.6 μg of gRNA plasmid (s) (*see* **Note 2**) to the transfection mix and swirl bottle to mix reagents together.
4. Incubate this reagent mix for 15 min at room temperature.
5. Add the reagent mix to 1 l of fresh packaging medium.
6. Remove medium from cells and replace with packaging medium containing the reagent mix.
7. Return cells to 37 °C/5% CO_2 incubator.

3.5.3 Day [1]: Medium Change

1. Remove medium from cells and replace with 335 ml of fresh packaging medium.
2. Return cells to 37 °C/5% CO_2 incubator.

3.5.4 Day [4]: Viral Supernatant Harvest

1. Harvest viral supernatant from cells.
2. Filter the supernatant through a 0.45 μM cellulose acetate filter to remove any cells that may be in the supernatant.
3. Determine viral titers using the LentiX qPCR kit according to manufacturer's instructions. Titers are typically in the range of 5.0×10^6 TU/ml.
4. Aliquot the newly filtered virus in 1 ml aliquots and store at −80 °C.

3.6 Determination of Multiplicity of Infection Parameters

This method describes a flow cytometry-based determination of MOI (multiplicity of infection) by using serial dilutions of virus (*see* **Note 3**). This is required to determine the amount of virus needed to set the MOI to 0.5 for the genome-wide profiling experiment.

3.6.1 Day [−1]: Cell Seeding

1. Seed cells into four 6-well plates at an appropriate seeding density (for HCT116 seed 1.0×10^5 cells/well) into 2 ml/well of high nutrient concentration tissue culture medium.
2. Incubate cells in 37 °C/5% CO_2 incubator for 24 h.

3.6.2 Day [0]: Infection

1. Remove an aliquot of your virus stock from −80 °C and quickly thaw in a 37 °C water bath.
2. Prepare a 10 ml master mix consisting of 8 ml of high nutrient concentration tissue culture medium and 5 μg/ml polybrene (10 mg/ml stock) and 2 ml of virus in a 15 ml conical tube. This tube will contain the highest concentration of virus (400 μl virus per 2 ml medium). This tube will be labeled as "A."
3. Label additional 15 ml conical tubes (B-L) and fill them with 5 ml high nutrient concentration tissue culture medium and 5 μl polybrene (10 mg/ml stock).
4. Perform serial dilutions (mix well before transferring to the next tube) by transferring 5 ml of solution from tube "A" to "B" to "C" up to "K". "L" is the no-virus control.
5. Incubate the serial dilutions at room temperature for 5 min.
6. Take the cells seeded into the four 6-well plates from the incubator, split into two plate-sets, aspirate medium from each well, and per plate set, replace with the 2 ml of medium from the serial dilution series.
7. Incubate cells in 37 °C/5% CO_2 incubator for 24 h.

3.6.3 Day [1]: Medium Change and Antibiotic Selection

1. Aspirate medium from all wells. In one serial dilution plate set, replace with 2 ml of fresh high nutrient concentration tissue culture medium. In the other plate set, replace with 2 ml of fresh high nutrient concentration tissue culture medium and the corresponding antibiotic at the concentration determined as described in Subheading 3.2.
2. Incubate cells in 37 °C/5% CO_2 incubator for 72 h.

3.6.4 Day [4]: Flow Cytometry Analysis

1. Remove cells from 37 °C/5% CO_2 incubator and aspirate medium from each well.
2. Dispense 250 μl of cell dissociation solution to each well and incubate for 5–10 min or until all cells have detached.
3. Dispense 750 μl of high nutrient concentration tissue culture medium to each well.
4. Mix well by repeat pipetting and immediately transfer 200 μl to an FACS compatible tube or plate.
5. Analyze samples on flow cytometer.
6. Determine the amount of virus required to get 50% fluorescent reporter-positive cells.

3.7 Chemogenomic Profiling

The length of the chemogenomic profiling experiment can be varied based on the scientific question. As demonstrated with the examples of NAMPT and ER signal peptidase inhibitors [12, 13], analysis of early time-points (day 7, 10, 14) is favorable for identification of hypersensitive hits outlining the target of the compound, whereas later time-points (day 18, 21 and beyond) result in clearer separation of hits that lead to drug resistance (Fig. 1). The splitting paradigm of the experiment allows the harvest of samples at different time-points which, in principle, allows for analysis of multiple time-points. Extension of experiments beyond 3 weeks is however not recommended as genetic drift over time between compound and DMSO treated samples will hamper analysis.

3.7.1 Day [−6] to Day [−1]: Expansion and Selection of Cas9 Expressing Cells

Aim is to expand a HCT116-Cas9 clone/population (*see* **Note 8**) and plate into five-layer cellstack flasks. During the experiment, the aim is to keep coverage above 1000 cells per gRNA construct. The number of cells to be plated will thus vary based on complexity of the gRNA library. If a 100,000 element library is to be screened at an MOI of 0.5, 2×10^8 cells need to be transduced (100,000 gRNAs × 1000 cells × 1/0.5 MOI). As cells are plated 1 day ahead of infection, they will double until infection and thus only half the number of cells will need to be seeded (1×10^8 cells). It is recommended not to seed more than 3×10^7 HCT116-Cas9 cells per five-layer cellstack (*see* **Note 9**) flask. If more cells are necessary, the cell number can be split into the appropriate number of flasks. As a reference, the cell number in this protocol reflect an experiment with a 100,000 gRNA element library split into two cell stack flasks with a complexity of 50,000 gRNAs each at day of transduction.

1. On day [−6]: Thaw and seed 3×10^7 Cas9-expressing cells into a five-layer cellstack flask, add 700 ml of high nutrient concentration tissue culture medium, and containing the appropriate antibiotic for the selection marker for the Cas9 construct at the concentration determined as described in Subheading 3.2.
2. Incubate cells at 37 °C/5% CO_2 for 72 h.
3. On day [−3] refresh medium (same as above) to maintain antibiotic selection. Incubate cells at 37 °C/5% CO_2.
4. On day [−1] it will be necessary to subculture the cells. Seed two five-layer cellstacks with 5×10^7 cells each in 500 ml high nutrient concentration tissue culture medium (no antibiotic selection), and incubate at 37 °C/5% CO_2 for 24 h.

3.7.2 Day [0]: Lentiviral Transduction of gRNA Pool

1. Per five-layer cellstack flask, prepare 500 ml of high nutrient concentration tissue culture medium, add 250 μl of polybrene (to a final concentration of 5 μg/ml) and add the appropriate

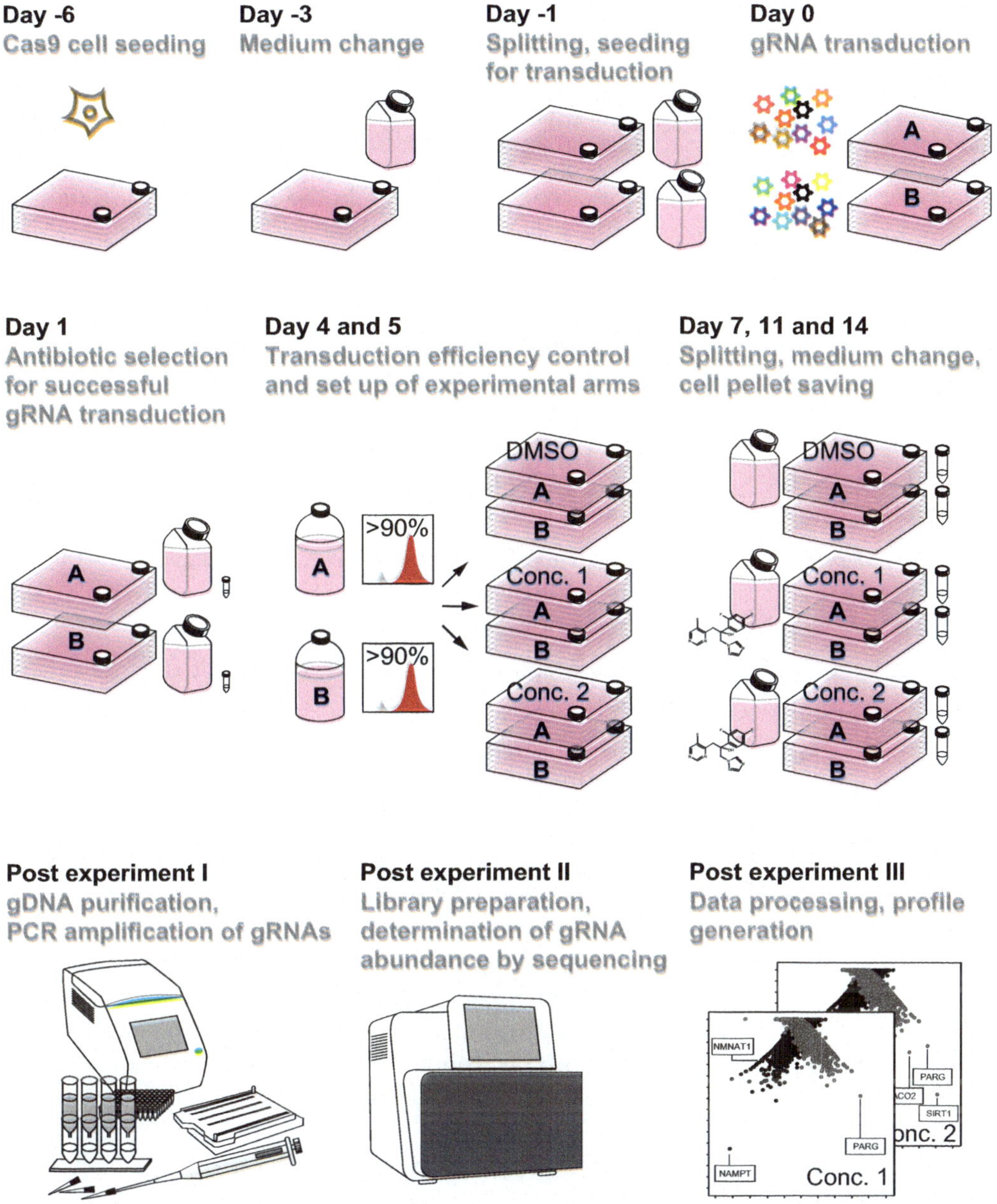

Fig. 1 Schematic representation of the chemogenomic profiling procedure described in Subheadings 3.7–3.10. The scheme depicts an example experiment comprising a 100,000 gRNA element library split into two five-layer cellstack flasks (50,000 elements each) and tested against two different test compound concentrations (e.g., IC_{30}, IC_{50}) and a no-drug control

amount of virus to reach an MOI of 0.5 as determined in Subheading 3.6. Mix well.

2. Incubate at room temperature for 5 min.
3. Decant medium (*see* **Note 10**) from five-layer cellstacks and replace with 500 ml of virus/polybrene containing medium.

4. Wipe cap neck with 70% ethanol.
5. Equilibrate evenly across the five layers of the cellstack.
6. Return cells to 37 °C/5% CO_2.

3.7.3 Day [1]: Medium Change and Antibiotic Selection

1. Per five-layer cellstack, prepare 1 l of high nutrient concentration tissue culture medium and the appropriate antibiotic concentration as determined in Subheading 3.2.
2. Carefully decant the virus-containing medium from the five-layer cellstack and replace with 1 l of high nutrient concentration tissue culture medium and antibiotic.
3. Immediately inactivate and wipe any spills, wipe cap neck with 70% ethanol and replace with new, sterile cap.
4. Equilibrate evenly across the five layers of the cellstack.
5. Return cells to 37 °C/5% CO_2 for 72 h.

3.7.4 Day [4]: Pool Collection and Quality Control

1. Per five-layer cellstack, decant medium, wash with 200 ml PBS and add 100 ml dissociation solution, equilibrate across all layers and pour off cell dissociation solution.
2. Incubate at 37 °C/5% CO_2 for 10 min or until all cells have detached.
3. Per five-layer cellstack, add 200 ml high nutrient concentration tissue culture medium.
4. Transfer cells to a 250 or 500 ml sterile bottle.
5. Count cell numbers and if a plasmid with a fluorescent protein reporter was used, take an aliquot and test for % of positive cells by flow cytometry analysis.
6. If the population is of good purity (*see* **Note 11**) seed 3×10^7 cells per cellstack and add 700 ml of high nutrient concentration tissue culture medium (*see* **Note 12**). Be aware that you will need the corresponding number of cellstacks per test regimen (compound concentration 1, compound concentration 2, DMSO control).
7. Return cells to 37 °C/5% CO_2 for 24 h (*see* **Note 13**).

3.7.5 Day [5]: Test Compound Addition

1. Per five-layer cellstack and experimental branch add the appropriate concentration of test compound as determined in Subheading 3.1 add same volume of DMSO to the control arm of the experiment. Mix well.
2. Return cells to 37 °C/5% CO_2 for 48 h.

3.7.6 Day [7, 11, and 14]: Test Compound Competition, Subculture, and Harvesting

1. Subculture cells at indicated days, seed 3×10^7 cells per flask into test compound containing 700 ml high nutrient concentration tissue culture medium (*see* **Note 12**).

2. On day 14 (but preferably also on earlier days), for each experimental arm collect twice $1x10^8$ cells (one will be processed for gDNA extraction, the other will be kept at −80 °C as back up).

3.8 Genomic DNA Extraction

1. Extract genomic DNA (gDNA) using the Whole Blood QIAamp DNA Blood Maxi Kit (Centrifugation Protocol) or equivalent gDNA extraction kit according to manufacturer's instructions.

3.9 Determination of gRNA Abundance by Next Generation Sequencing

1. Quantify the genomic DNA using Quant-iT PicoGreen following the manufacturer's recommendations.
2. For generation of Illumina sequencing libraries split the genomic DNA (gDNA) into multiple PCR reactions. Per 2000 gRNAs use one PCR reaction (in the example presented here with a 100,000 gRNA library, the gDNA would be split into 50 PCR reactions).
3. Add 4 μg of gDNA per tube.
4. Per tube, prepare the following PCR mix: 0.5 μM primer P1, 0.5 μM primer P2, 0.5 mM dNTPs, 1× Titanium Taq DNA polymerase, and buffer. Adjust volume to 100 μl.
5. Run the following reaction on a thermocycler 1× 98 °C for 5 min; 28× 95 °C for 15 s, 65 °C for 15 s, 72 °C for 30 s; 1× 72 °C for 5 min.
6. Purify the PCR product using a cleanup kit according to manufacturer's instructions.
7. Quantify product amount and quality by qPCR following the Sequencing Library qPCR Quantification Guide by Illumina (*see* **Note 5**).
8. Add 4 μg of DNA per lane of the Illumina flow-cell surface.
9. Sequence library with 1x30 base reads, using sequencing primer A and a 1×11b index read, using the standard Illumina indexing primer on an Illumina 2500 HiSeq instrument according to manufacturer's instructions (*see* **Note 14**).

3.10 Data Analysis

1. The Illumina HiSeq 2500 machine outputs BCL files that must be demultiplexed and converted to FASTQ format. This can be accomplished, along with adapter trimming, by using Illumina provided bcl2fastq software. See documentation for more options.

```
bcl2fastq -R <RUN FOLDER> -o <OUTPUT FOLDER> --sample-sheet <SAMPLE SHEET>
```

2. Isolate gRNA sequence from NGS reads. The specifics of this will depend on the design of the input library, but for reference the protocol here describes a 100,000-element library designed

with unique gRNAs of 20 base pairs length at the beginning of the read.

```
fastx_trimmer -l 20 -i <INFILE> -i <OUTFILE>
```

3. Align reads to input gRNA library using bowtie. This requires building a reference database of all gRNA sequences of the input library (consult bowtie documentation for specifics). Sequences are aligned allowing no mismatches and reads that do not align are discarded.

```
bowtie -q -v 0 --norc <REFERENCE DB> <FASTQ FILES> >> <OUTPUT FILE>
```

4. Aggregate and count each gRNA sequence using your programming language of choice, e.g., Python. This will provide the raw count data for all downstream statistics.
5. NGS-based counts can be modeled using the negative binomial distribution and there are multiple software packages to do this. Here we suggest using the DESeq2 package in the Bioconductor software suite coded in the R programming language. Consult the package documentation for specifics of implementation. Contrasts should be constructed to generate fold changes between compound treatment at both doses and DMSO treatment. Additionally, contrasts should be generated between DMSO treatment and input library plasmid to look for control cytotoxic genes dropping out as a quality control check for CRISPR/Cas9 activity.
6. Given a library design of multiple reagents targeting an individual gene, in the case of this protocol five gRNAs per gene, gene level statistics can be generated using the RSA algorithm [14] which calculates the cumulative minimum hypergeometric score for each reagent given the distribution of the entire library. This results in a log10 based p-value for each gene which should then be corrected for multiple hypothesis testing using the Benjamini–Hochberg procedure.
7. Screen quality can be assessed using multiple metrics including NGS read quality, read depth, read distribution, replicate correlation, performance of library controls (known cytotoxic genes, nontargeting controls, positive/negative controls), and principal component analysis.
8. Hit calling can be performed in multiple ways depending on the design of the experiment. In this protocol, bypass and resistance mechanisms can be identified by examining gRNAs that induce proliferation (positive fold change) in the comparison of compound-treated samples to DMSO. Targets and pathways directly affected by compound treatment can be

identified by examining hits that synergize with compound treatment at early time points by reducing proliferation (negative fold change) in the comparison to DMSO.

4 Notes

1. There exist many Cas9 encoding lentiviral vectors from commercial and noncommercial distributors. The following resource provides a good overview over available constructs https://www.addgene.org/crispr/cut/. In our experience EF1a promoter is superior over CMV. Silencing of CMV has been observed after retesting of banked samples.
2. There are several commercial and noncommercial distributors for gRNA libraries with different genomic coverage and species selectivity. The following resource provides a good first overview https://www.addgene.org/crispr/libraries/.
3. If the gRNA-encoding vector does not contain a fluorescent reporter but only an antibiotic selection marker the same protocol can be used with slight modifications: Instead of flow-cytometric analysis apply antibiotic selection pressure at day 1 (use the appropriate concentration of antibiotic as determined in Subheading 3.2) and in combination with a cell counter determine the amount of virus required to render 50% of cells antibiotic-resistant.
4. Denote a 10 base PCR-sample specific barcode used for data demultiplexing following sequencing.
5. If you have no access to a next generation sequencing instrument the following resource provides a list of commercial Illumina sequencing service providers: https://genohub.com/showcased-providers/.
6. Start with a low cell passage.
7. FLAER (Alexa Fluor 488 proaerolysin, FL2S, Cedarlanelabs) refers to a nontoxic variant of fluorescently labeled aerolysin that binds to GPI anchored proteins and consequently stains cells. Upon inactivation of the PIG-A gene involved in GPI anchor biosynthesis by a gRNA containing the following 20 base-pair homology to the PIG-A gene following 5-′-TGGCGTGGAAGAGAGCATCA-3′, FLAER fails to bind and reduced cellular staining can be monitored by flow-cytometric analysis. This protocol is an adaptation from a clinical diagnostic test used to diagnose mutations in the GPI biosynthesis pathway [15] and allows to verify homozygous inactivation and thus functional integrity of the CRISPR/Cas system in cells.

8. Practical experience led us to favor working cells derived from a single Cas9 clone rather than a Cas9 pool. Although unwanted background mutations can hamper the experiment when working with a single clone, we noticed much better signal-to-noise ratios for loss of function (hypersensitive) hits.
9. Cell number might need adjustment based on size and growth characteristics of cell line in question. If unsure, conduct pilot experiment in T-175 flaks and scale up results by a factor of 20.
10. Always treat virus-contaminated solutions with appropriate inactivating agent and consider local regulations concerning compound and/or biological waste disposal.
11. Expect a purity of around 95% of fluorescence positive cells. If a harvested population is substantially below this value, seed 1.4×10^8 cells/5-layer cellstack and select for an additional 72 h on antibiotic selection.
12. Once the cells are taken off antibiotic selection, the selective pressure is not reapplied. If the purity of fluorescence-positive cells drops between Day 4 and Day 14 consider forfeiting and repeating the experiment.
13. Our experiments are based on time-matched controls. We thus are not checking for confluency and are not splitting cells accordingly. The DMSO control should be more/over confluent at the indicated subculturing days compared to the highest test compound concentration. If a generation-time-matched control is desired for each experimental arm (control, concentration 1, concentration 2, concentration n), a T-175 flask needs to be seeded and propagated in parallel as confluency of cells in five-layer cellstack vessels cannot be determined by microscopy. The scaling factor from a T-175 flask to a 5-layer cellstack is 20.
14. 50–60 million reads were generated per 100,000 gRNA element sample, resulting in an average of approximately 500 reads per gRNA.

Acknowledgments

We thank Nadire Ramadan Cochran, David Estoppey, Malini Varadarajan, and Claudia Agarinis for their help setting up the described protocols and careful proofreading of the manuscript.

References

1. Vane JR (1971) Inhibition of prostaglandin synthesis as a mechanism of action for aspirin-like drugs. Nat New Biol 231(25):232–235
2. Huang SM, Mishina YM, Liu S, Cheung A, Stegmeier F, Michaud GA, Charlat O, Wiellette E, Zhang Y, Wiessner S, Hild M,

Shi X, Wilson CJ, Mickanin C, Myer V, Fazal A, Tomlinson R, Serluca F, Shao W, Cheng H, Shultz M, Rau C, Schirle M, Schlegl J, Ghidelli S, Fawell S, Lu C, Curtis D, Kirschner MW, Lengauer C, Finan PM, Tallarico JA, Bouwmeester T, Porter JA, Bauer A, Cong F (2009) Tankyrase inhibition stabilizes axin and antagonizes Wnt signalling. Nature 461 (7264):614–620. https://doi.org/10.1038/nature08356

3. Heitman J, Movva NR, Hall MN (1991) Targets for cell cycle arrest by the immunosuppressant rapamycin in yeast. Science 253 (5022):905–909
4. Winzeler EA, Shoemaker DD, Astromoff A, Liang H, Anderson K, Andre B, Bangham R, Benito R, Boeke JD, Bussey H, Chu AM, Connelly C, Davis K, Dietrich F, Dow SW, El Bakkoury M, Foury F, Friend SH, Gentalen E, Giaever G, Hegemann JH, Jones T, Laub M, Liao H, Liebundguth N, Lockhart DJ, Lucau-Danila A, Lussier M, M'Rabet N, Menard P, Mittmann M, Pai C, Rebischung C, Revuelta JL, Riles L, Roberts CJ, Ross-MacDonald P, Scherens B, Snyder M, Sookhai-Mahadeo S, Storms RK, Veronneau S, Voet M, Volckaert G, Ward TR, Wysocki R, Yen GS, Yu K, Zimmermann K, Philippsen P, Johnston M, Davis RW (1999) Functional characterization of the S. cerevisiae genome by gene deletion and parallel analysis. Science 285(5429):901–906
5. Giaever G, Shoemaker DD, Jones TW, Liang H, Winzeler EA, Astromoff A, Davis RW (1999) Genomic profiling of drug sensitivities via induced haploinsufficiency. Nat Genet 21(3):278–283. https://doi.org/10.1038/6791
6. Lee AY, St Onge RP, Proctor MJ, Wallace IM, Nile AH, Spagnuolo PA, Jitkova Y, Gronda M, Wu Y, Kim MK, Cheung-Ong K, Torres NP, Spear ED, Han MK, Schlecht U, Suresh S, Duby G, Heisler LE, Surendra A, Fung E, Urbanus ML, Gebbia M, Lissina E, Miranda M, Chiang JH, Aparicio AM, Zeghouf M, Davis RW, Cherfils J, Boutry M, Kaiser CA, Cummins CL, Trimble WS, Brown GW, Schimmer AD, Bankaitis VA, Nislow C, Bader GD, Giaever G (2014) Mapping the cellular response to small molecules using chemogenomic fitness signatures. Science 344 (6180):208–211. https://doi.org/10.1126/science.1250217
7. Hoepfner D, Helliwell SB, Sadlish H, Schuierer S, Filipuzzi I, Brachat S, Bhullar B, Plikat U, Abraham Y, Altorfer M, Aust T, Baeriswyl L, Cerino R, Chang L, Estoppey D, Eichenberger J, Frederiksen M, Hartmann N, Hohendahl A, Knapp B, Krastel P, Melin N, Nigsch F, Oakeley EJ, Petitjean V, Petersen F, Riedl R, Schmitt EK, Staedtler F, Studer C, Tallarico JA, Wetzel S, Fishman MC, Porter JA, Movva NR (2014) High-resolution chemical dissection of a model eukaryote reveals targets, pathways and gene functions. Microbiol Res 169(2-3):107–120. https://doi.org/10.1016/j.micres.2013.11.004
8. Roemer T, Xu D, Singh SB, Parish CA, Harris G, Wang H, Davies JE, Bills GF (2011) Confronting the challenges of natural product-based antifungal discovery. Chem Biol 18(2):148–164. https://doi.org/10.1016/j.chembiol.2011.01.009
9. Whitehurst AW, Bodemann BO, Cardenas J, Ferguson D, Girard L, Peyton M, Minna JD, Michnoff C, Hao W, Roth MG, Xie XJ, White MA (2007) Synthetic lethal screen identification of chemosensitizer loci in cancer cells. Nature 446(7137):815–819. https://doi.org/10.1038/nature05697
10. Bassik MC, Kampmann M, Lebbink RJ, Wang S, Hein MY, Poser I, Weibezahn J, Horlbeck MA, Chen S, Mann M, Hyman AA, Leproust EM, McManus MT, Weissman JS (2013) A systematic mammalian genetic interaction map reveals pathways underlying ricin susceptibility. Cell 152(4):909–922. https://doi.org/10.1016/j.cell.2013.01.030
11. Matheny CJ, Wei MC, Bassik MC, Donnelly AJ, Kampmann M, Iwasaki M, Piloto O, Solow-Cordero DE, Bouley DM, Rau R, Brown P, McManus MT, Weissman JS, Cleary ML (2013) Next-generation NAMPT inhibitors identified by sequential high-throughput phenotypic chemical and functional genomic screens. Chem Biol 20(11):1352–1363. https://doi.org/10.1016/j.chembiol.2013.09.014
12. Estoppey D, Hewett JW, Guy CT, Harrington E, Thomas JR, Schirle M, Cuttat R, Waldt A, Gerrits B, Yang Z, Schuierer S, Pan X, Xie K, Carbone W, Knehr J, Lindeman A, Russ C, Frias E, Hoffman GR, Varadarajan M, Ramadan N, Reece-Hoyes JS, Wang Q, Chen X, McAllister G, Roma G, Bouwmeester T, Hoepfner D (2017) Identification of a novel NAMPT inhibitor by CRISPR/Cas9 chemogenomic profiling in mammalian cells. Sci Rep 7:42728. https://doi.org/10.1038/srep42728
13. Estoppey D, Lee CM, Janoschke M, Lee BH, Wan KF, Dong H, Mathys P, Filipuzzi I, Schuhmann T, Riedl R, Aust T, Galuba O, McAllister G, Russ C, Spiess M, Bouwmeester T, Bonamy GMC, Hoepfner D

(2017) The natural product cavinafungin selectively interferes with Zika and dengue virus replication by inhibition of the host signal peptidase. Cell Rep 19(3):451–460. https://doi.org/10.1016/j.celrep.2017.03.071

14. Konig R, Chiang CY, Tu BP, Yan SF, DeJesus PD, Romero A, Bergauer T, Orth A, Krueger U, Zhou Y, Chanda SK (2007) A probability-based approach for the analysis of large-scale RNAi screens. Nat Methods 4 (10):847–849. https://doi.org/10.1038/nmeth1089

15. Brodsky RA (2009) How I treat paroxysmal nocturnal hemoglobinuria. Blood 113 (26):6522–6527. https://doi.org/10.1182/blood-2009-03-195966

Chapter 10

Exome Sequencing of Drug-Resistant Clones for Target Identification

Ting Han and Deepak Nijhawan

Abstract

Many small molecule compounds with anticancer activities are discovered through phenotype-based screens. However, discovering the targets of these small molecules has been challenging. The gold standard for target identification requires the discovery of mutations in the target protein that block the effects of small molecules in vitro as well as in vivo. Here we describe the procedures for isolating drug resistant clones using the colorectal cancer cell line HCT-116 followed by whole-exome sequencing to identify recurrent mutations associated with compound resistance. Together with downstream in vitro and in vivo validation experiments, this strategy enables rapid target discovery for cytotoxic compounds.

Key words Forward genetics, Phenotype-based screens, Anticancer toxins, Compound resistant mutations, Mismatch repair deficiency, Whole-exome sequencing, Target Identification

1 Introduction

The discovery of cancer drugs depends on the identification of chemicals that inhibit proteins essential to cancer. Unfortunately, many of the proteins genetically implicated as important for the proliferation of cancer cells have no known chemical inhibitors and often, no assay is available that can be used to isolate an inhibitor [1]. In fact, exhaustive efforts to identify chemical inhibitors for bona fide cancer genes have failed, resigning these proteins to be labeled "undruggable" [2]. Nonetheless, numerous small molecules have been identified as anticancer toxins in phenotypic screens [3, 4]. For many of these toxins, the protein target is either unknown or ambiguous. The identification of the proteins influenced by these orphan toxins offers a path to find new targets that are, by necessity, "druggable."

Despite its potential, target discovery of orphan toxins is seldom accomplished and it remains technically challenging. One approach to target identification depends on identifying a significant correlation between mRNA signatures derived from cells

Slava Ziegler and Herbert Waldmann (eds.), *Systems Chemical Biology: Methods and Protocols*, Methods in Molecular Biology, vol. 1888, https://doi.org/10.1007/978-1-4939-8891-4_10,

treated with either a toxin or a gene-specific RNAi [5]. More recently, recessive genetic systems including pooled genome-wide shRNA or CRISPR screens as well as mammalian haploid cell lines have successfully been used to discover the genes required for toxin activity [6–8]. Yet while each of these approaches guides us to the affected pathways, none of them is designed to identify the direct target.

Indeed, the identification of the direct target of an orphan compound is a critical step in drug discovery. Biochemical isolation of proteins that bind a toxin can be used to determine the direct target [9]. This approach, however, usually requires potent compounds that form relatively stable noncovalent interactions with their targets. As a result, each toxin first requires an extensive effort to optimize potency while including chemical groups that are necessary for affinity purification. Furthermore, toxins likely bind several proteins of which only one is the direct target. Therefore, an additional challenge to this approach is reconciling which of the many proteins that bind a toxin is relevant to its activity.

An unambiguous way of discovering the direct chemical target is through a mutation that prevents compound binding and renders cells resistant to the toxin. Forward genetics is an unbiased strategy to discover such compound-resistant alleles and has successfully been used to identify chemical targets in classic genetic systems such as yeast, worms, and bacteria [10]. These approaches had been limited to the few chemicals that elicit a selectable phenotype in one of the aforementioned organisms. Kapoor and colleagues have overcome this limitation by using a forward genetic strategy in a human cancer cell line, HCT-116, to identify chemical targets in a proof of principle study [11]. HCT-116 is a diploid cell line with a high nucleotide substitution rate (10^{-5}) [12]. Therefore, unlike the aforementioned recessive genetic systems, HCT-116 resistance is primed to be the result of heterozygous mutations in the target (Fig. 1a). The identification of such mutant alleles in multiple independently derived resistant clones strongly implicates the corresponding proteins as biologically relevant targets of the toxin.

We have developed a pipeline to systematically screen orphan compounds for resistance to HCT-116 cells, after which, we used whole-exome sequencing of resistant clones to discover their direct targets. Compound-resistant alleles are usually nonsynonymous mutations that have little impact on the normal functions of proteins, but are able to reduce their affinity for a toxin. In addition, the alleles are likely clustered around the drug-binding pocket and are genetically dominant over the wild-type allele in rendering drug resistance (Fig. 1a).

The discovery of a causal mutation underlying drug resistance relies on identifying independently acquired recurrent mutations. To efficiently isolate independent mutation events, we use a "barcoded" population of HCT-116 cells (Fig. 1b). In Subheading 3.1,

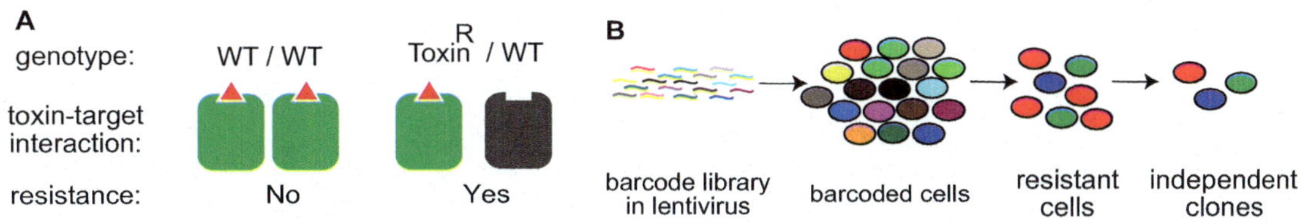

Fig. 1 Target identification of cytotoxic compounds in a population of barcoded HCT-116 cells. (**a**) Mutations that compromise compound binding to its target result in compound resistance (represented by an R superscript). (**b**) Efficient isolation of independent mutation events using a "barcoded" population of HCT-116 cells. Lines with different colors indicate different 20mer sequences

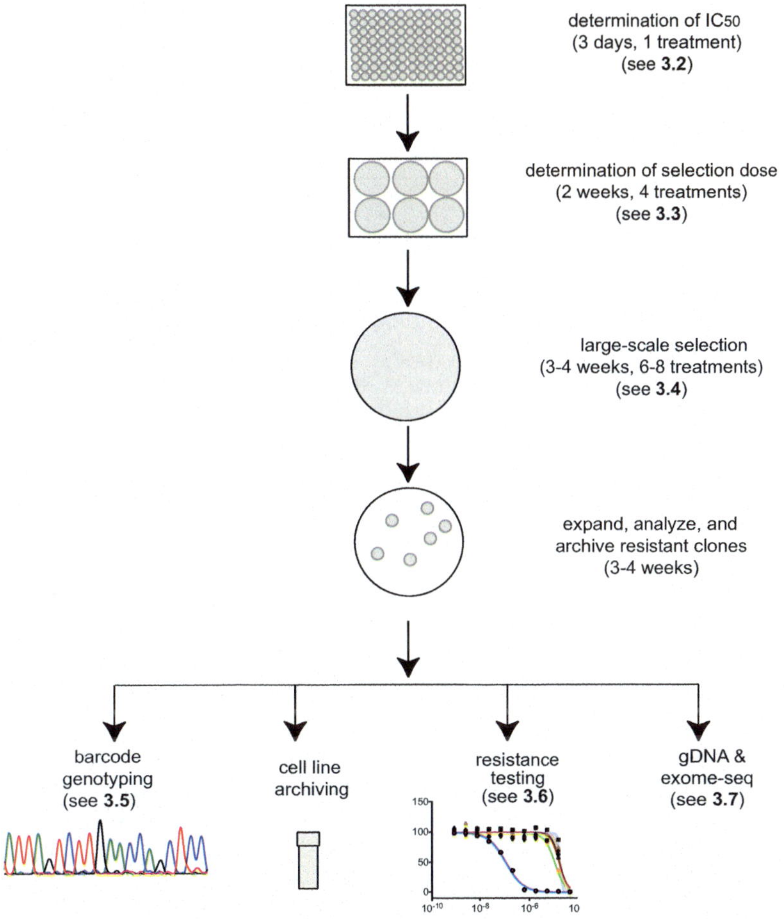

Fig. 2 The experimental pipeline for selecting resistant clones to cytotoxic compounds

we describe the procedures for generating a lentiviral library containing ~1000 independent random 20 mer oligonucleotides to infect HCT-116 cells. In Subheadings 3.2, 3.3, and 3.4, we outline the methods for selecting compound resistant clones using optimized compound concentrations (Fig. 2). In Subheading 3.5, we

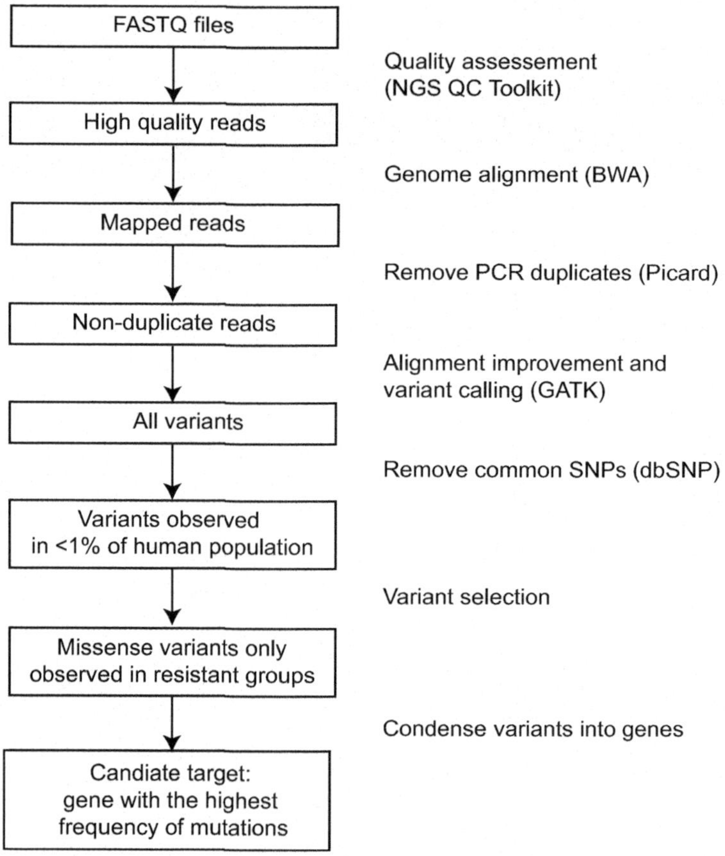

Fig. 3 The computational pipeline for analyzing exome-seq data

provide a detailed protocol for analyzing the barcode for each clone by Sanger sequencing a PCR product amplified from genomic DNA with primers that encompass the 20 mer sequence. Since the diversity of our barcode library is ~10^3, clones with different barcode sequences are virtually independent. In Subheading 3.6, we validate compound resistance of isolated clones and examine their sensitivity to the unrelated toxin paclitaxel. In Subheading 3.7, we isolate genomic DNA from six independent resistant clones and subject them to whole-exome sequencing. In Subheading 3.8, we briefly outline the bioinformatic procedures for mutation discovery using the whole-exome sequencing data (Fig. 3). These analyses typically result in a single candidate gene for further causality test and functional studies. Throughout this protocol, we use the compound indisulam as an example. Using methods described above, we discovered that mutations in the gene encoding the RBM39 RNA binding protein cause indisulam resistance [13]. In addition, we have successfully used this suite of methodologies to discover the target of another orphan toxin CD437 as DNA polymerase α [14].

2 Materials

2.1 Cell Culture

1. HCT-116 cells (ATCC) (*see* **Note 1**).
2. Lenti-X 293T cells (Clontech) (*see* **Note 1**).
3. Cell culture medium: Dulbecco's Modified Eagle Medium (DMEM) supplemented with 10% (v/v) fetal bovine serum (FBS), and 2 mM L-glutamine. For 500 mL of DMEM, add 50 mL of FBS and 5 mL of 200 mM L-glutamine.
4. Phosphate buffered saline (PBS).
5. 0.25% Trypsin–0.02% EDTA solution.
6. Cell culture vessels: tissue culture treated 6-well plates and 10 cm^2 plates.
7. Fugene HD transfection reagent (Promega).

2.2 Molecular Biology Reagents

1. Oligonucleotide sequences (*see* **Note 2**)
   oligo1: 5′-ggaaaggacgaaacACCGGTT-3′
   oliog2: 5′-cgagaattcNNNNNNNNNNNNNNNNNNNNNAAAAAACCGGTgtttcgtcctttcc-3′
   oligo3: 5′-ggaggcttggtaggtttaagaa-3′
   oligo4: 5′-ggatctctgctgtccctgtaat-3′
   oligo5: 5′-ggaggcttggtaggtttaagaa-3′.
2. Plasmids
   pLKO.1 (Addgene 10878)
   psPAX2 (Addgene 12260)
   pMD2.G (Addgene 12259).
3. Platinum Taq DNA polymerase kit (Sigma-Aldrich).
4. 3 M sodium acetate () dissolved in molecular biology grade water.
5. 100% ethanol (200 proof). Prepare 75% ethanol by mixing 3 volumes of 100% ethanol with 1 volume of molecular biology grade water.
6. Restriction enzymes and buffer: AgeI-HF, EcoRI-HF, and CutSmart buffer.
7. 0.8% agarose gel with 0.2 μg/mL of ethidium bromide.
8. DNA gel extraction kit.
9. Phenol–chloroform–isoamyl alcohol solution (25:24:1) (Sigma-Aldrich).
10. T4 DNA ligase with T4 DNA ligase buffer.
11. Stbl3 chemically competent cells.
12. Plasmid midiprep kit.
13. Lysis buffer: 10 mM Tris–HCl, pH 8.0, 50 mM KCl, 0.1 mg/mL gelatin, 0.45% NP-40, 0.45% Tween 20, and 0.2 mg/mL

of Proteinase K. Prepare the lysis buffer with molecular biology grade water, and filter-sterilize before use.

14. Taq DNA polymerase with standard Taq buffer.
15. DNA clean and concentrator −5 (Zymo Research).
16. QiaAmp DNA mini kit (Qiagen).

2.3 Chemicals

1. 8 mg/mL Polybrene dissolved in molecular biology grade water. Filter-sterilize the solution before use.
2. 2 mg/mL puromycin dissolved in molecular biology grade water. Filter-sterilize the solution before use.
3. 10 mM indisulam (MedKoo Biosciences) dissolved in dimethyl sulfoxide (DMSO).
4. 10 mM Paclitaxel dissolved in DMSO.

2.4 Sterile Supplies for Harvesting Cell Colonies

1. Medium Forceps (wrapped in aluminum foil and autoclaved).
2. 60 mm glass petri dish (Corning 3160-60) containing cloning cylinders (Corning 3166-10) (wrapped in aluminum foil and autoclaved).
3. 60 mm glass petri dish containing Dow Corning 976 V silicone high vacuum grease (wrapped in aluminum foil and autoclaved).

2.5 Cell Viability Assays

1. White 96-well assay plates.
2. CellTiter-Glo Luminescent Cell Viability Assay kit (Promega).
3. Luminescence plate reader.

2.6 Software and Tools

1. GraphPad Prism.
2. NGS QC Toolkit (v2.3.1).
3. Burrows-Wheeler Aligner (BWA, v0.7.5a).
4. Picard (v1.99).
5. Genome Analysis Toolkit (GATK, v2.7-2).

3 Methods

Standard aseptic tissue culture techniques are used for culturing HCT-116 and lenti-X 293T cells.

3.1 Generation of Barcoded HCT-116 Cells (See Note 3)

1. Assemble the hybridization and primer extension reaction on ice by mixing 6 μL of 10 μM oligo1, 6 μL of 10 μM oligo2, 10 μL of 10× PCR buffer without Mg^{2+}, 3 μL of 50 mM $MgCl_2$, 2 μL of 10 mM dNTP mix, 0.5 μL of Platinum Taq DNA polymerase, and 72.5 μL of H_2O. Incubate the reaction in a thermocycler at 94 °C for 5 min, cool down gradually to

55 °C (1% ramp), and then at 72 °C for 30 min. This step uses Taq DNA polymerase to perform a primer extension reaction to create double stranded DNAs harboring 20 nucleotide random sequences.

2. Precipitate DNA duplexes generated in the previous reaction by adding 12.5 μL of 3 M sodium acetate, 250 μL of 100% ethanol, and incubate for 2 h at −80 °C (*see* **Note 4**). Collect precipitated DNA by centrifugation for 20 min at 20,000 × *g* at room temperature, followed by two washes with cold 75% ethanol. Air-dry pellet briefly and resuspend in 40 μL of H_2O.
3. Assemble the vector digestion reaction on ice by mixing 2 μg of pLKO.1 vector, 10 μL of CutSmart buffer, 3 μL of Age-HF and 3 μL of EcoRI-HF in a 100 μL reaction. Assemble the insert digestion reaction by mixing 40 μL of DNA duplexes, 20 μL of CutSmart buffer, 3 μL of Age-HF, 3 μL of EcoRI-HF, and 134 μL of H_2O. Incubate both reactions at 37 °C for 4 h.
4. Run out the digestion reaction of pLKO.1 on an ethidium bromide-stained 0.8% agarose gel. Visualize the digested bands using a standard gel imager. Excise the digested vector backbone and extract DNA with a DNA gel extraction kit according to the manufacturer's instructions. Elute in 30 μL of H_2O.
5. Mix the digestion reaction of DNA duplexes with 200 μL of H_2O and 400 μL of phenol–chloroform–isoamyl alcohol solution. Vortex and spin for 5 min at 20,000 × *g*. Transfer the liquid phase into a new tube, mix with 40 μL of 3 M sodium acetate, 1 mL of 100% ethanol, and precipitate for 2 h at −80 °C. Collect precipitated DNA by centrifugation for 20 min at 20,000 × *g* at room temperature followed by two washes with cold 75% ethanol. After brief air-drying, resuspend pellet in 10 μL of H_2O.
6. Assembly ligation reactions on ice by mixing 25 ng of digested pLKO.1, 1 ng of digested DNA duplexes, 1 μL of 10× T4 DNA ligase buffer, and 1 μL of T4 DNA ligase in a 10 μL reaction. Incubate at 16 °C overnight. Perform a vector-only control by omitting digested DNA duplexes from the reaction.
7. For each reaction, transform 1 μL of ligation product into 50 μL of One Shot Stbl3 chemically competent cells following the manufacturer's instructions. Plate 89%, 10%, and 1% of transformed *E. coli* onto LB-ampicillin plates. Incubate the plates overnight at 37 °C.
8. Assess the number of colonies from both vector-only control and insert-containing ligation reactions. If there are >1000 colonies from insert containing ligation reaction, and <10 colonies on vector-only control, proceed by scraping all the colonies off the plate with 99% of transformants from the

insert-containing reaction and extract plasmids using a plasmid midiprep kit. The resulting plasmids are the pLKO-barcode library.

9. Package pLKO-barcode library into lentivirus by cotransfecting the pLKO-barcode library with psPAX2 and pMD2.G into lenti-X 293T cells. Per well in a 6-well plate, seed 0.5 million lenti-X 293T cells. Let the cells attach overnight, and transfect them with 1 μg of pLKO-barcode library, 1 μg of psPAX2, and 0.1 μg of pMD2.g mixed with 6 μL of Fugene HD. Collect the medium after 72 h post transfection. Filter the virus containing medium through a 0.45 μm filter.
10. Use a different amount of medium collected from transfected lenti-X 293T cells to infect HCT-116 cells together with 8 μg/mL polybrene and identify the viral dose that results in a low multiplicity of infection (MOI) to achieve ~30% cell survival after 3 days of selection with 2 μg/mL puromycin (*see* **Note 5**).
11. Passage the low MOI lentivirally transduced cells for 2 weeks (*see* **Note 6**) and expand them to twenty 10 cm^2 plates. Freeze cells in FBS supplemented with 10% DMSO (v/v) at a density of 5 million cells/mL using standard cryopreservation methods. These cells are henceforth called HCT-116BC cells.

3.2 Estimation of the IC_{50} of the Compound

1. Prepare a stock solution of indisulam in DMSO to a final concentration of 10 mM, make 50 μL aliquots and store at −20 °C.
2. Use the standard method to dissociate a 10 cm^2 plate of HCT-116BC cells with trypsin–EDTA and dilute the cell suspension to 40,000 cells/mL. Aliquot 100 μL per well in a white 96 well assay plate (*see* **Note 7**). Allow cells to attach overnight in the incubator.
3. Make eleven threefold serial dilutions of indisulam in DMSO, starting from the 10 mM stocks. The 12th sample is DMSO control with no indisulam added. Mix 2 μL of each dilution with 400 μL of medium. Aspirate old medium from the 96-well plate and add 100 μL of medium containing indisulam. Three wells are used per indisulam dose.
4. Seventy two hours following indisulam addition, measure cell viability using the CellTiter-Glo Luminescent Cell Viability Assay kit following the manufacturer's instructions. Record luminescence by a luminescence plate reader. Plot triplicate data with GraphPad Prism using baseline correction (by normalizing to DMSO control). IC_{50} of indisulam is estimated by the asymmetric (five parameter) equation and least squares fit.

3.3 Optimization of Selection Dose

1. Dissociate a 10 cm^2 plate of HCT-116BC cells by trypsin–EDTA and dilute the cell suspension to 50,000 cells/mL. Aliquot 2.5 mL of cell suspension per well in a 6-well plate. Allow cells to attach overnight in the incubator.
2. Dilute indisulam in DMSO to 200×, 250×, 300×, 350×, 400×, and 450× of IC_{50} measured in Subheading 3.2. Add 12.5 μL of diluted indisulam per well in the 6-well plate (*see* **Note 8**).
3. Every 3–4 days, replace medium with 2.5 mL fresh medium and add 12.5 μL of diluted indisulam.
4. Two weeks after the first indisulam treatment, examine the plate to identify the lowest concentration of indisulam that is sufficient to clear up the plate. This concentration will be used for isolating indisulam resistant clones.

3.4 Large-Scale Selection of Independent Compound-Resistant Clones

1. Plate six 10 cm^2 plates of HCT-116BC cells with 1 million cells in 8 mL of medium per plate. Allow cells to attach overnight.
2. Add indisulam to the final concentration as estimated in Subheading 3.3.
3. Every 3–4 days, replace medium with 8 mL of fresh medium and supplement with the same concentration of indisulam as estimated in Subheading 3.3.
4. After 3 weeks, indisulam-resistant colonies appear. Use a microscope with 4× or 10× objective to examine plates. Use a marking pen to circle ~20 well isolated colonies.
5. Remove medium and wash the cells once with PBS. Use sterile forceps to pick up a cloning cylinder and coat its bottom with a thin layer of silicone high vacuum grease. Place the cylinder over a colony and gently press down the cylinder evenly to glue it to the plate. Repeat the process for additional colonies.
6. Add 0.1 mL of trypsin–EDTA to the interior of each cloning cylinder. Incubate the plate at 37 °C for about 5 min. Prepare several 6-well plates with 2 mL of medium per well. Examine cells incubated with trypsin–EDTA under the microscope. When the cells begin to round up, add 0.1 mL of medium to the cylinder, pipette up and down a few times, and transfer cells to a well in the 6-well plate.
7. When cells become confluent on the 6-well plates, dissociate cells by digestion with trypsin–EDTA. Transfer 10% of dissociated cells into an Eppendorf tube and centrifuge at 180 × *g* for 3 min. Rinse the cell pellets once with PBS and store them in an −80 °C freezer. These cell pellets will be used for barcode sequencing (*see* Subheading 3.5). Plate the remaining 90% of cells onto three 10 cm^2 plates (*see* Subheading 3.6).

3.5 Barcode Amplification and Sequencing to Identify Independent Clones

1. Resuspend frozen pellets of isolated resistant clones in 50 μL of lysis buffer. Incubate at 55 °C for 2 h, followed by incubation at 95 °C for 15 min to inactivate Proteinase K. Lysates are directly used as the template for PCR amplification of barcode sequences.
2. Assembly the following PCR reactions on ice by mixing 2 μL of lysate, 0.5 μL of 10 μM oligo3, 0.5 μL of 10 μM oligo4, 2.5 μL of 10× Standard Taq buffer, 2 μL of 2.5 mM dNTP mix, and 17.2 μL of H_2O (*see* **Note 9**). Amplify reactions in a thermocycler using 35 cycles of the following sequence: 95 °C for 30 s, 60 °C for 30 s, and 72 °C for 30 s.
3. Clean up PCR products with the DNA clean and concentrator kit and submit the clean PCR products to a core facility for Sanger sequencing with oligo5.
4. Based on the barcode sequences, identify ≥6 clones with unique barcode sequences that represent independent clones.
5. For the independent clones, when they become confluent on three 10 cm^2 plates prepared in **step 8** of Subheading 3.4, freeze down cells from one plate in two cryopreservation vials. Scrape cells off from the second plate, spin down at 180 × *g* for 3 min, rinse the cell pellets with PBS, and freeze the pellets in −80 °C freezer for genomic DNA preparation (Subheading 3.7). Proceed with the third plate of cells for validation of indisulam resistance and testing of paclitaxel cross-resistance (Subheading 3.6).

3.6 Validation of Indisulam Resistance and Testing of Paclitaxel Cross-Resistance (See Note 10)

1. Prepare serial dilutions of paclitaxel as described in Subheading 3.2.
2. Dilute dissociated HCT-116BC and each resistant clone to 40,000 cells/mL. Aliquot 100 μL per well in a white 96-well assay plate. Plate six rows per clone. Allow cells to attach overnight in the incubator.
3. Mix 2 μL of each dilution of indisulam or paclitaxel with 400 μL of medium. Aspirate old medium from the 96-well plate and add 100 μL of medium containing paclitaxel.
4. Seventy two hours following indisulam or paclitaxel addition, measure cell viability using the CellTiter-Glo Luminescent Cell Viability Assay kit as described in Subheading 3.2. Clones, whose indisulam IC_{50} values are at least twofold higher than parental cells, and paclitaxel IC_{50} values are comparable to parental cells, are subjected to further analysis.

3.7 Genomic DNA Isolation and Whole-Exome Sequencing

1. Extract genomic DNA from HCT-116BC and six indisulam-resistant but paclitaxel-sensitive clones using DNA mini kit following manufacturer's instructions.

2. For each sample, submit ~10 μg of isolated genomic DNAs to a next generation sequencing core facility for whole-exome sequencing analysis.

3.8 Bioinformatic Analysis of Whole-Exome Sequencing Data

1. Assess the quality of sequencing with NGS QC Toolkit (v2.3.1) [15].
2. Align the high-quality reads to the human reference genome (hg19) using Burrows-Wheeler Aligner (BWA, v0.7.5a) [16].
3. Use Picard (v1.99) (http://picard.sourceforge.net) to remove PCR duplicates.
4. Use Genome Analysis Toolkit (GATK, v2.7-2) to improve the quality of the alignment, base quality recalibration and local realignment around indels [17, 18]. Perform variant calling and joint genotyping for all samples using HaplotypeCaller of GATK. Filter the variants in Variant Call Format (VCF) by applying the following criteria: AN (Total number of alleles in called genotypes) <24, DP (Approximate read depth) <10, QD (Variant Confidence/Quality by Depth) <5, GQ (Genotype confidence quality) <20.
5. Identify candidate variants using the following criteria: (1) missense; (2) recurrent in resistant clones and absent in parental HCT-116BC; (3) not listed in Common SNPs in the UCSC Genome Browser [19], a subset consisting of uniquely mapped variants that appear in at least 1% of the population.

4 Notes

1. ATCC provides a "certificate of analysis" for every lot of the HCT-116 cancer cell line. This analysis includes a STR (short tandem repeat) analysis which will verify the identity of the cell line as well as testing to confirm that they are not infected by Mycoplasma. HCT-116 cell lines can be used from an alternative source as long as both of these tests are also provided. Periodically monitor culture medium for mycoplasma contamination using a PCR-based mycoplasma detection kit. Discard any cell culture that is contaminated with mycoplasma.
2. Oligonucleotides can be ordered from a commercial source. Standard desalting is sufficient.
3. A population of HCT-116 cells is expected to contain several resistant founders representing individual mutational events. The barcoding strategy outlined in this protocol facilitates the identification of independent founder events. Resistant clones with the same barcode sequence are likely descendants from the same founder clone. Clones with different barcode sequences are likely derived from independent founders.

4. Clean up DNA duplexes by precipitation instead of binding to a silicon column. The length of DNA duplexes is too short to bind efficiently to the silicon column.
5. Low MOI infection ensures that the majority of transduced cells harbor a single barcode.
6. HCT-116BC cells are passaged for 2 weeks to accumulate enough mutations.
7. White plates reflect light and will maximize light output signal, and therefore are used in luminescence-based assays.
8. Six compound concentrations (1×, 1.25×, 1.5×, 1.75×, 2×, 2.25×) are tested initially. If 2.25× is still insufficient to clear up the plates, higher concentrations can be tested.
9. Use barrier tips for setting up PCR reactions to prevent contamination of PCR reagents. Always run a no-template control to make sure no contamination from previous PCR products can be detected.
10. Lack of cross-resistance to the chemically unrelated toxin, paclitaxel, reduces the likelihood that indisulam resistance is the result of nonspecific increase in compound metabolism or efflux.

Acknowledgment

T.H. is a Howard Hughes Medical Institute Fellow of the Life Sciences Research Foundation. D.N. is supported by a Harold C. Simmons Cancer Center Startup Award, a Disease-Oriented Clinical Scholar award, a Damon Runyon Clinical Investigator award (CI-68-13), a grant from the Welch Foundation (I-1879), and a grant from the Harrington Discovery Institute.

References

1. Vogelstein B, Papadopoulos N, Velculescu VE, Zhou S, Diaz LA Jr, Kinzler KW (2013) Cancer genome landscapes. Science 339 (6127):1546–1558. https://doi.org/10.1126/science.1235122
2. Spiegel J, Cromm PM, Zimmermann G, Grossmann TN, Waldmann H (2014) Small-molecule modulation of Ras signaling. Nat Chem Biol 10(8):613–622. https://doi.org/10.1038/nchembio.1560
3. Garnett MJ, Edelman EJ, Heidorn SJ, Greenman CD, Dastur A, Lau KW, Greninger P, Thompson IR, Luo X, Soares J, Liu Q, Iorio F, Surdez D, Chen L, Milano RJ, Bignell GR, Tam AT, Davies H, Stevenson JA, Barthorpe S, Lutz SR, Kogera F, Lawrence K, McLaren-Douglas A, Mitropoulos X, Mironenko T, Thi H, Richardson L, Zhou W, Jewitt F, Zhang T, O'Brien P, Boisvert JL, Price S, Hur W, Yang W, Deng X, Butler A, Choi HG, Chang JW, Baselga J, Stamenkovic I, Engelman JA, Sharma SV, Delattre O, Saez-Rodriguez J, Gray NS, Settleman J, Futreal PA, Haber DA, Stratton MR, Ramaswamy S, McDermott U, Benes CH (2012) Systematic identification of genomic markers of drug sensitivity in cancer cells. Nature 483(7391):570–575. https://doi.org/10.1038/nature11005
4. Barretina J, Caponigro G, Stransky N, Venkatesan K, Margolin AA, Kim S, Wilson CJ, Lehar J, Kryukov GV, Sonkin D, Reddy A, Liu M, Murray L, Berger MF,

Monahan JE, Morais P, Meltzer J, Korejwa A, Jane-Valbuena J, Mapa FA, Thibault J, Bric-Furlong E, Raman P, Shipway A, Engels IH, Cheng J, Yu GK, Yu J, Aspesi P Jr, de Silva M, Jagtap K, Jones MD, Wang L, Hatton C, Palescandolo E, Gupta S, Mahan S, Sougnez C, Onofrio RC, Liefeld T, MacConaill L, Winckler W, Reich M, Li N, Mesirov JP, Gabriel SB, Getz G, Ardlie K, Chan V, Myer VE, Weber BL, Porter J, Warmuth M, Finan P, Harris JL, Meyerson M, Golub TR, Morrissey MP, Sellers WR, Schlegel R, Garraway LA (2012) The Cancer Cell Line Encyclopedia enables predictive modelling of anticancer drug sensitivity. Nature 483(7391):603–607. https://doi.org/10.1038/nature11003

5. Lamb J, Crawford ED, Peck D, Modell JW, Blat IC, Wrobel MJ, Lerner J, Brunet JP, Subramanian A, Ross KN, Reich M, Hieronymus H, Wei G, Armstrong SA, Haggarty SJ, Clemons PA, Wei R, Carr SA, Lander ES, Golub TR (2006) The Connectivity Map: using gene-expression signatures to connect small molecules, genes, and disease. Science 313(5795):1929–1935. https://doi.org/10.1126/science.1132939

6. Brummelkamp TR, Fabius AW, Mullenders J, Madiredjo M, Velds A, Kerkhoven RM, Bernards R, Beijersbergen RL (2006) An shRNA barcode screen provides insight into cancer cell vulnerability to MDM2 inhibitors. Nat Chem Biol 2(4):202–206. https://doi.org/10.1038/nchembio774

7. Carette JE, Guimaraes CP, Varadarajan M, Park AS, Wuethrich I, Godarova A, Kotecki M, Cochran BH, Spooner E, Ploegh HL, Brummelkamp TR (2009) Haploid genetic screens in human cells identify host factors used by pathogens. Science 326(5957):1231–1235. https://doi.org/10.1126/science.1178955

8. Wang T, Wei JJ, Sabatini DM, Lander ES (2014) Genetic screens in human cells using the CRISPR-Cas9 system. Science 343 (6166):80–84. https://doi.org/10.1126/science.1246981

9. Harding MW, Galat A, Uehling DE, Schreiber SL (1989) A receptor for the immunosuppressant FK506 is a cis-trans peptidyl-prolyl isomerase. Nature 341(6244):758–760. https://doi.org/10.1038/341758a0

10. Carroll PM, Dougherty B, Ross-Macdonald P, Browman K, FitzGerald K (2003) Model systems in drug discovery: chemical genetics meets genomics. Pharmacol Ther 99 (2):183–220

11. Wacker SA, Houghtaling BR, Elemento O, Kapoor TM (2012) Using transcriptome sequencing to identify mechanisms of drug action and resistance. Nat Chem Biol 8 (3):235–237. https://doi.org/10.1038/nchembio.779

12. Glaab WE, Tindall KR (1997) Mutation rate at the hprt locus in human cancer cell lines with specific mismatch repair-gene defects. Carcinogenesis 18(1):1–8

13. Han T, Goralski M, Gaskill N, Capota E, Kim J, Ting TC, Xie Y, Williams NS, Nijhawan D (2017) Anticancer sulfonamides target splicing by inducing RBM39 degradation via recruitment to DCAF15. Science 356(6336):eaal3755. https://doi.org/10.1126/science.aal3755

14. Han T, Goralski M, Capota E, Padrick SB, Kim J, Xie Y, Nijhawan D (2016) The antitumor toxin CD437 is a direct inhibitor of DNA polymerase alpha. Nat Chem Biol 12 (7):511–515. https://doi.org/10.1038/nchembio.2082

15. Patel RK, Jain M (2012) NGS QC Toolkit: a toolkit for quality control of next generation sequencing data. PLoS One 7(2):e30619. https://doi.org/10.1371/journal.pone.0030619

16. Li H, Durbin R (2009) Fast and accurate short read alignment with Burrows-Wheeler transform. Bioinformatics 25(14):1754–1760. https://doi.org/10.1093/bioinformatics/btp324

17. McKenna A, Hanna M, Banks E, Sivachenko A, Cibulskis K, Kernytsky A, Garimella K, Altshuler D, Gabriel S, Daly M, DePristo MA (2010) The genome analysis toolkit: a MapReduce framework for analyzing next-generation DNA sequencing data. Genome Res 20 (9):1297–1303. https://doi.org/10.1101/gr.107524.110

18. DePristo MA, Banks E, Poplin R, Garimella KV, Maguire JR, Hartl C, Philippakis AA, del Angel G, Rivas MA, Hanna M, McKenna A, Fennell TJ, Kernytsky AM, Sivachenko AY, Cibulskis K, Gabriel SB, Altshuler D, Daly MJ (2011) A framework for variation discovery and genotyping using next-generation DNA sequencing data. Nat Genet 43(5):491–498. https://doi.org/10.1038/ng.806

19. Karolchik D, Barber GP, Casper J, Clawson H, Cline MS, Diekhans M, Dreszer TR, Fujita PA, Guruvadoo L, Haeussler M, Harte RA, Heitner S, Hinrichs AS, Learned K, Lee BT, Li CH, Raney BJ, Rhead B, Rosenbloom KR, Sloan CA, Speir ML, Zweig AS, Haussler D, Kuhn RM, Kent WJ (2014) The UCSC Genome Browser database: 2014 update. Nucleic Acids Res 42(Database issue): D764–D770. https://doi.org/10.1093/nar/gkt1168

Chapter 11

The Use of Large-Scale Chemically-Induced Transcriptome Data Acquired from LINCS to Study Small Molecules

Michio Iwata and Yoshihiro Yamanishi

Abstract

Identification of the modes of action of bioactive compounds is an important issue in chemical systems biology. In this chapter we review a recently developed data-driven approach using large-scale chemically induced transcriptome data acquired from the Library of Integrated Network-based Cellular Signatures to elucidate the modes of action of bioactive compounds. First, we present a method for pathway enrichment analyses of regulated genes to reveal biological pathways activated by compounds. Next, we present a method using the pre-knowledge on chemical–protein interactome for predicting potential target proteins, including primary targets and off-targets, with transcriptional similarity. Finally, we present a method based on the target proteins for predicting new therapeutic indications for a variety of diseases. These approaches are expected to be useful for mode-of-action analysis, drug discovery, and drug repositioning.

Key words Bioactive compound, Drug repositioning, Drug target, Modes of action, Transcriptome, Pathway activity

1 Introduction

Diseases are caused by dysfunction of the human biological system consisting of genes, proteins, and their pathways. Most drugs are small compounds that modulate the activity of biological systems for treatment of diseases. Thus, identification of the modes of action of bioactive small compounds is an important issue in the chemical systems biology of diseases. However, the mechanisms of many bioactive compounds, including approved drugs, remain unknown. Drugs interact with target proteins implicated in a disease of interest, but they may interact not only with their primary target proteins but also with other proteins (off-targets), which sometimes leads to unexpected side effects [1–3]. Side effects related to off-target interactions are generally undesired, but they occasionally lead to new therapeutic indications [4–6]. Therefore, it is crucial to understand the complex responses of the human

Slava Ziegler and Herbert Waldmann (eds.), *Systems Chemical Biology: Methods and Protocols*, Methods in Molecular Biology, vol. 1888, https://doi.org/10.1007/978-1-4939-8891-4_11,

biological system to bioactive compounds in medical and pharmaceutical research.

A promising unbiased approach for exploring the modes of action of bioactive compounds is to analyze transcriptional responses of human cell lines to chemical treatment on a genome-wide scale. Owing to recent development of public databases, chemically induced gene expression data have become available for some compounds on various human cell lines. As a toxicogenomics project in Japan, the Toxicogenomics Project-Genomics Assisted Toxicity Evaluation System (TG-GATEs) in which 170 compounds were used in vitro was established [7]. The Connectivity Map (CMap) database in which 1309 compounds were used for perturbations of five cancer cell lines was constructed by the Broad Institute in the United States [8]. These databases make it possible to relate compound activities to gene expression patterns.

These transcriptome resources are useful for pharmaceutical applications. In particular, the CMap data have been used for drug repositioning. For example, a search for compounds whose gene expression patterns are inversely correlated with those of a disease of interest has been proposed for drug repositioning [9–11]. An evaluation of the transcriptional similarity of drugs has been performed [12], and the use of transcriptional similarity has been proposed to predict the targets of drugs [13, 14]. Extraction of drug-induced transcriptional modules has been performed by using a module detection approach in which gene expression patterns in each module were conserved among different cell lines as well as among different organisms [15]. The use of a probabilistic approach has been proposed for the information retrieval problem in CMap [16]. Various normalization methods and gene selection criteria have been developed to handle the batch effects in the gene expression data in CMap [17–20]. However, previous methods have heavily depended on the coverage of compounds in CMap, which makes it difficult to perform large-scale analyses. In addition, the cell-specificity in gene expression profiles has not been considered in the previous studies.

Recently, the Library of Integrated Network-based Cellular Signatures (LINCS) was established by the Broad Institute in the United States [21]. The LINCS L1000 database stores a huge number of gene expression profiles that represent the transcriptomic responses of 77 human cell lines to various perturbations such as chemical treatment, gene knockdown, and gene overexpression. For chemical treatment, gene expression profiles are available for 20,413 compounds and 72 human cell lines in LINCS. Note that the numbers of compounds and human cell lines in LINCS are much larger than those in CMap are. Thus, the LINCS resources are expected to be useful for medical and pharmaceutical research, and there are some recent reports on the use of LINCS. For example, a normalization method was proposed for raw data within

LINCS L1000 data [22], a large-scale analysis was performed to analyze the correlations among chemical structures and gene expression profiles [23], and a series of statistical methods were proposed to elucidate the modes of actions of bioactive compounds in a cell-specific manner [24].

In this chapter, we review a developed data-driven approach using large-scale chemically induced transcriptome data acquired from the LINCS to elucidate the modes of action of bioactive compounds [24]. First, we present a method for pathway enrichment analyses of regulated genes to reveal biological pathways activated by compounds. Next, we present a method using pre-knowledge on the chemical–protein interactome for predicting potential target proteins, including primary targets and off-targets, with transcriptional similarity. Finally, we present a method for predicting new therapeutic indications for a variety of diseases based on the target proteins. The approaches are expected to be useful for the mode-of-action analysis, drug discovery, and drug repositioning.

2 Materials

2.1 Chemically Induced Gene Expression Data from LINCS

Gene expression profiles in LINCS were obtained from the Broad Institute's website (http://www.lincscloud.org). LINCS stores gene expression profiles on 77 human cell lines and various cellular perturbations, including chemical treatment, gene knockdown, gene overexpression, and several more interventions. The gene expression data from the 68 cell lines of chemical treatment experiments were used in this study. Gene expression levels were measured by using flow cytometry [25], and test samples were prepared by using 384-well plates. LINCS provides 978 landmark genes that are referred to as "L1000 genes." The expression levels of the L1000 were experimentally measured [21]. The expression levels of the remaining genes (approximately 21,000 genes) were computationally estimated on the basis of the Gene Expression Omnibus [26]. The gene expression data for the L1000 genes were used in this study.

Gene expression levels were measured at 6, 6.4, 24, 24.4, and 48 h after chemical treatment. Each gene expression profile was filed by using its "distil_id." The total number of profiles was 1,328,098. In the first step, 663,594 chemical treatment profiles (denoted as "trt_cp" in the information) and 28,557 control profiles (denoted as "ctl_vehicle") were selected. The correspondence between the chemical treatment profiles and the control profiles was determined by comparing their "distil_ids." The gene expression profiles that lacked corresponding control profiles were not used. The name of each compound was converted into its corresponding InChIKey (http://www.iupac.org/home/

publications/e-resources/inchi.html) via the perturbation ID by using the information provided in LINCS. The gene expression profiles measured at 6 (6.4) h of 68 cell lines treated with 16,268 bioactive compounds were used in this study.

2.2 Chemically Induced Gene Expression Data from CMap

Gene expression profiles in CMap (build 02) were obtained from the Broad Institute's website (http://www.broadinstitute.org/cmap). CMap contains 7056 profiles based on five cell lines treated with 1309 compounds. Gene expression levels were measured at 6 h after chemical treatment. The gene expression profiles of 22,277 genes were experimentally measured by using microarray technology. Some profiles were measured by using the HT_HG-U133A platform for the human cell lines MCF7 (breast adenocarcinoma), PC3 (prostate cancer), and HL60 (promyelocytic leukemia). The name of each compound was converted into its corresponding InChIKey via ChemBank [27]. The process yielded the gene expression profiles of 1133 chemical compounds. Gene expression profiles consisting of the 978 L1000 genes from LINCS were also prepared to make comparisons between CMap and LINCS.

2.3 Chemically Induced Gene Expression Data from TG-GATEs

Gene expression profiles in TG-GATEs were obtained from the Toxicogenomics Project of Japan (http://toxico.nibio.go.jp/english/). TG-GATEs stores 170 compounds and primary hepatocytes in vitro (rat or human). Gene expression data were measured at three time points (i.e., 2, 8, 24 h) after chemical treatment. The name of each compound was converted into its corresponding InChIKey. This process yielded the gene expression signatures of 157 compounds.

2.4 Comparison of TG-GATEs, CMap, and LINCS

A comparison of the statistics of the chemically induced gene expression profile data was made among three different databases: TG-GATEs, CMap, and LINCS. Table 1 shows the numbers of cell lines and compounds stored in the three databases. In LINCS, gene expression data measured at 6 h after chemical treatment were used. The numbers of cell lines stored in TG-GATEs, CMap, and LINCS were 1, 5, and 68, respectively. MCF7, PC3, and HL60 are common human cancer cell lines used in both CMap and LINCS, and primary human hepatocytes is present only in TG-GATEs. The numbers of compounds stored in TG-GATEs, CMap, and LINCS were 157, 1133, and 16,268, respectively. The numbers of cell lines and compounds in LINCS were much larger than those used in TG-GATEs and CMap were. These observations indicate that the LINCS resources are suitable for large-scale applications.

Control gene expression profiles measured at 6 (6.4) h of the 68 human cell lines in LINCS were analyzed by performing hierarchical clustering. Figure 1 shows the distribution of 68 human cell lines in LINCS, most of which are cancer cell lines derived from

Table 1
Number of cell lines and chemical compounds present in the three databases containing chemically induced gene expression profiles: TG-GATEs, CMap, and LINCS

Database	Number of cell lines	Number of compounds
TG-GATEs	1	157
CMap	5	1133
LINCS	68	16268

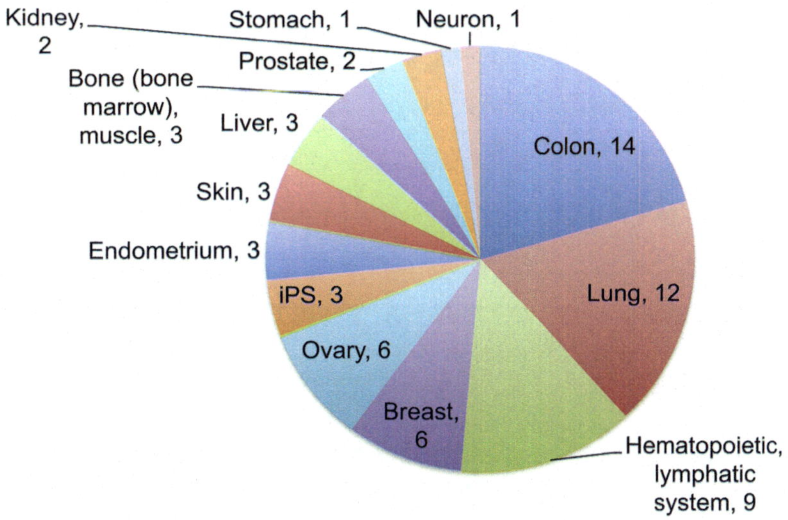

Fig. 1 Tissue distribution of 68 cell lines in Library of Integrated Network-based Cellular Signatures

various tissues such as colon, lung, breast, ovary, skin, liver, and prostate. Figure 2a shows the clustering dendrogram of 957 control gene expression profiles in which multiple control profiles under different conditions were distinguished and analyzed for each human cell line. In contrast, Fig. 2b shows the clustering dendrogram of 68 control gene expression profiles in which multiple control gene expression profiles were averaged into a single profile for each human cell line. We observed that the same cell lines and similar cell lines from the same tissues tended to be closely located, whereas the cell lines from different tissues tended to be separately and distantly located. For example, the cell lines in CMap (i.e., MCF7, PC3, and HL60) were scattered and distantly located. These observations suggest that gene expression profiles are highly variable among different cell lines and that it is important to consider cell specificity in gene expression profiles.

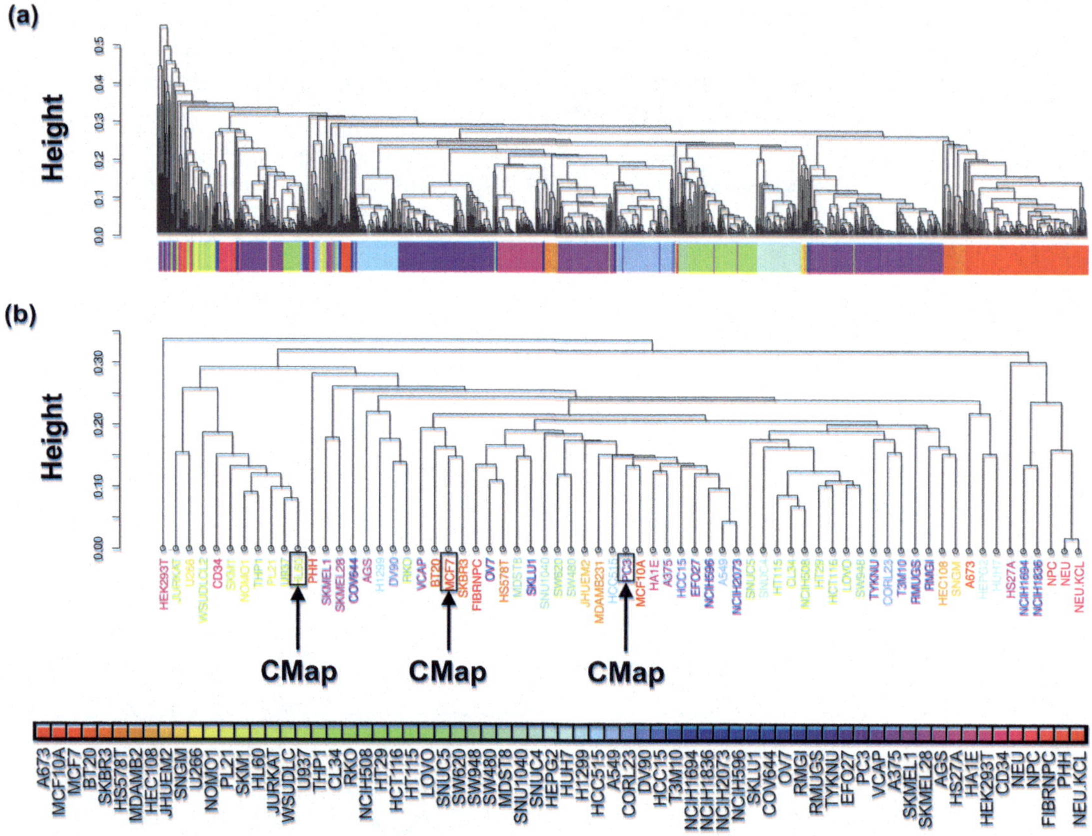

Fig. 2 Clustering of cell lines performed by using transcriptional similarity scores between control profiles in Library of Integrated Network-based Cellular Signatures. Individual cell lines are indicated by different colors, and cell lines obtained from the same tissue are indicated by similar colors. Dendrogram (**a**) shows the result of clustering 957 control profiles. Dendrogram (**b**) shows the result of clustering 68 cell lines, for which multiple control profiles are averaged over the same cell line. Black arrows indicate cell lines also present in Connectivity Map

3 Methods

3.1 Construction of Gene Expression Signatures

Chemically induced gene expression profiles are constructed and referred to as "gene expression signatures" or "signatures" for short, as described below (*see* **Note 1**).

1. Represent the gene expression signature of each compound by a feature vector as $\boldsymbol{x} = (x_1, x_2, \ldots, x_d)^{\mathrm{T}}$, where d is the number of features and is identical to the number of genes. Each compound on the same cell line has multiple signatures based on various concentrations and different time points. In this study, multiple signatures are integrated into a single signature by averaging the signatures, and 6.4 and 24.4 h are regarded as 6 and 24 h, respectively.

2. Prepare two kinds of gene expression profiles in LINCS: "Level 3" and "Level 4". Level 3 data correspond to a set of profiles generated on the basis of invariant-set scaling and quantile normalization. Level 4 data correspond to a set of profiles generated on the basis of robust *z*-scores relative to population controls by using the average over all other wells on the 384-well plate. The difference is similar to that of two types of normalization procedures: "biological control" and "mean centering".
3. Construct a gene expression signature that is based on the logarithmic ratio of the chemical treatment profile to the biological control profile in the "biological control" normalization [8].
4. Center the signature to have a mean of 0 and scale it to have standard deviation of 1. For the CMap data, the Robust Multi-array Average [28] method is applied to the original data in Affymetrix CEL file format. In LINCS, a similar approach is applied to Level 3 data generation.
5. Prepare a control profile by averaging chemical treatment profiles measured under the same biological control conditions, which is referred to as the averaged control profile in the "mean centering" normalization. The mean centering normalization procedure is known to reduce batch effects in CMap [17].
6. Construct a gene expression signature based on the logarithmic ratio of the chemical treatment profile against the averaged control profile.
7. Center the signature to have mean 0 and scale it to have a standard deviation of 1. In LINCS, a similar approach is applied to Level 4 data generation.

3.2 An Overview of the Approach

Here, we present an overview of a recently developed data-driven approach to analyze a set of chemically induced gene expression signatures [24]. To elucidate the mode of action of a query compound, the approach consists of three procedures: (1) identification of active pathways, (2) prediction of potential target proteins, and (3) prediction of new therapeutic indications. Figure 3 illustrates each procedure in the approach.

1. Map upregulated and downregulated genes in the signatures onto many biological pathway maps, and evaluate the statistical significance of the pathway enrichment of the regulated genes in each pathway by using a hypergeometric test. The procedure for identifying biological pathways that are activated or inactivated by a query compound is shown in Fig. 3a.
2. Evaluate transcriptional similarities by calculating correlation coefficients between gene expression signatures in a cell-specific manner. If the query compound is transcriptionally similar to interactome compounds of known target proteins,

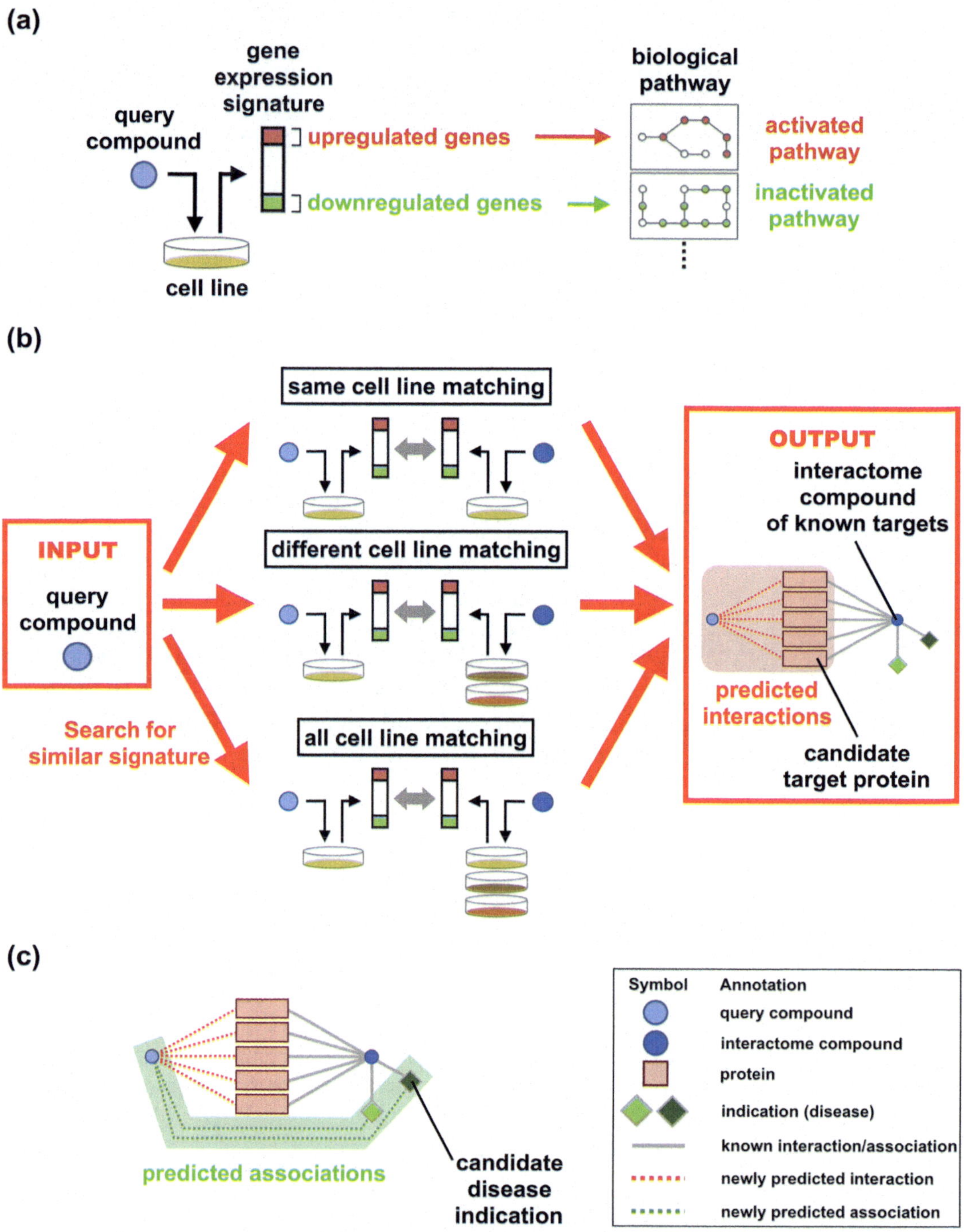

Fig. 3 Overview of the approach. (**a**) Identification of active pathways, (**b**) prediction of potential target proteins, and (**c**) prediction of new therapeutic indications

the query compound is predicted to have the same target proteins as the similar compounds. The procedure for predicting potential target proteins of a query compound is shown in Fig. 3b.

3. If the query compound is transcriptionally similar to interactome compounds with known indications and shares the same target proteins with the similar compounds, the query compound is predicted to have the same therapeutic indications as those of similar compounds. The procedure for predicting new therapeutic indications (i.e., applicable diseases) of a query compound is shown in Fig. 3c.

3.3 Pathway Enrichment Analyses

Pathway enrichment analyses are performed for upregulated and downregulated genes by chemical treatments. Note that the same procedure was used in previous studies for different biological problems [29, 30].

1. Select upregulated genes (e.g., top 5%) or downregulated genes (e.g., bottom 5%) from a signature induced by a compound (*see* **Note 2**). Let *Gcompound* denote a set of upregulated or downregulated genes in a signature induced by a compound.
2. Prepare the information on biological pathway maps. Let *Gpathway* denote a set of genes in a pathway map.
3. Let $r = |Gcompound|$, $k = |Gpathway|$, $z = |Gcompound \cap Gpathway|$ and l equals the total number of genes in the entire dataset ($l = 978$). Then, z is assumed to follow a hypergeometric distribution.
4. Compute the probability of observing an intersection of size z between the *Gpathway* and the *Gcompound* as follows:

$$P(G_{\text{pathway}}, G_{\text{compound}}) = \sum_{i=z}^{\min(k,r)} \frac{\binom{k}{i}\binom{l-k}{r-i}}{\binom{l}{r}}. \quad (1)$$

5. Correct the resulting *P*-values by using the false discovery rate [31].

For example, 163 biological pathway maps in the KEGG PATHWAY database are used in which the pathways belonged to the following categories: Metabolism (except for Global and overview maps), Environmental Information Processing (except for Membrane Transport and Signaling Molecules and Interaction), Cellular Processes (except for Transport and Catabolism), and Organismal Systems. Figure 4 shows the distribution of biological pathways that were detected by the pathway enrichment analysis at

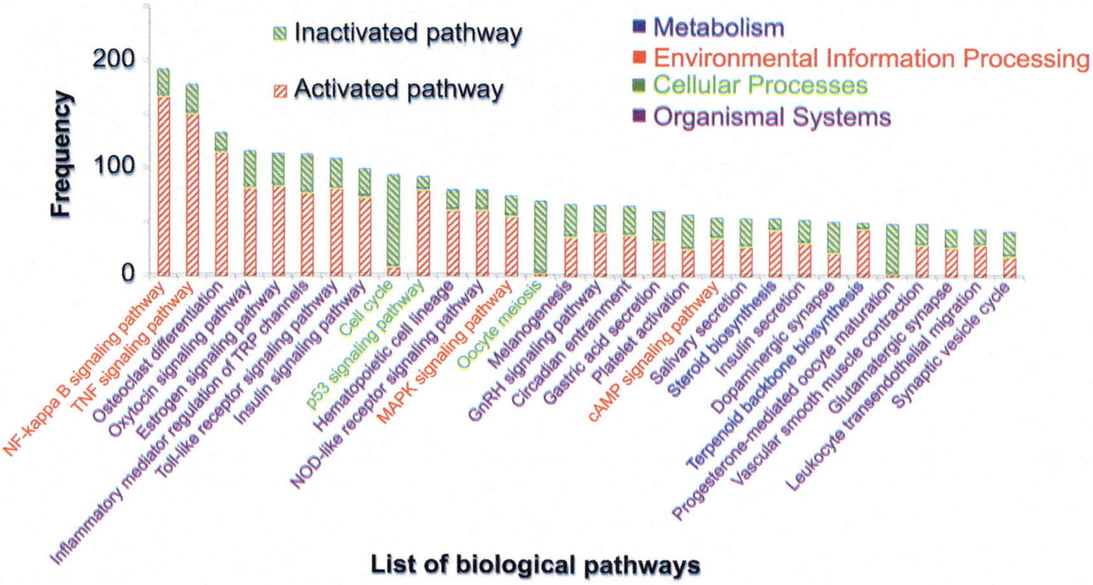

Fig. 4 Distribution of the pathways detected for all compounds. Red bars indicate the numbers of activated pathways enriched with upregulated genes. Green bars indicate the numbers of inactivated pathways enriched with downregulated genes. The pathways in the Metabolism, Environmental Information Processing, Cellular Processes, and Organismal Systems categories in the KEGG PATHWAY database are colored in blue, red, green, and purple, respectively

a statistically significant level ($P < 0.05$) for all compounds in LINCS.

3.4 Target Protein Prediction

Potential target proteins of a query compound are predicted by a similarity search on the basis of a large set of gene expression signatures by using preknowledge on chemical–protein interactome data. The details of the similarity search are described in the next subsection.

1. Obtain compound–protein interactome data from databases such as ChEMBL [32], MATADOR [33], DrugBank [34], the Psychoactive Drug Screening Program Ki [35], KEGG [36], the BindingDB [37], and the Therapeutic Target Database [38] (*see* **Note 3**). Compounds and proteins included in the chemical–protein interactome data are referred to as interactome compounds and interactome proteins, respectively.
2. Represent a query compound by the feature vector $\boldsymbol{x}^{(\text{query})} = (x_1, x_2, \ldots, x_d)^{\text{T}}$.
3. Represent each interactome compound by the feature vector $\boldsymbol{x}^{(\text{inter})} = (x'_1, x'_2, \ldots, x'_d)^{\text{T}}$, where d is the number of genes.
4. Represent each interactome compound by a target protein interaction profile $\boldsymbol{y}^{(\text{inter})} = (y_1, y_2, \ldots, y_p)^{\text{T}}$, where y_k indicates the presence or absence of an interaction with the kth protein

by 1 or 0 ($k = 1, 2, \cdots, p$), and p is the number of interactome proteins.

5. Calculate the transcriptional similarity scores between a given query compound and all the interactome compounds in your chemical–protein interactome data.
6. Out of the interactome compounds of known interactions with the kth protein ($k = 1, 2, \cdots, p$), select an interactome compound with the highest similarity to the query compound and use the corresponding similarity score as a prediction score to assess the possibility that the query compound interacts with the kth protein.
7. Repeat the above procedure for all p interactome proteins and assign the prediction scores for compound–protein pairs (pairs between the query compound and all interactome proteins).
8. Predict high-scoring compound–protein pairs as candidates for interaction pairs.

3.5 Cell-Based Similarity Search

The transcriptional similarity between two gene expression signatures $\boldsymbol{x}$ and $\boldsymbol{x}'$ is calculated by using the Pearson correlation coefficient as follows:

$$S(\boldsymbol{x}, \boldsymbol{x}') = \frac{\sum_{i=1}^{d} (x_i - \overline{x})(x'_i - \overline{x}')}{\sqrt{\sum_{i=1}^{d} (x_i - \overline{x})^2} \sqrt{\sum_{i=1}^{d} (x'_i - \overline{x'})^2}}, \quad (2)$$

where $\overline{x}$ and $\overline{x}'$ represent the average of the gene expression values in $\boldsymbol{x}$ and $\boldsymbol{x}'$, respectively.

A cell-based similarity search is performed by considering the cell specificity in the gene expression profiles and by using the following strategies (*see* **Note 4**).

1. Calculate the transcriptional similarity score S_{same} between a query compound $\boldsymbol{x}^{(\text{query})}$ and an interactome compound $\boldsymbol{x}^{(\text{inter})}$ on the basis of the same cell lines as follows:

$$\begin{aligned} S_{\text{same}}(\boldsymbol{x}^{(\text{query})}, \boldsymbol{x}^{(\text{inter})}) = \max\{ & S(\boldsymbol{x}_{\text{cell }1}^{(\text{query})}, \boldsymbol{x}_{\text{cell }1}^{(\text{inter})}), \\ & S(\boldsymbol{x}_{\text{cell }2}^{(\text{query})}, \boldsymbol{x}_{\text{cell }2}^{(\text{inter})}), \\ & \cdots, \\ & S(\boldsymbol{x}_{\text{cell }n}^{(\text{query})}, \boldsymbol{x}_{\text{cell }n}^{(\text{inter})})\}, \end{aligned} \quad (3)$$

where $\max\{\cdot, \cdot, \cdots\}$ is an operation to take the maximum value and n is the number of cell lines. This strategy is referred to as the same cell line-matching strategy. It is supposed that gene expression data of compounds are generated from the same cell line.

2. Calculate the transcriptional similarity score S_{diff} between a query compound $\boldsymbol{x}^{(\text{query})}$ and an interactome compound $\boldsymbol{x}^{(\text{inter})}$ on the basis of different cell lines as follows:

$$S_{\text{diff}}(\boldsymbol{x}^{(\text{query})}, \boldsymbol{x}^{(\text{inter})}) = \max\{S(\boldsymbol{x}^{(\text{query})}_{\text{cell 1}}, \boldsymbol{x}^{(\text{inter})}_{\text{cell 2}}), S(\boldsymbol{x}^{(\text{query})}_{\text{cell 1}}, \boldsymbol{x}^{(\text{inter})}_{\text{cell 3}}), \ldots, S(\boldsymbol{x}^{(\text{query})}_{\text{cell }(n-1)}, \boldsymbol{x}^{(\text{inter})}_{\text{cell } n})\}, \quad (4)$$

This strategy is referred to as the different cell line-matching strategy. Gene expression data are not always available from the same cell line. It is supposed that gene expression data of compounds are generated from different cell lines.

3. Calculate the transcriptional similarity score S_{all} between a query compound $\boldsymbol{x}^{(\text{query})}$ and an interactome compound $\boldsymbol{x}^{(\text{inter})}$ on the basis of the integration of the same cell lines and different cell lines as follows:

$$S_{\text{all}}(\boldsymbol{x}^{(\text{query})}, \boldsymbol{x}^{(\text{inter})}) = \max\{S_{\text{same}}(\boldsymbol{x}^{(\text{query})}, \boldsymbol{x}^{(\text{inter})}), S_{\text{diff}}(\boldsymbol{x}^{(\text{query})}, \boldsymbol{x}^{(\text{inter})})\}. \quad (5)$$

This strategy is referred to as the all cell line-matching strategy. In practice, some compounds have gene expression data generated from the same cell line, whereas other compounds have gene expression data generated from different cell lines. It is supposed that gene expression data of compounds are generated from the same cell lines and different cell lines.

3.6 Therapeutic Indication Prediction

New therapeutic indications for query compounds are predicted on the basis of the results of the compound–target interaction predictions.

1. Prepare drug–disease association data from medical books [39] and KEGG DISEASE [36] by regarding therapeutic indications as drug–disease associations. Diseases defined in the International Classification of Diseases version 10 were used in this chapter. In total, 5606 drug–disease associations involving 2266 drugs and 461 diseases were obtained.
2. If the query compound has high transcriptional similarity with interactome compounds of known indications and shares the same target proteins with the interactome compounds, the query compound is predicted to have the same therapeutic indications as those of the most similar compound. The prediction scores for compound–disease associations are the same as those of the associated compound–protein interactions.

4 Notes

1. Experimental noise in gene expression data is an obstacle in the analysis of gene expression signatures and can influence biological interpretations. For this problem to be minimized, the use of the expression ratios only for top-ranked and bottom-ranked genes has been proposed in previous studies [12, 13, 17–20]. The top 50 and bottom 50 ranked genes or the top 100 and bottom 100 ranked genes are often used, and the expression ratios of the remaining genes are assigned values of 0. Note that the results depend on the choice of a threshold value for the rank.
2. The pathway enrichment analysis can suggest pathway activities triggered by compound–protein interactions, but the performance of the pathway enrichment analysis depends heavily on the quality and coverage of biological pathways. Upregulated or downregulated genes by chemical perturbation were selected, and significantly enriched pathways with upregulated and downregulated genes were identified. The performance also depends on the threshold values for determining upregulated and downregulated genes. In this study, the top 5% and bottom 5% of genes were extracted as upregulated and downregulated genes, respectively. Another approach would be to choose a threshold value for the gene expression ratio values rather than their ranks.
3. In the target protein prediction, compound–protein interactome data were obtained from databases such as ChEMBL [32], MATADOR [33], DrugBank [34], the Psychoactive Drug Screening Program Ki [35], KEGG [36], the BindingDB [37], and the Therapeutic Target Database [38]; and compound–protein interaction pairs annotated as active interactions or with binding affinities of < 30 μM (e.g., IC50) were used. Note that the performance of target protein prediction depends on the quality and quantity of compound–protein interactions in the databases, and the performance also depends on the threshold value for the interaction strength.
4. LINCS has enabled large-scale analysis of cell-specific gene expression signatures, because it covers a variety of cell lines (68 cell lines in this study). However, there remains much room for improvement in the statistical analysis of cell specificity. There is a strong incentive to develop statistical methods to appropriately analyze cell-specific gene expression patterns.

Aknowledgements

This work is supported by JST PRESTO Grant Number JPMJPR15D8.

References

1. Whitebread S, Hamon J, Bojanic D, Urban L (2005) Keynote review: in vitro safety pharmacology profiling: an essential tool for successful drug development. Drug Discov Today 10:1421–1433
2. Blagg J (2006) Structure–activity relationships for in vitro and in vivo toxicity. Annu Rep Med Chem 41:353–368
3. Lin SF, Xiao KT, Huang YT, Chiu CC, Soo VW (2010) Analysis of adverse drug reactions using drug and drug target interactions and graph-based methods. Artif Intell Med 48:161–166
4. Ashburn TT, Thor KB (2004) Drug repositioning: identifying and developing new uses for existing drugs. Nat Rev Drug Discov 3:673–683
5. Chong CR, Sullivan DJ (2007) New uses for old drugs. Nature 448:645–646
6. Novac N (2013) Challenges and opportunities of drug repositioning. Trends Pharmacol Sci 34:267–272
7. Igarashi Y, Nakatsu N, Yamashita T, Ono A, Ohno Y, Urushidani T, Yamada H (2015) Open TG-GATEs: a large-scale toxicogenomics database. Nucleic Acids Res 43: D921–D927
8. Lamb J, Crawford ED, Peck D, Modell JW, Blat IC, Wrobel MJ, Lerner J, Brunet JP, Subramanian A, Ross KN, Reich M, Hieronymus H, Wei G, Armstrong SA, Haggarty SJ, Clemons PA, Wei R, Carr SA, Lander ES, Golub TR (2006) The connectivity map: using gene-expression signatures to connect small molecules, genes, and disease. Science 313:1929–1935
9. Dudley JT, Sirota M, Shenoy M, Pai RK, Roedder S, Chiang AP, Morgan AA, Sarwal MM, Pasricha PJ, Butte AJ (2011) Computational repositioning of the anticonvulsant topiramate for inflammatory bowel disease. Sci Transl Med 3:96ra76
10. Kosaka T, Nagamatsu G, Saito S, Oya M, Suda T, Horimoto K (2013) Identification of drug candidate against prostate cancer from the aspect of somatic cell reprogramming. Cancer Sci 104:1017–1026
11. van Noort V, Schölch S, Iskar M, Zeller G, Ostertag K, Schweitzer C, Werner K, Weitz J, Koch M, Bork P (2014) Novel drug candidates for the treatment of metastatic colorectal cancer through global inverse gene-expression profiling. Cancer Res 74:5690–5699
12. Iorio F, Tagliaferri R, di Bernardo D (2009) Identifying network of drug mode of action by gene expression profiling. J Comput Biol 16:241–251
13. Iorio F, Bosotti R, Scacheri E, Belcastro V, Mithbaokar P, Ferriero R, Murino L, Tagliaferri R, Brunetti-Pierri N, Isacchi A, di Bernardo D (2010) Discovery of drug mode of action and drug repositioning from transcriptional responses. Proc Natl Acad Sci USA 107:14621–14626
14. Hizukuri Y, Sawada R, Yamanishi Y (2015) Predicting target proteins for drug candidate compounds based on drug-induced gene expression data in a chemical structure-independent manner. BMC Med Genomics 8:82
15. Iskar M, Zeller G, Blattmann P, Campillos M, Kuhn M, Kaminska KH, Runz H, Gavin AC, Pepperkok R, van Noort V, Bork P (2013) Characterization of drug-induced transcriptional modules: towards drug repositioning and functional understanding. Mol Syst Biol 9:662
16. Parkkinen JA, Kaski S (2014) Probabilistic drug connectivity mapping. BMC Bioinformatics 15:113
17. IIskar M, Campillos M, Kuhn M, Jensen LJ, van Noort V, Bork P (2010) Drug-induced regulation of target expression. PLoS Comput Biol 6:e1000,925
18. Wang K, Sun J, Zhou S, Wan C, Qin S, Li C, He L, Yang L (2013) Prediction of drug–target interactions for drug repositioning only based on genomic expression similarity. PLoS Comput Biol 9:e1003,315
19. Cheng J, Xie Q, Kumar V, Hurle M, Freudenberg JM, Yang L, Agarwal P (2013) Evaluation of analytical methods for connectivity map data. Pac Symp Biocomput 2013:5–16
20. Cheng J, Yang L, Kumar V, Agarwal P (2014) Systematic evaluation of connectivity map for disease indications. Genome Med 6:540

21. Duan Q, Flynn C, Niepel M, Hafner M, Muhlich JL, Fernandez NF, Rouillard AD, Tan CM, Chen EY, Golub TR, Sorger PK, Subramanian A, Ma'ayan A (2014) LINCS Canvas Browser: interactive web app to query, browse and interrogate LINCS L1000 gene expression signatures. Nucleic Acids Res 42: W449–W460
22. Liu C, Su J, Yang F, Wei K, Ma J, Zhou X (2015) Compound signature detection on LINCS L1000 big data. Mol Biosyst 11:714–722
23. Chen B, Greenside P, Paik H, Sirota M, Hadley D, Butte AJ (2015) Relating chemical structure to cellular response: an integrative analysis of gene expression, bioactivity, and structural data across 11,000 compounds. CPT Pharmacomet Syst Pharmacol 4:576–584
24. Iwata M, Sawada R, Iwata H, Kotera M, Yamanishi Y (2017) Elucidating the modes of action for bioactive compounds in a cell-specific manner by large-scale chemically-induced transcriptomics. Sci Rep 7:40164
25. Peck D, Crawford ED, Ross KN, Stegmaier K, Golub TR, Lamb J (2006) A method for high-throughput gene expression signature analysis. Genome Biol 7:R61
26. Barrett T, Troup DB, Wilhite SE, Ledoux P, Rudnev D, Evangelista C, Kim IF, Soboleva A, Tomashevsky M, Edgar R (2007) NCBI GEO: mining tens of millions of expression profiles--database and tools update. Nucleic Acids Res 35:D760–D765
27. Seiler KP, George GA, Happ MP, Bodycombe NE, Carrinski HA, Norton S, Brudz S, Sullivan JP, Muhlich J, Serrano M, Ferraiolo P, Tolliday NJ, Schreiber SL, Clemons PA (2008) ChemBank: a small-molecule screening and cheminformatics resource database. Nucleic Acids Res 36:D351–D359
28. Irizarry RA, Bolstad BM, Collin F, Cope LM, Hobbs B, Speed TP (2003) Summaries of Affymetrix GeneChip probe level data. Nucleic Acids Res 31:e15
29. Mizutani S, Pauwels E, Stoven V, Goto S, Yamanishi Y (2012) Relating drug–protein interaction network with drug side effects. Bioinformatics 28:i522–i528
30. Hung JH (2013) Gene set/pathway enrichment analysis. Methods Mol Biol 939:201–213
31. Benjamini Y, Hochberg Y (1995) Controlling the false discovery rate: a practical and powerful approach to multiple testing. J R Stat Soc Ser B Methodol 57:289–300
32. Gaulton A, Bellis LJ, Bento AP, Chambers J, Davies M, Hersey A, Light Y, McGlinchey S, Michalovich D, Al-Lazikani B, Overington JP (2012) ChEMBL: a large-scale bioactivity database for drug discovery. Nucleic Acids Res 40:D1100–D1107
33. Günther S, Kuhn M, Dunkel M, Campillos M, Senger C, Petsalaki E, Ahmed J, Urdiales EG, Gewiess A, Jensen LJ, Schneider R, Skoblo R, Russell RB, Bourne PE, Bork P, Preissner R (2008) SuperTarget and Matador: resources for exploring drug–target relationships. Nucleic Acids Res 36:D919–D922
34. Knox C, Law V, Jewison T, Liu P, Ly S, Frolkis A, Pon A, Banco K, Mak C, Neveu V, Djoumbou Y, Eisner R, Guo AC, Wishart DS (2011) DrugBank 3.0: a comprehensive resource for 'omics' research on drugs. Nucleic Acids Res 39:D1035–D1041
35. Roth BL, Lopez E, Patel S, Kroeze WK (2000) The multiplicity of serotonin receptors: uselessly diverse molecules or an embarrassment of riches? Neuroscientist 6:252–262
36. Kanehisa M, Goto S, Furumichi M, Tanabe M, Hirakawa M (2010) KEGG for representation and analysis of molecular networks involving diseases and drugs. Nucleic Acids Res 38: D355–D360
37. Liu T, Lin Y, Wen X, Jorissen RN, Gilson MK (2007) BindingDB: a web-accessible database of experimentally determined protein–ligand binding affinities. Nucleic Acids Res 35: D198–D201
38. Zhu F, Shi Z, Qin C, Tao L, Liu X, Xu F, Zhang L, Song Y, Liu X, Zhang J, Han B, Zhang P, Chen Y (2012) Therapeutic target database update 2012: a resource for facilitating target-oriented drug discovery. Nucleic Acids Res 40:D1128–D1136
39. Papadakis MA, McPhee SJ, Rabow MW (2014) Current medical diagnosis and treatment 2014. McGraw Hill Medical, New York

Chapter 12

Integrated Analysis of Drug Sensitivity and Selectivity to Predict Synergistic Drug Combinations and Target Coaddictions in Cancer

Alok Jaiswal, Bhagwan Yadav, Krister Wennerberg, and Tero Aittokallio

Abstract

High-throughput drug sensitivity testing provides a powerful phenotypic profiling approach to identify effective drug candidates for individual cell lines or patient-derived samples. Here, we describe an experimental-computational pipeline, named target addiction scoring (TAS), which mathematically transforms the drug response profiles into target addiction signatures, and thereby provides a ranking of potential therapeutic targets according to their functional importance in a particular cancer sample. The TAS pipeline makes use of drug polypharmacology to integrate the drug sensitivity and selectivity profiles through systems-wide interconnection networks between drugs and their targets, including both primary protein targets as well as secondary off-targets. We show how the TAS pipeline enables one to identify not only single-target addictions but also combinatorial coaddictions among targets that often underlie synergistic drug combinations.

Key words Precision oncology, Drug sensitivity testing, Drug polypharmacology, Drug–target interactions, Target addictions, Target deconvolution, Drug combinations

1 Introduction

Recent advances in high-throughput genetic and molecular profiling techniques have enabled an extensive exploration of the genomic landscapes of human tumors, leading to an in-depth understanding of the plethora of genetic alterations present in various cancer types [1–4]. However, these genetic profiling studies have also revealed the massive complexity and heterogeneity across tissue types and even within the same tumor [5]. Furthermore, our limited understanding of the functional consequences of these genetic alterations has posed significant challenges for translating this knowledge into clinically actionable targeted treatment strategies [6]. Cancer cells often exhibit a selective dependence on a few genes or pathways, a phenomenon referred to as "oncogenic" addiction, which opens up the possibility of exploiting these

Slava Ziegler and Herbert Waldmann (eds.), *Systems Chemical Biology: Methods and Protocols*, Methods in Molecular Biology, vol. 1888, https://doi.org/10.1007/978-1-4939-8891-4_12,

dependencies for therapeutic developments [7]. However, the extensive genetic heterogeneity results in alterations of multiple molecular pathways, and therefore makes it difficult to pinpoint the specific addictions in particular cancer cells. Complementary strategies are thus needed to identify selective vulnerabilities and cellular addictions of cancer cells toward developing targeted drug combinations that can inhibit the multiple cancer-survival or resistance-driving pathways.

High-throughput compound sensitivity screens that make use of cell-based viability assays to profile the activity of a library of small bioactive molecules (referred hereafter as to "drugs") have facilitated the identification of promising drug candidates for individual cancer patients (so-called functional precision oncology) [8–12]. The direct advantage of functional drug sensitivity screening assays, compared to genomic and transcriptomic-based profiling, is that they provide a complementary phenotypic information of cancer cells, which is often more clinically actionable. Furthermore, such drug response phenotypic profiles can be used to identify molecular targets or pathway-level addictions based on the knowledge of primary or 'on-targets' of the drugs combined with their functional response profiles [13–18], a process called 'target deconvolution'. However, most drugs are promiscuous and lead to a polypharmacological or "off-target" effects [19], which in turn complicate the identification of target addictions directly from the measured drug response phenotypes. Since it is experimentally and economically infeasible to exhaustively explore the full space of all drug–target combinations, there is a need for systematic computational approaches that enable the identification of target addiction patterns from drug sensitivity data.

Here, we describe our experimental-computational target deconvolution pipeline, named target addiction scoring (TAS) [20, 21], which makes use of drug polypharmacology to model the complex interconnections between drug responses and target interaction profiles that consider both primary protein targets as well as secondary off-targets. The TAS approach is individualized in the sense that it uses the drug sensitivity profile from a given sample (e.g., cancer cell line or patient-derived sample), screened against a library of bioactive compounds, and then transforms the observed phenotypic profile into a sample-specific target addiction profile, through the information encoded in system-wide drug–target interactions that connects the compounds to their complete spectrum of protein targets (i.e., target selectivity profiles). Systematic mapping of such target addiction signatures provides a ranking of the therapeutic targets according to their functional importance in a given cancer sample. In addition to demonstrating how the TAS approach enables the identification of single-target addictions, we also show here how it can be used to infer combinatorial coaddictions among therapeutic targets, such as kinase pairs, by means of combinatorial TAS approach.

We have previously demonstrated the applicability of TAS by profiling a panel of triple negative breast cancer (TNBC) cell line responses to 239 kinase inhibitors [20, 22], in which case the approach was called kinase inhibition sensitivity scoring (KISS). We observed that the KISS profiles of TNBC cells closely resembled the patient-derived TNBC subtypes, but also revealed novel druggable addictions [20]. However, our approach is not only applicable to cancer cell lines but also to patient-derived cancer samples amenable to high-throughput drug screening [8–10, 12]. Moreover, the TAS approach extends naturally to other target classes beyond kinase targets, although the accuracy and coverage of the addiction profiles is critically dependent on the availability of drug–target bioactivity profiles. Drug–target bioactivity data are generally sparse and notoriously heterogeneous due to various experimental assays, and often biased toward specific target classes, such as kinases, and the use of different bioactivity end-points make the implementation of target deconvolution approaches difficult, despite the curated and standardized drug/target databases such as ChEMBL [23]. Therefore, we additionally demonstrate how to use the bioactivity profiles for the purposes of target deconvolution and prediction of drug–target combinations.

2 Materials

The approach makes use of two input data sources, drug sensitivity (Subheading 2.1) and target selectivity (Subheading 2.2) profiles. Drug sensitivity profiling data makes the TAS signatures context-specific for a given cancer cell line or patient sample, whereas the underlying drug–target bioactivity profiles provide the system-wide interactions between drugs and targets that map the potential activity space, independent of the cellular context.

2.1 Drug Sensitivity Profiling Data

1. A library of targeted compounds is screened against cancer cell lines or patient-derived samples using a high-throughput drug sensitivity profiling assay (*see* **Note 1**). At FIMM, our oncology collection contains both FDA/EMA-approved drugs as well as investigational compounds. Each compound is tested in five increasing concentrations, across a 10,000-fold range, allowing for the establishment of dose–response curves.
   (a) We previously tested 21 breast cancer cell lines against 239 anticancer compounds [20, 22], using the FIMM drug sensitivity and resistance testing (DSRT) platform [8, 12].

   (b) We have also screened patient-derived cells from acute myeloid leukemia (AML) patients, and used healthy bone marrow samples as controls in the DSRT platform [8, 24].

2. One can also use publicly available cancer cell line profiling datasets, such as those from the Sanger institute's Genomics of Drug Sensitivity in Cancer (GDSC) project [14, 15], using which we applied the TAS pipeline on a subset of 107 pan-cancer cell lines tested against 138 targeted compounds that included also other drugs than kinase inhibitors. The Broad Institute's Cancer Cell Line Encyclopedia (CCLE) [13] and Cancer Therapeutic Research Portal (CTRP) [16, 18] are also useful resources of drug sensitivity data across cancer cell lines, along with their comprehensive genomic and molecular profiles.

2.2 Drug–Target Bioactivity Profiles

1. List of primary "on-targets" for each compound can be obtained from various resources, such as the targets listed Sanger GDSC (http://www.cancerrxgene.org/); Broad CTRP (http://portals.broadinstitute.org/ctrp/?page=#ctd2BodyHome); DrugBank (https://www.drugbank.ca/) or Drug Gene Interaction Database (DGIdb, http://dgidb.genome.wustl.edu/).
2. For more extensive and quantitative information of drug–target bioactivity profiles for primary, secondary, and downstream "off-target" effects, which are required for the systems-level application of TAS, one can use data from large-scale target selectivity profiling assays, including those for kinase inhibitors [25–27].
3. Comprehensive bioactivity data for many target classes can be retrieved from our publicly available Drug Target Commons (DTC) database (https://drugtargetcommons.fimm.fi/), a crowdsourced platform for bioactivity data annotation, integration, standardization, and curation, which also includes assay and bioactivity data from ChEMBL [28, 29].

2.3 Software Tools/R-Packages

1. DSS R-package is available as open-source R implementation at https://bitbucket.org/BhagwanYadav/drug-sensitivity-score-dss-calculation
2. TAS R-package is available as open-source R implementation at https://bitbucket.org/BhagwanYadav/target-addiction-score-tas-calculation.
3. SynergyFinder is available both as web implementation (https://synergyfinder.fimm.fi) and R-package (http://bioconductor.org/packages/release/bioc/html/synergyfinder.html).

3 Methods

The following steps detail the computational methodologies for the TAS approach, both for single-target addiction scoring (Subheading 3.3) and combinatorial TAS (Subheading 3.4).

3.1 Drug Sensitivity Scoring (DSS)

1. Calculate quantitative DSS values by fitting the dose–response curves using four-parameter logistic model [24]. DSS is a closed-form integration of the area under the dose–response curve (AUC), which effectively summarizes the complex dose–response relationship into a single metric (*see* **Note 2**). An open-source R implementation of DSS package is made available for this purpose (*see* Subheading 2.3).
2. Calculate differential DSS (dDSS) by comparing the DSS of cancer sample to that of the controls (*see* **Note 3**). Differential DSS allows one to estimate the selective response of a given cancer sample to the particular drug, compared to the control background response distribution, which is eventually useful when quantifying the selective target addictions of the particular cancer samples.

3.2 Drug–Target Interactions

1. Extract the quantitative drug–target interaction bioactivity profiles of compounds from the DTC web portal (https://drugtargetcommons.fimm.fi/).
2. Define the potent targets of a particular compound by using all the available bioactivity data for the compound and a suitable activity threshold:
   (a) If there are enough dose-response bioactivity end-points (K_d, K_i, or IC_{50}) for the compound (say, for more than 20 distinct proteins), define the potent targets as those with less than 50-fold bioactivity compared to the smallest bioactivity value over all the proteins (this typically corresponds to the 'on target'). In case of limited number of data points, define potent targets as those with K_d, K_i or $IC_{50} \leq 100$ nM (*see* **Note 4**).
   (b) For the activity measurements (activity%, residual activity %, %inhibition, or potency), often resulting from assays with only single or a few concentration points, define a more stringent threshold: ≤10% residual activity (≥90% inhibition) for the test concentration of ≤1000 nM and ≤20% for a test concentration of ≤500 nM in biochemical assays. For cell-based assays, set the threshold to ≤50% residual activity for the test concentration of ≤1000 nM and ≤10% for the test concentration ≤10,000 nM.

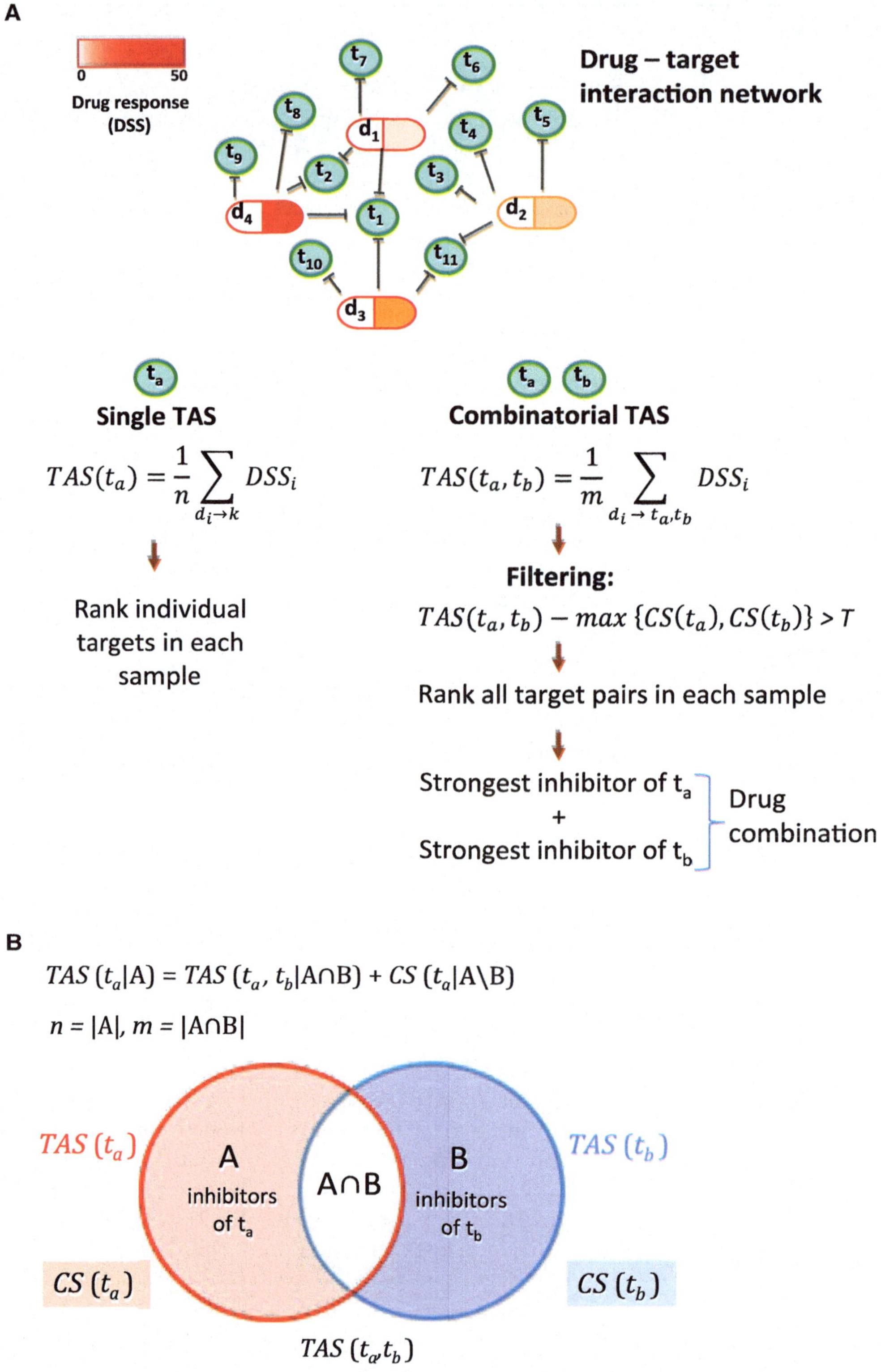

Fig. 1 Schematic illustration of the single and combinatorial target addiction scoring (TAS). (**a**) Left: Single TAS approach ranks each target (t) in the context of a given drug–target network based on the average drug

3.3 Target Addiction Scoring (TAS)

The TAS value provides an estimate of the sensitivity of a cell to the inhibition of a particular target. More specifically, for a given target t, TAS_t is calculated by averaging the observed drug response (e.g., DSS_i) (*see* **Note 2**) over all those n_t inhibitors (i) that target the protein t (Fig. 1a). Mathematically, TAS defines a transformation between the spaces spanned by the compounds and their targets, which maps observed drug responses to the underlying target addictions:

$$\mathrm{TAS}_t = \sum_{i=1}^{n_t} \frac{\mathrm{DSS}_i}{n_t}$$

1. Calculate the single-target TAS for each sample separately and sort the targets based on the increasing TAS values, which enables one to prioritize the pharmacologically actionable target addictions in individual cancer samples for further preclinical validation (*see* **Notes 5** and **6**). An open-source R implementation of TAS package is made available for this purpose (*see* Subheading 2.3).
2. Determine the statistical significance of a TAS value empirically using permutation testing; based on a vector of inhibitors per target, randomly select a given number of inhibitors, and then average their DSS values in a given sample. The permutation procedure is repeated, simulating at least 10,000 random TAS values in the given sample. The empirical p-value is defined by the percentage of the permuted TAS values above or equal to the observed one.

3.4 Prediction of Synergistic Target Pairs

1. For all those target pairs (t_1, t_2) that are inhibited by the same compound, calculate a combinatorial TAS score by averaging the drug response (e.g., DSS) over the set of common inhibitors that target both t_1 and t_2. Target pairs that have no common inhibitors in the library are excluded from the combinatorial TAS analysis.

Fig. 1 (continued) sensitivity score (DSS) over the subset of its potent inhibitors (n). Single TAS enables one to prioritize pharmacologically actionable kinase signal addictions in individual cancer cell samples for experimental validation. Right: The TAS concept was extended to ranking target pairs (t_a, t_b) based on their average combinatorial effect over the subset of inhibitors targeting both of the kinases (m). Combinatorial TAS enables one to identify synthetic lethal type of target pairs, which may correspond to synergistic drug combinations between their inhibitors. Experimental validation of the combinatorial TAS predictions can be carried out by testing the most potent inhibitors of t_a and t_b in combination assays. (**b**) Relationships between the single and combinatorial TAS defined using set-theoretic operations among the set of inhibitors. Using this notation, the cardinality of the set A is n and the cardinality of the intersection between A and B is m. The complement score (CS) of t_a or t_b is defined as the average drug response over the inhibitors that belong to the difference between sets A and B or B and A, respectively (the shaded portions) (modified from ref. 20 with permission from Elsevier)

2. Calculate complement scores for each target in a target pair. The complement score (CS) of t_1 or t_2 is defined as the average response (e.g., DSS) over the inhibitors that inhibit only t_1 or t_2, respectively, but not both of the targets (*see* Fig. 1b for a schematic illustration of the complement score concept).
3. Exclude those target pairs for which the difference between the combinatorial TAS and the complement scores is below a selected cutoff value (e.g., $T = 4$). Rank the remaining target pairs by the magnitude of their combinatorial TAS, and map back to drug pairs by selecting the strongest inhibitors of targets t_1 and t_2 alone, excluding their common inhibitors, using the lowest bioactivity values.
4. Experimentally validate the top coaddicted target pairs predicted by the combinatorial TAS, e.g., by testing the inhibitors in an 8×8 dose-matrix format covering seven increasing concentrations of each drug, along with all their pairwise combinations [20, 30] (*see* Fig. 2). Evaluate the degree of synergy between the pairwise combination effects using the synergy scoring models implemented in the SynergyFinder package [32] (*see* **Note 7**).

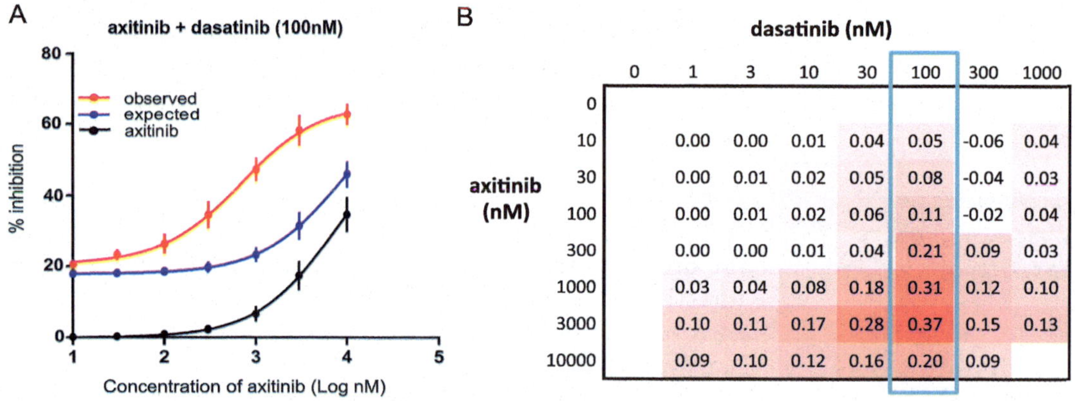

Fig. 2 Combinatorial TAS predicts synergistic drug combinations and coaddicted target pairs. (**a**) Dose–response curves for the predicted synergistic effect between axitinib and dasatinib on cell viability in HCC1937 human breast cancer cells. The red dose–response curve shows the observed combination effect between axitinib (at multiple concentrations, *y*-axis) and dasatinib (at fixed 100 nM concentration; see part **b**, highlighted column). The black curve represents the dose response to axitinib only. The blue curve represents the expected dose response of axitinib (at multiple concentrations) combined with dasatinib (at fixed 100 nM concentration). The expected combinatorial effects were calculated based on the Bliss independence model [31]. Each combination was tested in two to four replicates. Points and error bars represent the mean and its standard error (SEM), respectively, and the solid curve is the logistic function fit. (**b**) An example of the Bliss excess scores matrix for the synergistic combination effects between axitinib and dasatinib (see part **a**). The values in the matrix are synergy scores calculated using the Bliss independence model at each tested combinatorial concentration of the two drugs (the first row and column of the matrix show the single-drug dose–response curves, where the other drug is at zero concentration). The highlighted column (dasatinib 100 nM) corresponds to the red curve in part **a** (reproduced from ref. 20 with permission from Elsevier)

4 Notes

1. The DSRT platform uses of a library of compounds dissolved in 100% dimethyl sulfoxide (DMSO) and dispensed on tissue culture treated 384-well plates using an acoustic liquid handling device. The compounds are plated at five different concentrations in tenfold dilutions covering a 10,000-fold concentration range, centered around a compound-specific relevant cellular activity concentration (e.g., 1–10,000 nM for a compound with an on-target cellular half-maximal effect of about 100 nM). The predrugged plates are stored in pressurized storage pods filled with inert nitrogen gas. For the assay, the compounds are first dissolved with 5 μl of culture medium while shaking for 30 min, 20 μl of single cell suspension (10,000 cells) is then seeded to each well using a MultiDrop Combi peristaltic dispenser. The plates are incubated in a humidified environment at 37 °C and 5% CO_2 and after 72 h cell viability was measured using CellTiter-Glo luminescent assay (Promega), according to manufacturer's instructions. The data are normalized to negative control (DMSO only) and positive control wells (100 μM benzethonium chloride).
2. In addition to the drug sensitivity score (DSS), a number of other metrics, such as the area under the drug response (AUC) and half-maximal inhibitory concentration (IC_{50}), are also frequently used for summarizing dose–response curves and quantifying drug sensitivities. In our studies, we have mainly used DSS as the response metric, as it has been shown to provide robust and reproducible drug response profiles [24, 33].
3. High-throughput drug sensitivity screening is amenable to profiling of a compendium of patient-derived cells or cancer cell lines, making it possible also to study the variability of drug response across different samples to the same drug. Whenever possible, we calculate the differential DSS (dDSS) values in each patient sample by subtracting the mean DSS for a drug across all healthy controls from the observed DSS of a given cancer sample. However, sometimes it is difficult to define the control responses for cancer cell lines; in our previous work on breast cancer cell lines [20], we used the DSS of MCF10A cells as a control to calculate the differential DSS. Alternatively, we have also used the average DSS across all the samples screened against a given drug for calculating dDSS [21, 22].
4. Defining potent on- and off-targets for a compound may not be straightforward, especially in cases where only a few bioactivity values are available from target selectivity profiling studies. Further challenges originate from differences in the bioactivity assays. For instance, biochemical assays typically

generate higher potency levels than cell-based assays, and therefore require more stringent cutoff values. Dose-response bioactivity end-points are often more reliable than those based on single or a few concentration points, and therefore K_d, K_i, and IC_{50} end-points are preferred [34]. In cases where there are multiple bioactivity values for the same compound–target pair, originating from different studies or other data resources, one can take median bioactivity value that is more robust to outliers than the mean value.

5. We manually filter out targets involved in cellular metabolism, such as cytochrome (CYP) and ATP-binding cassette (ABC) genes, from the set of potential drug targets when analyzing in vitro drug responses, since those are unlikely to contribute to the drug's mode of action in cell line drug testing.
6. It is also possible to incorporate information from biological pathways and signaling networks for identifying target addictions of proteins that are upstream or downstream of a given drug target. Often, one will find drugs that show selective drug response in a cancer sample, even if their direct targets might not show selective addiction. Therefore, additional molecular profiling datasets such as transcriptomic, proteomic, and genome-wide loss-of-function RNAi or CRISPR screens, that provide complementary target validation information, can be used to identify targets that mechanistically explain the observed drug responses and also serves as a preclinical validation of the hits from the TAS approach.
7. For drug combination synergy scoring, we have previously made available the SynergyFinder web implementation (https://synergyfinder.fimm.fi) and R-package (http://bioconductor.org/packages/release/bioc/html/synergyfinder.html) to calculate the drug synergy scores using different reference models, including Zero Interaction Potency (ZIP), Bliss and Loewe models, as well as the Highest Single Agent (HSA) [32]. Sometimes it may be convenient to calculate the synergy scores using various models to come up with the most reliable determination of the degree of synergy for a given drug pair [35].

Acknowledgments

This work was supported by the Academy of Finland (grants 272437, 269862, 279163, 292611, 295504, 310507); the Cancer Society of Finland (TA, KW); the Integrative Life Science Doctoral Program at the University of Helsinki (AJ).

References

1. Vogelstein B, Papadopoulos N, Velculescu VE, Zhou S, Diaz LA, Kinzler KW (2013) Cancer genome landscapes. Science 339(6127):1546
2. Garraway Levi A, Lander Eric S (2013) Lessons from the cancer genome. Cell 153(1):17–37
3. The Cancer Genome Atlas Research N (2017) Integrated genomic and molecular characterization of cervical cancer. Nature 543 (7645):378–384
4. The Cancer Genome Atlas Research N (2017) Integrated genomic characterization of oesophageal carcinoma. Nature 541 (7636):169–175
5. Marusyk A, Almendro V, Polyak K (2012) Intra-tumour heterogeneity: a looking glass for cancer? Nat Rev Cancer 12(5):323–334
6. Yi S, Lin S, Li Y, Zhao W, Mills GB, Sahni N (2017) Functional variomics and network perturbation: connecting genotype to phenotype in cancer. Nat Rev Genet 18(7):395–410
7. Weinstein IB, Joe A (2008) Oncogene addiction. Cancer Res 68(9):3077
8. Pemovska T, Kontro M, Yadav B, Edgren H, Eldfors S, Szwajda A, Almusa H, Bespalov MM, Ellonen P, Elonen E, Gjertsen BT, Karjalainen R, Kulesskiy E, Lagström S, Lehto A, Lepistö M, Lundán T, Majumder MM, Lopez Marti JM, Mattila P, Murumägi A, Mustjoki S, Palva A, Parsons A, Pirttinen T, Rämet ME, Suvela M, Turunen L, Västrik I, Wolf M, Knowles J, Aittokallio T, Heckman CA, Porkka K, Kallioniemi O, Wennerberg K (2013) Individualized systems medicine (ISM) strategy to tailor treatments for patients with chemorefractory acute myeloid leukemia. Cancer Discov 3(12):1416–1429
9. Pemovska T, Johnson E, Kontro M, Repasky GA, Chen J, Wells P, Cronin CN, McTigue M, Kallioniemi O, Porkka K, Murray BW, Wennerberg K (2015) Axitinib effectively inhibits BCR-ABL1(T315I) with a distinct binding conformation. Nature 519(7541):102–105
10. Tyner JW, Yang WF, Bankhead A, Fan G, Fletcher LB, Bryant J, Glover JM, Chang BH, Spurgeon SE, Fleming WH, Kovacsovics T, Gotlib JR, Oh ST, Deininger MW, Zwaan CM, Den Boer ML, van den Heuvel-Eibrink MM, Hare T, Druker BJ, Loriaux MM (2013) Kinase pathway dependence in primary human leukemias determined by rapid inhibitor screening. Cancer Res 73(1):285
11. Friedman AA, Letai A, Fisher DE, Flaherty KT (2015) Precision medicine for cancer with next-generation functional diagnostics. Nat Rev Cancer 15(12):747–756
12. Malani D, Murumagi A, Yadav B, Kontro M, Eldfors S, Kumar A, Karjalainen R, Majumder MM, Ojamies P, Pemovska T, Wennerberg K, Heckman C, Porkka K, Wolf M, Aittokallio T, Kallioniemi O (2017) Enhanced sensitivity to glucocorticoids in cytarabine-resistant AML. Leukemia 31(5):1187–1195
13. Barretina J, Caponigro G, Stransky N, Venkatesan K, Margolin AA, Kim S, Wilson CJ, Lehar J, Kryukov GV, Sonkin D, Reddy A, Liu M, Murray L, Berger MF, Monahan JE, Morais P, Meltzer J, Korejwa A, Jane-Valbuena J, Mapa FA, Thibault J, Bric-Furlong E, Raman P, Shipway A, Engels IH, Cheng J, Yu GK, Yu J, Aspesi P, de Silva M, Jagtap K, Jones MD, Wang L, Hatton C, Palescandolo E, Gupta S, Mahan S, Sougnez C, Onofrio RC, Liefeld T, MacConaill L, Winckler W, Reich M, Li N, Mesirov JP, Gabriel SB, Getz G, Ardlie K, Chan V, Myer VE, Weber BL, Porter J, Warmuth M, Finan P, Harris JL, Meyerson M, Golub TR, Morrissey MP, Sellers WR, Schlegel R, Garraway LA (2012) The Cancer Cell Line Encyclopedia enables predictive modelling of anticancer drug sensitivity. Nature 483(7391):603–307
14. Garnett MJ, Edelman EJ, Heidorn SJ, Greenman CD, Dastur A, Lau KW, Greninger P, Thompson IR, Luo X, Soares J, Liu Q, Iorio F, Surdez D, Chen L, Milano RJ, Bignell GR, Tam AT, Davies H, Stevenson JA, Barthorpe S, Lutz SR, Kogera F, Lawrence K, McLaren-Douglas A, Mitropoulos X, Mironenko T, Thi H, Richardson L, Zhou W, Jewitt F, Zhang T, O/'Brien P, Boisvert JL, Price S, Hur W, Yang W, Deng X, Butler A, Choi HG, Chang JW, Baselga J, Stamenkovic I, Engelman JA, Sharma SV, Delattre O, Saez-Rodriguez J, Gray NS, Settleman J, Futreal PA, Haber DA, Stratton MR, Ramaswamy S, McDermott U, Benes CH (2012) Systematic identification of genomic markers of drug sensitivity in cancer cells. Nature 483 (7391):570–575
15. Iorio F, Knijnenburg TA, Vis DJ, Bignell GR, Menden MP, Schubert M, Aben N, Gonçalves E, Barthorpe S, Lightfoot H, Cokelaer T, Greninger P, van Dyk E, Chang H, de Silva H, Heyn H, Deng X, Egan RK, Liu Q, Mironenko T, Mitropoulos X, Richardson L, Wang J, Zhang T, Moran S, Sayols S, Soleimani M, Tamborero D, Lopez-Bigas N, Ross-Macdonald P, Esteller M, Gray NS, Haber DA, Stratton MR, Benes CH, Wessels LFA, Saez-Rodriguez J, McDermott U, Garnett MJ (2016) A landscape of

pharmacogenomic interactions in cancer. Cell 166(3):740–754

16. Basu A, Bodycombe Nicole E, Cheah Jaime H, Price Edmund V, Liu K, Schaefer Giannina I, Ebright Richard Y, Stewart Michelle L, Ito D, Wang S, Bracha Abigail L, Liefeld T, Wawer M, Gilbert Joshua C, Wilson Andrew J, Stransky N, Kryukov Gregory V, Dancik V, Barretina J, Garraway Levi A, Hon CS-Y, Munoz B, Bittker Joshua A, Stockwell Brent R, Khabele D, Stern Andrew M, Clemons Paul A, Shamji Alykhan F, Schreiber Stuart L (2013) An interactive resource to identify cancer genetic and lineage dependencies targeted by small molecules. Cell 154 (5):1151–1161
17. Rees MG, Seashore-Ludlow B, Cheah JH, Adams DJ, Price EV, Gill S, Javaid S, Coletti ME, Jones VL, Bodycombe NE, Soule CK, Alexander B, Li A, Montgomery P, Kotz JD, Hon CS-Y, Munoz B, Liefeld T, Dancik V, Haber DA, Clish CB, Bittker JA, Palmer M, Wagner BK, Clemons PA, Shamji AF, Schreiber SL (2016) Correlating chemical sensitivity and basal gene expression reveals mechanism of action. Nat Chem Biol 12(2):109–116
18. Seashore-Ludlow B, Rees MG, Cheah JH, Cokol M, Price EV, Coletti ME, Jones V, Bodycombe NE, Soule CK, Gould J, Alexander B, Li A, Montgomery P, Wawer MJ, Kuru N, Kotz JD, Hon CS-Y, Munoz B, Liefeld T, Dančík V, Bittker JA, Palmer M, Bradner JE, Shamji AF, Clemons PA, Schreiber SL (2015) Harnessing connectivity in a large-scale small-molecule sensitivity dataset. Cancer Discov 5(11):1210–1223
19. Hopkins AL (2008) Network pharmacology: the next paradigm in drug discovery. Nat Chem Biol 4(11):682–690
20. Szwajda A, Gautam P, Karhinen L, Jha Sawan K, Saarela J, Shakyawar S, Turunen L, Yadav B, Tang J, Wennerberg K, Aittokallio T (2015) Systematic mapping of kinase addiction combinations in breast cancer cells by integrating drug sensitivity and selectivity profiles. Chem Biol 22(8):1144–1155
21. Yadav B, Gopalacharyulu P, Pemovska T, Khan SA, Szwajda A, Tang J, Wennerberg K, Aittokallio T (2015) From drug response profiling to target addiction scoring in cancer cell models. Dis Model Mech 8(10):1255–1264
22. Gautam P, Karhinen L, Szwajda A, Jha SK, Yadav B, Aittokallio T, Wennerberg K (2016) Identification of selective cytotoxic and synthetic lethal drug responses in triple negative breast cancer cells. Mol Cancer 15(1):34
23. Hersey A, Chambers J, Bellis L, Patrícia Bento A, Gaulton A, Overington JP (2015) Chemical databases: curation or integration by user-defined equivalence? Drug Discov Today Technol 14:17–24
24. Yadav B, Pemovska T, Szwajda A, Kulesskiy E, Kontro M, Karjalainen R, Majumder MM, Malani D, Murumägi A, Knowles J, Porkka K, Heckman C, Kallioniemi O, Wennerberg K, Aittokallio T (2014) Quantitative scoring of differential drug sensitivity for individually optimized anticancer therapies. Sci Rep 4:5193
25. Davis MI, Hunt JP, Herrgard S, Ciceri P, Wodicka LM, Pallares G, Hocker M, Treiber DK, Zarrinkar PP (2011) Comprehensive analysis of kinase inhibitor selectivity. Nat Biotechnol 29(11):1046–1051
26. Metz JT, Johnson EF, Soni NB, Merta PJ, Kifle L, Hajduk PJ (2011) Navigating the kinome. Nat Chem Biol 7(4):200–202
27. Knapp S, Arruda P, Blagg J, Burley S, Drewry DH, Edwards A, Fabbro D, Gillespie P, Gray NS, Kuster B, Lackey KE, Mazzafera P, Tomkinson NCO, Willson TM, Workman P, Zuercher WJ (2013) A public-private partnership to unlock the untargeted kinome. Nat Chem Biol 9(1):3–6
28. Bento AP, Gaulton A, Hersey A, Bellis LJ, Chambers J, Davies M, Krüger FA, Light Y, Mak L, McGlinchey S, Nowotka M, Papadatos G, Santos R, Overington JP (2014) The ChEMBL bioactivity database: an update. Nucleic Acids Res 42(D1):D1083–D1090
29. Gaulton A, Bellis LJ, Bento AP, Chambers J, Davies M, Hersey A, Light Y, McGlinchey S, Michalovich D, Al-Lazikani B, Overington JP (2012) ChEMBL: a large-scale bioactivity database for drug discovery. Nucleic Acids Res 40(D1):D1100–D1107
30. He L, Kulesskiy E, Saarela J, Turunen L, Wennerberg K, Aittokallio T, Tang J (2016) Methods for high-throughput drug combination screening and synergy scoring. bioRxiv. https://doi.org/10.1101/051698
31. Bliss CI (1939) The toxicity of poisons applied jointly. Ann Appl Biol 26(3):585–615
32. Ianevski A, He L, Aittokallio T, Tang J (2017) SynergyFinder: a web application for analyzing drug combination dose–response matrix data. Bioinformatics 33(15):2413–2415. https://doi.org/10.1093/bioinformatics/btx168
33. Mpindi JP, Yadav B, Östling P, Gautam P, Malani D, Murumägi A, Hirasawa A, Kangaspeska S, Wennerberg K, Kallioniemi O, Aittokallio T (2016) Consistency in drug response profiling. Nature 540(7631):E5–E6

34. Santos R, Ursu O, Gaulton A, Bento AP, Donadi RS, Bologa CG, Karlsson A, Al-Lazikani B, Hersey A, Oprea TI, Overington JP (2017) A comprehensive map of molecular drug targets. Nat Rev Drug Discov 16 (1):19–34

35. Tang J, Wennerberg K, Aittokallio T (2015) What is synergy? The Saariselkä agreement revisited. Front Pharmacol 6:181

Chapter 13

Predicting Drug Interactions From Chemogenomics Using INDIGO

Sriram Chandrasekaran

Abstract

Designing effective antibiotic combination regimens is critical for countering drug resistance in pathogens. Yet the large combinatorial search-space makes the identification of effective combinations a significant challenge. There is a great need for computational approaches that can rapidly prioritize potential combination regimens based on the antagonistic and synergistic interactions among the constituent antibiotics. This protocol outlines the steps to predict antibiotic interactions from chemogenomics data using the INDIGO algorithm. INDIGO predicted novel drug–drug interaction outcomes quantitatively with high accuracy based on experimental evaluation of predictions in *E. coli* and *S. aureus*, and it overcomes several limitations of existing drug-interaction prediction algorithms. The INDIGO approach also expands the applicability of chemogenomic data from model organisms to a broader set of less-studied pathogens. INDIGO can predict drug-interaction outcomes in the bacterial pathogens *S. aureus* and *M. tuberculosis*, using chemogenomics data from *E. coli* by quantifying the degree of conservation of the drug–gene interaction network between different species. The INDIGO approach, which is demonstrated for *E. coli* and *S. aureus* in this protocol, can be applied easily to other organisms including pathogens.

Key words Drug synergy, Antibiotics, Drug resistance, Drug combinations, Chemogenomics, Machine learning, *Staphylococcus aureus*, *Mycobacterium tuberculosis*

1 Introduction

Each year 100,000 people die and a million others become sick by antibiotic-resistant bacteria in the USA. The recent discovery of a *E. coli* strain resistant to the drug of last resort—colistin, heralds the era of pandrug-resistant bacterial infections for which there are no antibiotics. Due to a dwindling supply of new antimicrobials, treatment of infections caused by antibiotic-resistant pathogens increasingly relies on the use of combinations of existing antibiotics [1, 2]. Multidrug combinations are the standard mode of treatment for tuberculosis (TB), which affects 9 million people and causes 1.4 million deaths worldwide annually [3]. Standard TB treatment takes months with a multidrug regimen, while treating drug-resistant strains may take 2 years. New combination regimens are needed

Slava Ziegler and Herbert Waldmann (eds.), *Systems Chemical Biology: Methods and Protocols*, Methods in Molecular Biology, vol. 1888, https://doi.org/10.1007/978-1-4939-8891-4_13,

to counter resistance and reduce treatment times for treating diseases like TB.

A well-designed drug combination regimen can simultaneously enhance potency and reduce risk of resistance [4]. Yet identifying promising multidrug regimens takes considerable time and effort. The current approach is based on trial and error, and involves screening different panels of drug combinations against a pathogen. Combining drugs in a cocktail can lead to surprising nonlinear effects; some drugs can enhance each other's action leading to higher potency (synergy), or drugs can interfere with their action leading to reduced potency (antagonism). Combinations with strong activity in vitro can guide clinical outcomes. For example, trimethoprim–sulfamethoxazole combination commonly used in the clinic is based on the synergistic interaction between the two drugs. In addition, bactericidal and bacteriostatic antibiotics are avoided in the clinic due to their strong antagonism [5]. Unfortunately, designing a four-drug combination, such as currently used for treating TB, from a set of just 100 drugs would require ~4 million drug interaction experiments. There is hence a huge unmet need for a rational basis by which to prioritize new drug combinations for treating multidrug-resistant pathogens [1]. Once an optimal composition of drugs in a combination is determined, their doses can be optimized using established tools such as feedback system control [6].

To address this challenge, a computational approach called INDIGO was developed to identify antibiotic combinations that interact synergistically or antagonistically in inhibiting bacterial growth [7] (Fig. 1). INDIGO makes use of chemogenomic

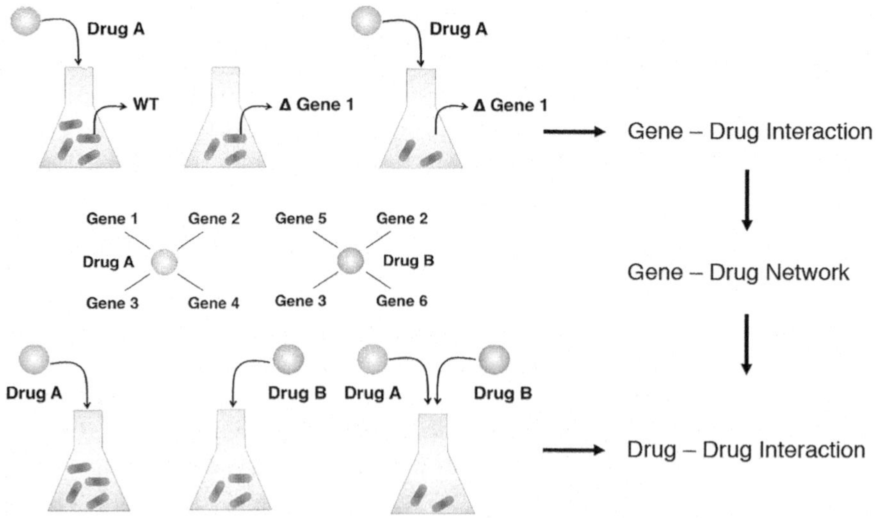

Fig. 1 Overview of the theoretical framework used by INDIGO. INDIGO uses gene-drug interaction data to predict drug–drug interactions

profiling, a high-throughput approach to identify genes that, when deleted, can enhance or reduce sensitivity to an individual drug [8]. INDIGO takes a systems approach to predict drug interactions; it quantifies the contribution of individual genes with a chemical–genetic interaction on the overall drug interaction outcome (Fig. 1). In contrast, current bioinformatics tools rely on simplistic "guilt-by-association" or "similarity is synergy" relationships to predict drug interactions. Algorithms based on drug similarity (or dissimilarity) are less effective in predicting interaction outcomes for new drugs classes and lack a model for antagonism [7]. Experimental data suggests that drugs with similar targets and chemogenomic profiles can have both synergistic and antagonistic outcomes.

INDIGO uses Boolean operations to simulate a joint chemogenomic profile of a drug combination. The sigma and delta scores, calculated by the algorithm by adding and subtracting the chemogenomic profiles respectively, capture the similarity and uniqueness in the mechanism of action of the individual drugs in a combination. A machine learning algorithm called random-forests is then used to build a predictive model that links interaction outcome of drug combinations to the joint chemogenomic profile (sigma and delta scores) of the drug pair (Fig. 2). INDIGO also identifies genes in the chemogenomics data that are most predictive of drug interactions in an unsupervised fashion by using the random forest algorithm.

Chemogenomic data compiled from the literature for the model organism *E. coli* was used as input to predict interaction outcomes for 2628 novel drug combinations using INDIGO.

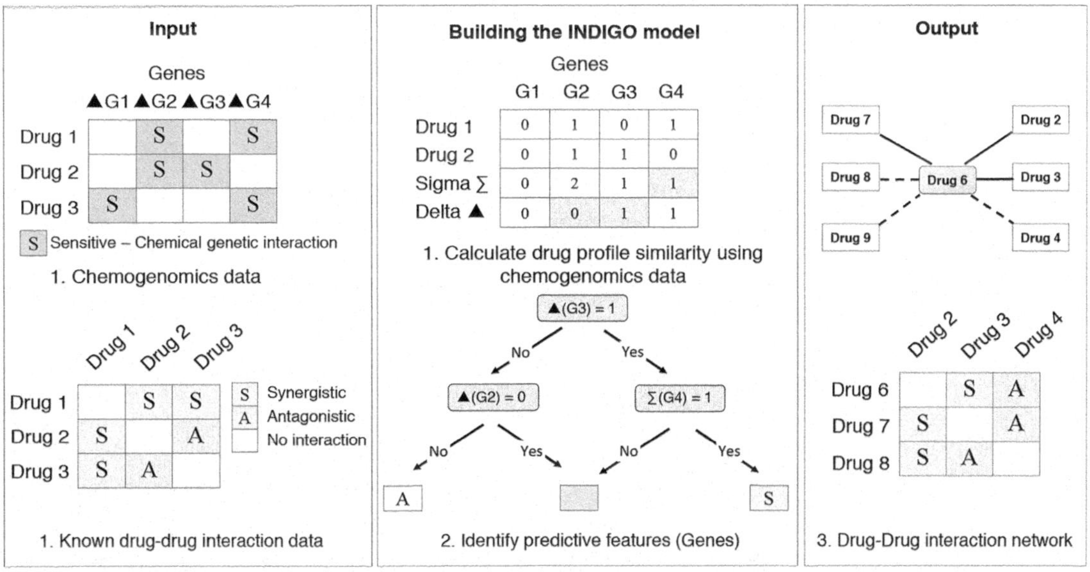

Fig. 2 Steps used by INDIGO to construct a drug interaction model

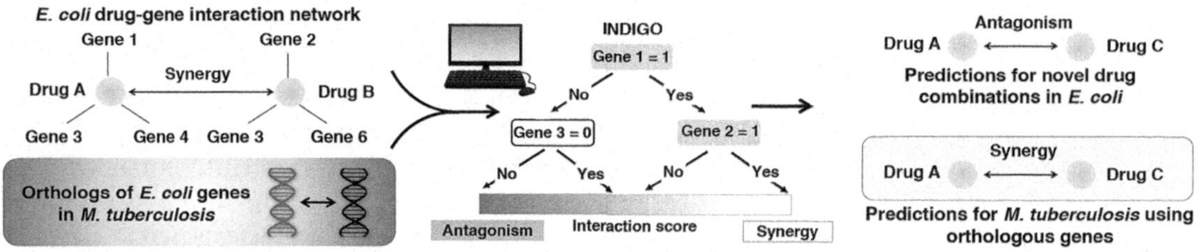

Fig. 3 Predicting drug interactions in pathogen species using model organism data by INDIGO. In this simplified example, Gene 3 is not conserved between *E. coli* and *M. tuberculosis*, therefore its contribution to the drug interaction model is "deleted" and the prediction from the *E. coli* INDIGO model without the contribution from this gene is used to infer drug interaction outcomes in *M. tuberculosis*

INDIGO was extensively tested experimentally and it predicted novel drug interaction outcomes with high accuracy in 66 validation experiments consisting of novel pairwise combinations of antibiotics (p-value $= 10^{-6}$). INDIGO also correctly predicted synergy between commonly used, clinically validated antibiotic combinations that required clinical validation over decades of study. Analysis of *E. coli* INDIGO model also revealed a core set of genes and pathways that are predictive of antibiotic interactions and provides insight into the biology that underpins drug interactions.

The INDIGO approach can also be applied to predict drug interaction outcomes in other bacterial pathogens by identifying orthologs of pathogen genes in the *E. coli* chemogenomics data (Fig. 3). This approach was based on the observation that top predictive genes in the *E. coli* INDIGO model were significantly enriched for genes orthologous between *E. coli*, a gram-negative bacterium, and the gram-positive microbe *S. aureus*. INDIGO assumes that interactions that depend on orthologous genes would be conserved between the two species, while those that are predicted to depend on nonorthologous genes would not be conserved. Quantification of the extent of conservation of the core drug-interaction genes in the *E. coli* INDIGO model accurately predicted the extent of conservation of drug-interaction outcomes in the bacterial pathogens *M. tuberculosis* and *S. aureus*. Novel predictions were tested experimentally for 45 drug pairs in *S. aureus* [7]. INDIGO provides the first example of leveraging omics data in model systems to facilitate the discovery of effective combination therapies in less-studied pathogens.

1.1 Key Terms

- *Random forests*—*Random forests* is a powerful machine learning algorithm that builds an ensemble of *decision trees* to create a predictive model for a process of interest. In a *decision tree* algorithm, information in the input data traverses down the tree like a flow chart and is binned to smaller branches based on decisions made at individual nodes (e.g., *see* Fig. 2). The random forest algorithm builds many diverse decision trees for a data of interest and outputs the average prediction from the ensemble of decision trees. While each decision tree can be a poor predictor, by combining many weak predictors, the random forest algorithm can provide an accurate prediction for a process of interest. Just like any other machine learning algorithm, Random forests requires a training data set to learn a predictive model, in this case, this data is known drug interaction outcomes. INDIGO then uses the random forest algorithm to build an ensemble of decision-trees using genes in the chemogenomics data and identifies those genes that are predictive of drug interaction outcomes using a training drug interaction data set.

- *Chemogenomics*—INDIGO uses chemogenomics data to comprehensively characterize the set of processes involved in drug action and interaction. A gene is predicted to have a chemical–-genetic interaction with a drug if a mutation or deletion leads to a statistically significant alteration in sensitivity to the drug of interest. A drug–gene interaction network is created by integrating chemogenomics data from hundreds of drugs. This static interaction network is then converted into a predictive model by the random forest algorithm.
- *Bliss drug interaction model*—Drug interactions can be quantified using a variety of approaches. The Bliss score is widely used due to its simplicity. The Bliss ε score is defined as

$$\varepsilon = W_{xy} - W_x \times W_y.$$

where W_x and W_y are the normalized growth rates caused by the addition of drugs x and y, and W_{xy} is the normalized growth rate caused by the addition of both drugs relative to no drug treatment. The Bliss score ε estimates the deviation of the observed combined effect of two drugs (W_{xy}) to an estimated value calculated based on the assumption that the two drugs act independently ($W_x \times W_y$). Interactions with negative ε are synergistic, while those that are positive are antagonistic. However, the Bliss score does not account for the dose of the drug and can incorrectly predict synergy or antagonism between identical (same) drugs.

- *Loewe's model*—Loewe additivity model defines a drug as additive or noninteracting with itself [9]. Antagonistic and synergistic interactions are inferred based on deviations from additivity. The Loewe model takes as input the growth of the bacteria at different dose combinations of the individual drugs to generate a quantitative interaction score (α score) between the drug pair, with low scores ($\alpha < -0.5$) corresponding to synergy and high scores ($\alpha > 1$) corresponding to antagonism. Synergy implies that the same amount of growth inhibition is achieved with a lower dose when both drugs are combined.

- *Orthologous genes*—Orthologs are genes in different organisms that have descend from a common ancestral gene [10]. Finding orthologous genes is a common step in comparative genomics to explore the evolution of genes. Genes that are orthologous in different species are hypothesized to have similar functions. This ortholog functional conservation hypothesis is the basis for many comparative genomics methods to infer gene functions across species. INDIGO uses this data to infer the conservation of drug interaction outcomes across species by finding the extent of conservation of drug interaction related genes (Fig. 3).

2 Materials

2.1 Software Prerequisites and Installation

1. MATLAB 7.0 or above (http://www.mathworks.com).
2. Random Forest toolbox for MATLAB https://code.google.com/p/randomforest-matlab/downloads/list.
3. INDIGO MATLAB code (http://www.sriramlab.org/software/) (*see* **Note 1**).

2.2 User Data

1. *Drug interaction training data*: A list of drug combinations and their interaction outcome. The interaction outcome can be discrete values (e.g., −1 for synergy, +1 for antagonism) or continuous values based on the Bliss, FIC (Fractional inhibitory concentration) or Loewe's models (*see* **Note 2**).
2. *Chemogenomics profiles*: Chemogenomics data should be provided for all the drugs that are part of a combination regimen. This should contain the set of genes that when inhibited or knocked off alters sensitivity to the drug of interest. This data can be quantitative or qualitative. If quantitative, the data should be z-normalized such that the rows have zero mean and unit variance. This enables uniform interpretation of the values across conditions. For examples, scores below −1 would suggest increased sensitivity with 1 standard deviation below the average observed change. By default, INDIGO uses those genes that have a z score below −2 to define a chemical genetic interaction with a drug.
3. *Orthologous genes for prediction in other organisms*: This data is necessary to enable drug interaction prediction in other bacteria. This consists of a list of genes that are shared between *E. coli* and the bacterium of interest (*see* **Note 3**).

3 Methods

3.1 Formatting Input Data for INDIGO

Example datasets for this tutorial is available at sriramlab.org/software. The following datasets are in the file.

1. ecoli_training.xlsx has the experimental drug interaction data (training set). The rows correspond to individual drug combination and the interaction outcome of that combination.
2. ecoli_phenotype_data_cell.xlsx has the chemogenomic data from Nichols et al. [8]. The rows correspond to individual gene knockout strains, while the columns correspond to different drug perturbations.
3. identifiers_match.xlsx—This file matches the drug identifiers used in the drug interaction data with the chemogenomic data. This file can be empty if the drug identifiers used in both the datasets are the same.

3.2 Training the INDIGO Model

Example:

The function (`indigo_train`) builds an INDIGO model using the drug interaction data and chemogenomics data provided as input (*See* Fig. 4 for a list of inputs and outputs).

```
>> [train_interactions, train_interaction_scores, chemogenomics_labels, indigo_model] = indigo_train( 'ecoli_training.xlsx', 'identifiers_match.xlsx','ecoli_phenotype_data_cell.xlsx');
```

Output:

1. `indigo_model`—trained INDIGO model
2. `train_interactions, train_interaction_scores`—These two variables have the interaction data used as input along with the drug names.

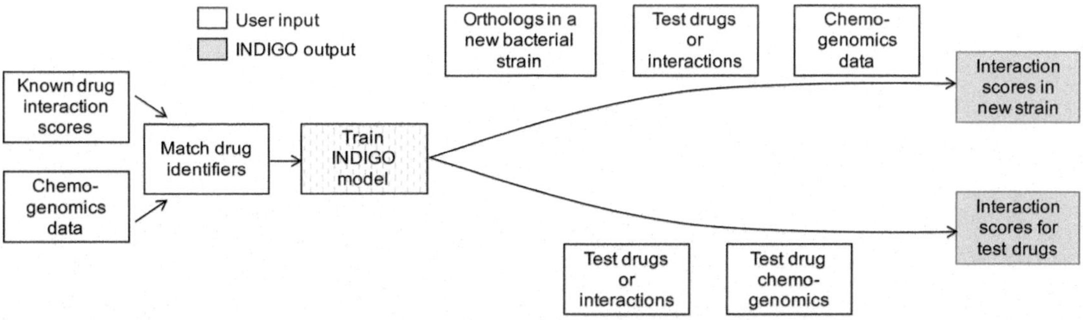

Fig. 4 Schematic of the input and output datasets for running INDIGO. INDIGO takes in drug interaction scores and chemogenomics data to train the random forest algorithm. The trained algorithm can be used to predict new drug interactions using chemogenomics data for the new drugs. The predictions can be made for both the model organism in which the experiments were done and in other strains if orthologous genes in the strain of interest are provided as additional input

3. `chemogenomics_labels` has the gene identifiers in the chemogenomic data.

The drug interactions and chemogenomics data can also be directly inputted from the MATLAB environment as follows.

```
>> [train_interactions, train_interaction_scores, chemogenomics_labels, indigo_model] = indigo_train(([], 'identifiers_match.xlsx', [], 2, phenotype_data, chemogenomics_labels, conditions, interaction_scores, interaction_pairs);
```

Input:

1. `interaction_pairs` is a cell array that contains the list of drug combinations.
2. `interaction_scores`—the corresponding interaction outcomes.

These variables can be imported into MATLAB from the Microsoft Excel file provided using the `xlsread` function. The function `xlsread` parses the first worksheet in the Microsoft Excel spreadsheet workbook named `'ecoli_training.xlsx'` and returns the data in a numeric matrix and a cell array.

```
>> [interaction_scores, interaction_pairs] = xlsread('ecoli_training.xlsx');
```

```
>> [train_interactions, train_interaction_scores, chemogenomics_labels, indigo_model] = indigo_train( [], 'identifiers_match.xlsx', 'ecoli_phenotype_data_cell.xlsx', [], [], [], [], interaction_scores, interaction_pairs);
```

The chemogenomics data can also be directly inputted from the MATLAB workspace (*see* **Note 4**). The function `process_chemgen` converts the chemogenomics data into a readable format for INDIGO.

```
>> [phenotype_data, chemogenomics_labels, conditions] = process_chemgen('ecoli_phenotype_data_cell.xlsx');
```

Output:

1. `chemogenomics_labels` contains the gene identifiers in the chemogenomic data.
2. `phenotype_data` has the corresponding chemogenomic interaction values.
3. `conditions` contains the drug names for each chemogenomic profile.

3.3 Predicting New Drug Interactions

The trained model (`indigo_model`) can be used to predict new drug interactions. The function (`indigo_predict`) takes as input the trained INDIGO model along with the drug of interest (e.g.,

Vancomycin ('VAN')) for which interaction has to be inferred. It also requires the list of drug identifiers for matching the drug interaction data with the chemogenomic data (*see* **Note 5**).

```
>>[test_interactions, testinteractions_scores]= indigo_predict(indigo_model, {'VA-
N'},1,'identifiers_match.xlsx', 'ecoli_phenotype_data_cell.xlsx');
```

One can also simultaneously predict interactions for multiple drugs.

```
>> testdrugs = {'VAN';'SPE'; 'FUS';'RIF'};
>> [test_interactions, testinteractions_scores] = indigo_predict(indigo_model,
testdrugs,1,'identifiers_match.xlsx', 'ecoli_phenotype_data_cell.xlsx');
```

Output:

1. `test_interactions`—This has the list of interactions predicted by INDIGO. The identifiers in the chemogenomic data is used to label the drugs. Interactions are ordered alphabetically.
2. `test_interactions_scores`—This has the interaction scores for the drug interactions predicted by INDIGO. Interactions are ordered alphabetically (same order as `test_interactions`).

INDIGO can also be used to predict interactions of specific combinations instead of all interactions for a drug. The parameter "`input_type`" can be set to 2 (for interaction prediction) or 1 (for a specific drug).

```
>> testdrugcombination = {'VAN','TOB';'NIT','OXA';'CHL','OXA';'AMK','CEF';'FUS','-
TOB'};
>> input_type = 2; % this value alerts INDIGO that the input is a list of interactions
>> [test_interactions, testinteractions_scores] = indigo_predict(indigo_model,
testdrugcombination, input_type, 'identifiers_match.xlsx','ecoli_phenotype_data_-
cell.xlsx');
```

3.4 Comparing Predictions with Experiment

The accuracy of INDIGO predictions can be assessed by comparison with experimentally measured interaction outcomes; in this case, interaction data for 66 novel drug combinations is used for validation (data from [7]). To perform this comparison, first load the experimental drug interaction data for the test set drugs (provided with the code)

```
>> [test_exp_interaction_scores, test_exp_interaction_pairs] = xlsread('ecoli_test.
xlsx');
>> [test_interactions, testinteractions_scores] = indigo_predict(indigo_model,
test_exp_interaction_pairs, input_type, 'identifiers_match.xlsx','ecoli_phenotype_-
data_cell.xlsx');
```

Next, find the correlation with experimental data.

```
>> corr(test_exp_interaction_scores, testinteractions_scores, 'type', 'spearman')
```

This should output ~0.55, the correlation between predicted and experimental interaction outcomes for the 66 drug combinations (*see* **Note 6**).

3.5 Orthology-Based Predictions

As an example of applying INDIGO to predict interaction outcomes in a species without chemogenomics data, the *E. coli* INDIGO model will be used to predict interaction outcomes in *S. aureus*. For predicting interaction outcomes in a different species than the one in which INDIGO was trained on, INDIGO requires the list of orthologous genes shared in the species of interest as an additional input (Fig. 3).

First, train INDIGO on the entire drug interaction data and get the *E. coli* INDIGO model.

Input:

1. `ecoli_fulldata.xlsx`—This Excel file has the entire drug interaction dataset.

```
>> [interaction_list, interaction_scores, chemogenomics_labels, indigo_model, sig-
ma_delta_scores, conditions] = indigo_train('ecoli_fulldata.xlsx', 'identifiers_-
match.xlsx', 'ecoli_phenotype_data_cell.xlsx');
```

Output:

1. `indigo_model` is the INDIGO model that is now trained on the entire dataset (both training and test data).
2. `interaction_scores`—The corresponding interaction scores for the entire dataset is stored in this variable.
3. `conditions`—The list of drugs or conditions for which chemogenomic data is available.
4. `sigma_delta_scores`—This has the sigma and delta scores calculated by INDIGO for each drug interaction pair.

Now load the list of orthologs between *E. coli* and *S. aureus*.

Input:

`ecoli_staph_orthologs.xlsx` is an Excel file that is provided with the code and contains the genes in *E. coli* that are also predicted to be present in *S. aureus*. Note that the gene identifiers used should match the corresponding gene IDs in the *E. coli* chemogenomics data.

```
>> [~, ecoli_staph_orth] = xlsread('ecoli_staph_orthologs.xlsx');
```

Output:

1. `ecoli_staph_orth`—A list of orthologous genes between *E. coli* and *S. aureus.*

Next, use INDIGO to estimate the interactions that would differ between the two species. INDIGO calculates this by deleting the contribution of nonorthologous genes in the random forest model.

Inputs:

1. `chemogenomics_labels`—gene identifiers in the chemogenomic data (output of the `indigo_train` function).
2. `ecoli_staph_orth`—orthologs between *E. coli* and *S. aureus.*
3. `sigma_delta_scores`—sigma and delta scores calculated by INDIGO for each drug interaction pair (output of `indigo_train`).
4. `indigo_model`—INDIGO model trained on the entire dataset (output of `indigo_train`).

```
>> deviations = indigo_orthology(chemogenomics_labels, ecoli_staph_orth, sigma_delta_scores, indigo_model);
```

Output:

1. `deviations`—predicted difference in interaction score between *E. coli* and *S. aureus.* Interactions are in the same order as in the variable `interaction_list`.

3.6 Comparing Orthology-Based Predictions with Experiment

To assess the accuracy of the INDIGO orthology model, the predicted interaction scores will be compared with experimentally measured interaction outcomes for 45 drug combinations in *S. aureus* (data from [7]). To perform this comparison, first load the experimental drug interaction data in *S. aureus* (provided with the code).

Inputs:

1. `'saureus_fulldata.xlsx'`—entire experimental drug interaction dataset for *S.aureus.* This file contains the difference between experimental *E. coli* interaction scores and *S. aureus* interaction scores.

```
>> [saureus_exp_interaction_scores, saureus_exp_interaction_pairs] = xlsread('saureus_fulldata.xlsx');
```

Using this list of interactions that were measured in *S. aureus* as input (stored in `saureus_exp_interaction_pairs`), INDIGO outputs the sigma and delta scores that were calculated from chemogenomics data for these drug combinations.

```
>> [test_interactions, testinteractions_scores, indigo_model, sigma_delta_scores_t-
est] = indigo_predict(indigo_model, saureus_exp_interaction_pairs, 2, 'identifiers_-
match.xlsx', [],[],phenotype_data, chemogenomics_labels, conditions);
```

Next, run INDIGO with the sigma and delta scores as input and predict the degree of interaction outcome change with *S. aureus*.

```
>> deviations = indigo_orthology(chemogenomics_labels, ecoli_staph_orth, sigma_del-
ta_scores_test, indigo_model);
```

Finally, find the correlation with experimental data.

```
>> corr(saureus_exp_interaction_scores, deviations, 'type', 'spearman')
```

This should output ~0.52, the correlation between predicted difference and observed experimental difference in the interaction outcome between the two species (*see* **Note 6**).

4 Notes

1. Ensure that all the provided files (MATLAB code, Microsoft Excel files) are in the same folder. There is no installation step for the random forest toolbox or INDIGO. Download and extract files to the MATLAB folder, add them to the MATLAB path. You are now ready to run INDIGO.
2. Irrespective of the method used for quantifying drug interactions, the same metric should be used consistently across all interaction data.
3. The list of orthologous genes should use the same identifiers used in the chemogenomics data.
4. In case both the Excel file and MATLAB variables are provided as input to INDIGO, the MATLAB data overrides the data input from the Microsoft Excel file.
5. If the names of the drugs in the test drug interaction data are the same as in the chemogenomic data, the contents of the file, `identifiers_match.xlsx` can be left blank.
6. Since the Random Forest algorithm has a stochastic element to it, the correlation with the experimental data will vary to some extent each time the algorithm is run. Hence the correlation value mentioned (0.52 or 0.55) is an average measure and the output might vary by approximately 10% in each run.

Acknowledgment

I thank Chen Li for critical reading of the manuscript.

References

1. Hopkins AL (2008) Network pharmacology: the next paradigm in drug discovery. Nat Chem Biol 4(11):682–690. https://doi.org/10.1038/nchembio.118
2. Mdluli K, Kaneko T, Upton A (2015) The tuberculosis drug discovery and development pipeline and emerging drug targets. Cold Spring Harb Perspect Med 5(6). https://doi.org/10.1101/cshperspect.a021154
3. Ramon-Garcia S, Ng C, Anderson H, Chao JD, Zheng X, Pfeifer T, Av-Gay Y, Roberge M, Thompson CJ (2011) Synergistic drug combinations for tuberculosis therapy identified by a novel high-throughput screen. Antimicrob Agents Chemother 55 (8):3861–3869. https://doi.org/10.1128/AAC.00474-11
4. Baym M, Stone LK, Kishony R (2016) Multidrug evolutionary strategies to reverse antibiotic resistance. Science 351(6268):aad3292. https://doi.org/10.1126/science.aad3292
5. Lobritz MA, Belenky P, Porter CBM, Gutierrez A, Yang JH, Schwarz EG, Dwyer DJ, Khalil AS, Collins JJ (2015) Antibiotic efficacy is linked to bacterial cellular respiration. Proc Natl Acad Sci 112(27):8173–8180. https://doi.org/10.1073/pnas.1509743112
6. Silva A, Lee B-Y, Clemens DL, Kee T, Ding X, Ho C-M, Horwitz MA (2016) Output-driven feedback system control platform optimizes combinatorial therapy of tuberculosis using a macrophage cell culture model. Proc Natl Acad Sci 113(15):E2172–E2179
7. Chandrasekaran S, Cokol-Cakmak M, Sahin N, Yilancioglu K, Kazan H, Collins JJ, Cokol M (2016) Chemogenomics and orthology-based design of antibiotic combination therapies. Mol Syst Biol 12(5):872. https://doi.org/10.15252/msb.20156777
8. Nichols RJ, Sen S, Choo YJ, Beltrao P, Zietek M, Chaba R, Lee S, Kazmierczak KM, Lee KJ, Wong A, Shales M, Lovett S, Winkler ME, Krogan NJ, Typas A, Gross CA (2011) Phenotypic landscape of a bacterial cell. Cell 144(1):143–156. https://doi.org/10.1016/j.cell.2010.11.052
9. Loewe S (1953) The problem of synergism and antagonism of combined drugs. Arzneimittelforschung 3(6):285–290
10. Whiteside MD, Winsor GL, Laird MR, Brinkman FS (2013) OrtholugeDB: a bacterial and archaeal orthology resource for improved comparative genomic analysis. Nucleic Acids Res 41 (Database issue):D366–D376. https://doi.org/10.1093/nar/gks1241

Chapter 14

Computational Analyses Connect Small-Molecule Sensitivity to Cellular Features Using Large Panels of Cancer Cell Lines

Matthew G. Rees, Brinton Seashore-Ludlow, and Paul A. Clemons

Abstract

We recently pioneered several analyses of small-molecule sensitivity data collected from large-scale perturbation of hundreds of cancer cell lines with hundreds of small molecules, with cell viability measured as a readout of compound sensitivity. We performed these studies using cancer cell lines previously annotated with cellular, genomic, and basal gene-expression features. By combining small-molecule sensitivity data with these other datasets, we identified new candidate biomarkers of sensitivity, gained insights into small-molecule mechanisms of action, and proposed candidate hypotheses for cancer dependencies (including candidate combination therapies). Nevertheless, given the size of these datasets, we expect that many connections between cellular features and small-molecule sensitivity remain underexplored. In this chapter, we provide a step-by-step account of foundational data-analysis methods underlying our published studies, including working MATLAB code applied to our own public datasets. These procedures will allow others to repeat analyses of our data with new parameters, in additional contexts, and to adapt our procedures to their own datasets.

Key words Computational biology, Chemical biology, Pharmacogenomics, Biomarkers, Cancer dependencies, Combination therapy, Public datasets, Data sharing, Reproducibility

1 Introduction

In this chapter, we provide a hands-on MATLAB walk-through of foundational data-analysis procedures underlying the results in the Cancer Therapeutics Response Portal (CTRP; https://portals.broadinstitute.org/ctrp/), which provides access to data and visualizations stemming from small-molecule sensitivity profiling of cancer cell lines [1–3]. When combined with large-scale genomic and transcriptomic characterization of cancer cell lines, such as those from the Cancer Cell Line Encyclopedia (CCLE) [4] and the Genomics of Drug Sensitivity in Cancer (GDSC) [5–7], small-molecule sensitivity data have been used to identify new candidate cancer dependencies [1–3, 8–14]. Their availability has also

Slava Ziegler and Herbert Waldmann (eds.), *Systems Chemical Biology: Methods and Protocols*, Methods in Molecular Biology, vol. 1888, https://doi.org/10.1007/978-1-4939-8891-4_14,

stimulated development of foundational analysis methods to find novel cancer dependencies in these or similar data, such as those emerging from genetic perturbation experiments in cancer cell lines [15–18].

Foundational analysis methods for CTRP include creating cell-line sensitivity profiles [1–3] based on area-under-curve (AUC) values computed from concentration-response curves; enrichment analysis for cellular (lineage or mutational) features among cell lines sensitive to small molecules [1, 2]; and cross-correlation analysis of small-molecule sensitivity with basal gene-expression patterns [3]. The data underlying CTRP have also been used in the development of several additional methods that integrate prior knowledge, harness new datasets, or result in more complex models. For example, we developed annotated cluster multidimensional enrichment (ACME) analysis [2], a novel method to combine clustering of small-molecule sensitivity data with prior knowledge (e.g., about compound targets or cellular pathways) to formulate new cancer-dependency hypotheses.

The emergence of these datasets has excited the computational biology community, as there is a need for novel method development to mine these data and uncover novel connections. Other recent methods that include a component of small-molecule sensitivity analysis include DAISY [19], DiSCoVER [20], EDDY [21], OncoGPS [22], and RWEN [23]; extensions to The Cancer Proteome Atlas (TCPA) [24, 25] and GDSC itself [26]; and a number of other related studies [27–32]. This rapid proliferation of tools, methods, and comparative analyses has also sparked an important line of critical reflection about the reproducibility of findings and their concordance across different data sources with different experimental designs [33–36]. These developments evince an appetite in the computational biology community to use small-molecule sensitivity profiling data in diverse and creative ways.

Accordingly, we present this chapter to carefully describe the foundational analyses of CTRP, showcase a MATLAB reference implementation of them, and share tips, caveats, and other considerations that we encountered during their development. We cover three procedures in detail:

- *Procedure 1*: Preparation of concentration-response areas-under-curve (AUCs) as a measure of small-molecule sensitivity [1–3].
- *Procedure 2*: Enrichment analysis of mutation features among cell lines of a particular lineage sensitive to individual compounds [1, 2].
- *Procedure 3*: Correlation analysis of basal gene-expression levels to compound sensitivity across panels of cell lines from a lineage [3].

In each case, the procedure selects at random an appropriate subset of available public data and is followed by a basic visualization, interpretable by users without a strong computational background, that illustrates the essential output, including statistical and quality-control measures we use to prioritize the most important connections. We focus on datasets that best illustrate the core steps in the analysis and that best highlight important tips and caveats. Our goal here is to present step-by-step instruction in the core methods, including working code, advice, and notes on the process, rather than highlight specific findings of potential biological interest. Indeed, our choice of including a random component into each analysis subset selection means not only that specific output figures will differ from run to run, but also that the code in principle could highlight a previously unappreciated biological connection. We strive to present the workflow in a manner that is logically modular, so that studies of reproducibility (by removing the random component), extensions to larger data subsets (at the cost of run-time), and the addition of other perturbations (e.g., new compounds, genetic perturbations) or alternative cellular features (e.g., proteomic, metabolomic, or epigenomic features), are straightforward for a user with baseline familiarity in MATLAB.

2 Materials

Conceptually, the analyses presented in the chapter are general enough to be carried out in any programming language. In our experience, scientific computing software, such as R (R Foundation for Statistical Computing; Vienna, Austria) or MATLAB (MathWorks, Inc.; Natick, MA), provides the appropriate balance of flexibility and the provision of built-in functions for common data manipulations. Therefore, to minimize environment configuration, and focus the discussion on the core principles of the analyses, we have based this narrative on a modular MATLAB reference implementation designed to operate on publicly available data and metadata files.

2.1 Supplementary ZIP File

The code to accompany this chapter is provided as a supplementary ZIP file (*implement.zip*) containing the main script (*master.m*) and three subservient procedural scripts (*code\scr* folder), plus all custom functions necessary to execute the procedures (*code\fun* folder and subfolders). The ZIP file also contains a preconfigured directory structure, with folders to house input data (*data* folder), input metadata (*meta* folder), and output tables (*out* folder). The supplementary ZIP file can be downloaded from GitHub (https://github.com/remontoire-pac/ctrp-reference; *see* **Note 1**). The ZIP file should be unpacked using a standard utility and placed in the local computing environment in a location that can be added to the MATLAB search path (*see* **Note 2**).

2.2 MATLAB Environment

The implementation described in this chapter was initially developed on a Dell OptiPlex 9020 (Intel i7-4790 CPU @ 2×3.60GHz, 32.0 GB RAM) running 64-bit Windows 7 Enterprise, Service Pack 1, and MATLAB 2014b. Development was completed, and main testing performed, on a virtual machine (Intel Xeon CPU E5-2695 v4 @ 2×2.10GHz, 32.0 GB RAM) running Windows Server 2016 Datacenter and MATLAB 2018a. Development was initiated in MATLAB 2014b, and this is the earliest version of MATLAB that can execute the complete set of procedures as written, due to the addition of *fishertest.m* in that release. The procedures require the MATLAB Statistics and Machine Learning Toolbox (called the Statistics Toolbox in MATLAB 2014b), along with the author-provided custom functions. A successful independent set of tests was completed on a Microsoft Surface Pro 4 (Intel i7-6650 CPU @ 2×2.20GHz, 16 GB RAM) running 64-bit Windows 10 Pro and MATLAB 2017b. In all three of these environments, the entire code package runs in 5–10 min and creates three figures, with some time variation per instantiation depending on the random component of data subset selection for analysis (*see* **Note 3**).

2.3 Downloading Public Datasets

All data and metadata required to run the analyses presented in this chapter are available for free public download from the National Cancer Institute (NCI) Office of Cancer Genomics (OCG) and were originally produced in our laboratories as part of work supported by the Cancer Target Discovery and Development (CTD^2) Network of research centers (*see* **Note 4**). To complete the download and deployment of the necessary input files for analysis, the following steps are sufficient.

1. Navigate a web browser to the OCG FTP server link ftp://caftpd.nci.nih.gov/pub/OCG-DCC/CTD2/Broad/.
2. Navigate down to the directory *CTRPv2.0_2015_ctd2_ExpandedDataset* and download the ZIP file *CTRPv2.0_2015_ctd2_ExpandedDataset.zip*.
3. Extract the file *v20.data.per_cpd_pre_qc.txt* and place it in the local *data* folder (created by unpacking the supplementary ZIP file *implement.zip* in the local environment; *see* Subheading 2.1).
4. Extract the files *v20.meta.per_cell_line.txt*, *v20.meta.per_compound.txt*, and *v20.meta.per_experiment.txt*, and place them in the local *meta* folder.
5. Return to the parent FTP directory, navigate down to the directory *CTRPv2.2_2015_pub_CancerDisc_5_1210*, and download the ZIP file *CTRPv2.2_2015_pub_CancerDisc_5_1210.zip*.

6. Extract the files *v22.data.auc_sensitivities.txt*, *v22.anno.ccl_anno_features.txt*, and *v22.anno.ccl_mut_features.txt*, and place them in the local *data* folder.
7. Extract the file *v22.meta.per_compound.txt* and place it in the local *meta* folder.
8. Return to the parent FTP directory, navigate down to the directory *CTRPv2.1_2016_pub_NatChemBiol_12_109*, and download the ZIP file *CTRPv2.1_2016_pub_NatChemBiol_12_109.zip*.
9. Extract the files *v21.data.auc_sensitivities.txt* and *v21.data.gex_avg_log2.txt*, and place them in the local *data* folder.
10. Extract the files *v21.meta.gex_features.txt* and *v21.meta.per_compound.txt*, and place them in the local *meta* folder.

3 Methods

Each of the following procedures features a specific analysis that assumes the user has downloaded and unpacked the author-provided files (*see* Subheading 2.1), configured an appropriate MATLAB environment (*see* Subheading 2.2), and downloaded the relevant public datasets (*see* Subheading 2.3).

3.1 Sensitivity Calculation from Small-Molecule Concentration-Response Data

1. Read input data and experiment metadata as DataFrames to prepare for curve-fitting analysis (*master.m*, lines 21–23; *see* **Note 5**).
2. Walk-through (**step 3**) or run (*skip to* **step 21**) the first subsidiary script (*procedure1CurveFit.m*) to perform curve-fit analysis (*master.m*, line 25; *see* **Note 6**).
3. Index input DataFrame with unique combinations of compound and experiment identifiers (*procedure1CurveFit.m*, lines 1–2; *see* **Note 7**).
4. Reconcile relationships between experiment identifiers and cell lines (*procedure1CurveFit.m*, lines 4–6; *see* **Note 8**).
5. Define a set of curves to be fit (*procedure1CurveFit.m*, lines 8–10; *see* **Note 9**).
6. Create an empty MATLAB structure to hold the curve-fitting results (*procedure1CurveFit.m*, lines 12–13).
7. Define concentration limits for area-under-curve (AUC) numeric integration (*procedure1CurveFit.m*, lines 15–17; *see* **Note 10**).
8. Initialize curve-fit parameters and options for nonlinear fits (*procedure1CurveFit.m*, lines 19–22; *see* **Note 11**).

9. Get data points for an individual curve fit and AUC integration (*procedure1CurveFit.m*, lines 28–34; *see* **Note 12**).
10. Seed curve-fit parameters by guessing $\log_2(EC_{50})$ from the data (*procedure1CurveFit.m*, lines 36–42; *see* **Note 13**).
11. Handle data censoring depending on the value of a precomputed quality-control type (*procedure1CurveFit.m*, lines 44–72; *see* **Note 14**).
12. Decide whether enough data points remain to perform a curve fit (*procedure1CurveFit.m*, lines 75–97; *see* **Note 15**).
13. Provisionally fit a three-parameter sigmoid curve to data points (*procedure1CurveFit.m*, lines 99–103).
14. Conditionally fit a two-parameter sigmoid curve depending on apparent $\log_2(EC_{50})$ of the provisional three-parameter fit (*procedure1CurveFit.m*, lines 105–112; *see* **Note 16**).
15. Append the current curve-fit results to the growing MATLAB structure (*procedure1CurveFit.m*, lines 116–134; *see* **Note 17**).
16. Create output DataFrames for fitted-curve per-point and per-curve data for all fit curves (*procedure1CurveFit.m*, lines 138–141).
17. Iterate over curve-fit results structure to build DataFrames for quality control and output (*procedure1CurveFit.m*, lines 143–218; *see* **Note 18**).
18. Define post curve-fit quality-control filters and apply to DataFrames (*procedure1CurveFit.m*, lines 220–241; *see* **Note 19**).
19. Resolve experiment identifiers to cell-line identifiers (*procedure1-CurveFit.m*, lines 243–245; *see* **Note 20**).
20. Write output files for downstream analysis and interpretation (*procedure1CurveFit.m*, lines 247–249).
21. Read additional metadata containing compound and cell-line information for interpretation and for figure labels (*master.m*, lines 27–29; *see* **Note 21**).
22. Select suitable curves to illustrate concentration-response curve fit and extract data (*master.m*, lines 31–43; *see* **Note 22**).
23. Plot both original data and predicted curves with error bars to produce a variant of Fig. 1 (*master.m*, lines 45–62; *see* **Note 23**).

3.2 Mutation Enrichment Among Small-Molecule Sensitive Cell Lines

1. Read input data and metadata as DataFrames to prepare for enrichment analysis (*master.m*, lines 66–69; *see* **Note 24**).
2. Walk-through (**step 3**) or run (*skip to* **step 16**) the second subsidiary script (*procedure2Enrichment.m*) to perform enrichment analysis (*master.m*, line 71; *see* **Note 25**).

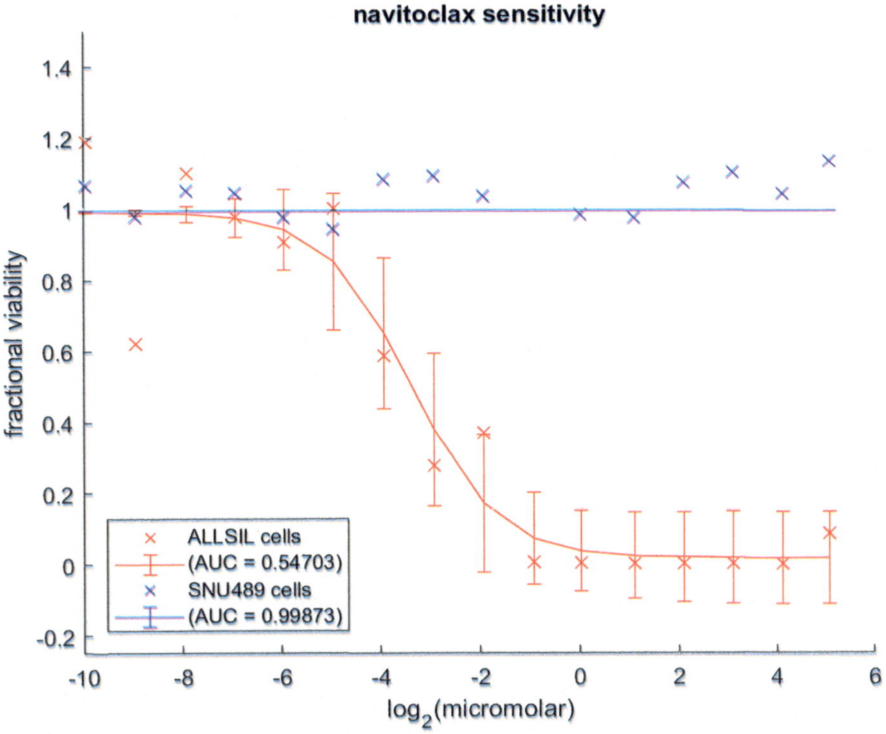

Fig. 1 Representative curve-fit visualization showing the differential sensitivity of two cell lines. Two plots are presented for each of two cell lines, one more sensitive (red) and one less sensitive (blue), to navitoclax, a compound annotated in the Cancer Therapeutics Response Portal (CTRP; http://portals.broadinstitute.org/ctrp) as an inhibitor of BCL2, BCL-xL, and BCL-W. Unconnected crosses represent the original data and are labeled in the MATLAB figure legend with the cell-line name. Line plots with error bars represent the corresponding fit curves and are labeled in the MATLAB figure legend with the computed area-under-curve (AUC)

3. Index input DataFrame with unique combinations of compound and experiment identifiers (*procedure2Enrichment.m*, lines 1–2; *see* **Note 26**).
4. Verify that indices for rows and columns match compounds and cell lines, respectively (*procedure2Enrichment.m*, lines 4–6).
5. Create a matrix of sensitivity values for compounds by cell lines (*procedure2Enrichment.m*, lines 8–12; *see* **Note 27**).
6. Create an indicator (i.e., binary) matrix of cell lineage and histology features by cell lines (*procedure2Enrichment.m*, lines 15–18; *see* **Note 28**).
7. Define a subset of cell lines to be tested and restrict compound data to this subset (*procedure2Enrichment.m*, lines 20–25; *see* **Note 29**).
8. Create an indicator (i.e., binary) matrix of cellular mutation features by cell lines and restrict this matrix to the same cell-line subset (*procedure2Enrichment.m*, lines 27–34; *see* **Note 30**).

9. Restrict analysis to mutation features with an appropriate number of examples (*procedure2Enrichment.m*, lines 36–39; *see* **Note 31**).
10. Define a set of compounds to be tested (*procedure2Enrichment.m*, lines 41–43; *see* **Note 32**).
11. Create an empty DataFrame to hold enrichment results (*procedure2Enrichment.m*, lines 45–48; *see* **Note 33**).
12. For a given compound to be tested, verify that enough cell lines were examined, and compute raw enrichment output using Fisher's exact tests (*procedure2Enrichment.m*, lines 52–58; *see* **Note 34**).
13. For the current compound, record cell lineage, compound, and mutation feature labels, and append labeled results to the growing result DataFrame (*procedure2Enrichment.m*, lines 60–66; *see* **Note 35**).
14. When all compounds are tested (by iterating over **steps 12–13**), define statistical and quality-control filters for all enrichment results (*procedure2Enrichment.m*, lines 70–77; *see* **Note 36**).
15. Apply filters to raw enrichment results and write output files for downstream analysis and interpretation (*procedure2Enrichment.m*, lines 79–83; *see* **Note 37**).
16. Read additional metadata containing compound information for interpretation and for figure labels (*master.m*, lines 73–74; *see* **Note 38**).
17. Select suitable results to illustrate enrichment analysis and extract data (*master.m*, lines 76–90; *see* **Note 39**).
18. Perform one-sided T-test as an additional statistical annotation for boxplots (*master.m*, lines 92–93).
19. Create labels for visualizations using appropriate metadata (*master.m*, lines 95–98; *see* **Note 40**).
20. Plot enrichment results as both heatmap and boxplot representations to produce a variant of Fig. 2 (*master.m*, lines 100–122; *see* **Note 41**).

3.3 Correlation of Small-Molecule Sensitivity with Basal Gene-Expression

1. Read input data as DataFrames to prepare for correlation analysis (*master.m*, lines 126–128; *see* **Note 42**).
2. Walk-through (**step 3**) or run (*skip to* **step 17**) the third subsidiary script (*procedure3Correlation.m*) to perform correlation analysis (*master.m*, line 130; *see* **Note 43**).
3. Index input DataFrame with unique combinations of compound and cell-line identifiers (*procedure3Correlation.m*, lines 1–2; *see* **Note 44**).

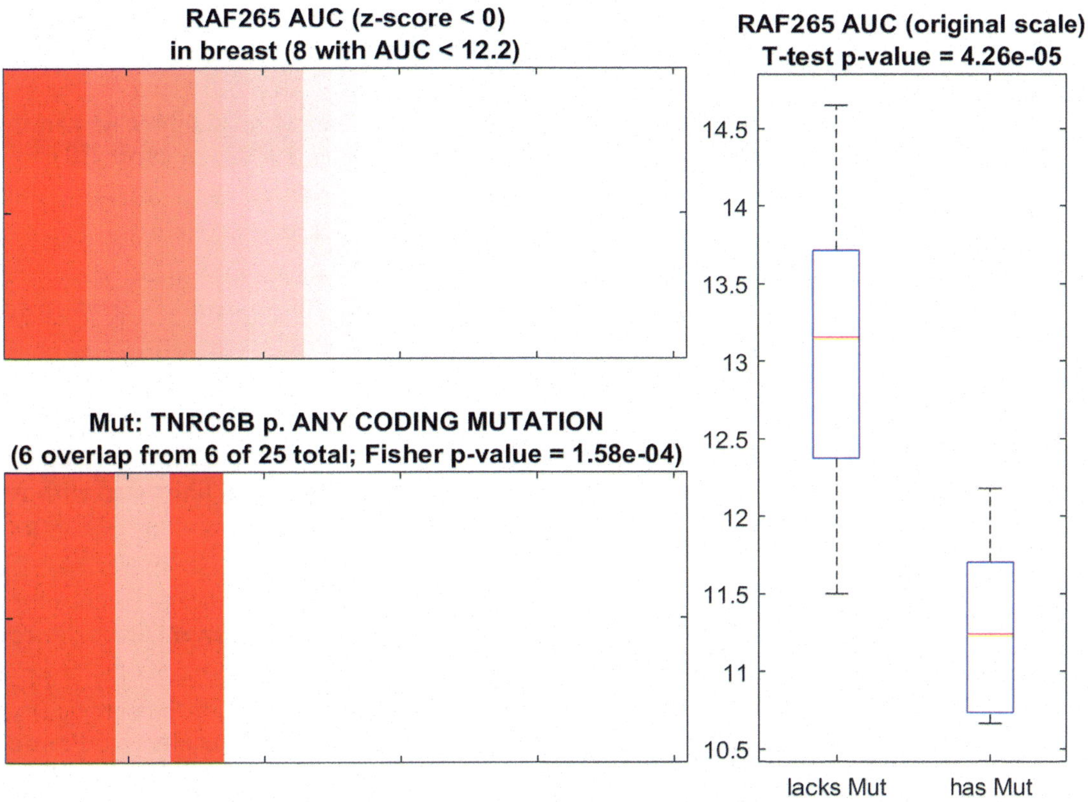

Fig. 2 Representative visualization of enrichment analysis for a single compound tested in multiple cell lines of the same type. In this case, 25 breast-derived cancer lines were tested with RAF265 (annotated in CTRP as an inhibitor of VEGFR2 and BRAF), and then sorted by area-under concentration-response curve (AUC) in the top left panel (increasing red color represents lower AUCs below the mean and therefore more sensitivity). Enrichment analysis resulted in an optimal cutoff of AUC < 12.2 which corresponds to eight total cell lines in the bottom left panel, of which six carry a coding mutation in *TNRC6B* (red = has mutation; pink = lacks mutation). These were the only six *TNRC6B* mutants in this subset of 25 breast cancer-derived cell lines. The right panel depicts an alternative representation (box-whisker plot) and statistical analysis (t-test) of the same information, showing the relative distribution of AUC values for cell lines with or without coding mutations in *TNRC6B*

4. Create a matrix of sensitivity values for compounds by cell lines (*procedure3Correlation.m*, lines 4–8; *see* **Note 45**).
5. Index gene-expression DataFrame with unique combinations of gene and cell-line identifiers (*procedure3Correlation.m*, lines 10–11; *see* **Note 46**).
6. Create a matrix of gene-expression features by cell lines (*procedure3Correlation.m*, lines 13–17; *see* **Note 47**).
7. Define a subset of cell lines to be tested (*procedure3Correlation.m*, lines 19–22; *see* **Note 48**).

8. Restrict cell lines considered to those with both sensitivity and expression data (*procedure3Correlation.m*, lines 24–27; *see* **Note 49**).
9. Restrict genes considered to those with adequate dynamic range (*procedure3Correlation.m*, lines 29–33; *see* **Note 50**).
10. Restrict compounds considered to those with differentially sensitive cell lines (*procedure3Correlation.m*, lines 35–37; *see* **Note 51**).
11. Define a set of compounds to be tested (*procedure3Correlation.m*, lines 39–40; *see* **Note 52**).
12. Compute raw correlation output using normalized Pearson correlation coefficients (*procedure3Correlation.m*, lines 42–43; *see* **Note 53**).
13. Compute *p*-values and index to statistically filter the output correlations (*procedure3Correlation.m*, lines 45–48; *see* **Note 54**).
14. Create a results DataFrame and append cell lineage labels, plus compound and gene identifiers (*procedure3Correlation.m*, lines 50–57; *see* **Note 55**).
15. Append p-values, correlation z-scores, correlation coefficients, and the numbers of participating cell lines to results DataFrame (*procedure3Correlation.m*, lines 59–63; *see* **Note 56**).
16. Define and apply correlation quality-control filters, then write output DataFrame (*procedure3Correlation.m*, lines 65–86; *see* **Note 57**).
17. Read additional metadata containing gene information for interpretation and for figure labels (*master.m*, lines 132–133).
18. Select suitable correlation results to illustrate correlation analysis and extract data (*master.m*, lines 135–142; *see* **Note 58**).
19. Create labels for visualizations using appropriate metadata (*master.m*, lines 144–147; *see* **Note 59**).
20. Plot correlation results as a scatterplot to produce a variant of Fig. 3 (*master.m*, lines 149–161; *see* **Note 60**).

4 Notes

1. We anticipate that the GitHub repository may grow over time, possibly including code updates, additional procedures, and other information. However, we will keep the original version corresponding exactly to this chapter available indefinitely.
2. To keep the distribution file size small, we include the complete directory structure, but do not redistribute the source data and metadata. Rather, we include instructions for downloading the

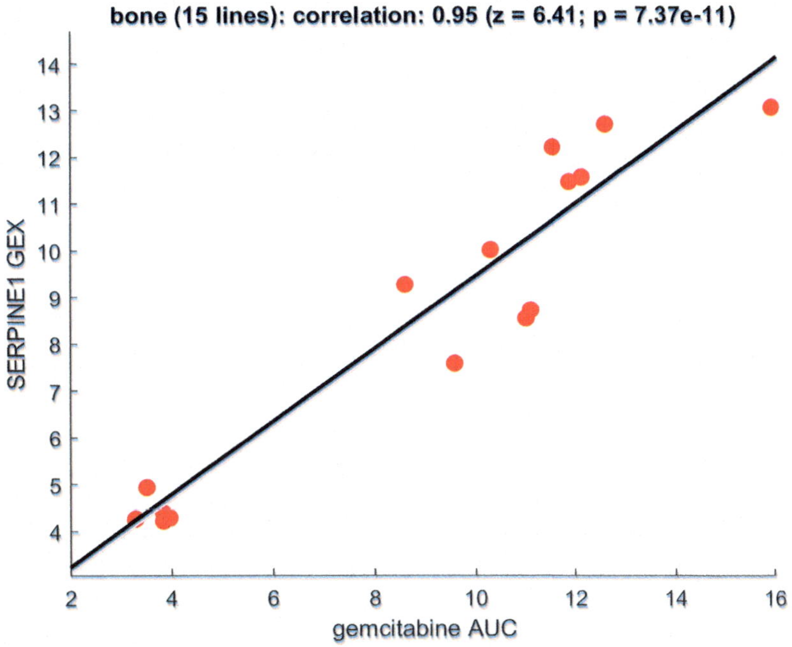

Fig. 3 Representative visualization of correlation analysis for a single compound tested in multiple cell lines of the same type. In this case, 15 bone-derived cancer lines were tested with gemcitabine (annotated in CTRP as an inhibitor of CMPK1, RRM1, TYMS). Sensitivity to gemcitabine (low AUC) is correlated with low expression of *SERPINE1* in these cell lines, and each of the AUC and gene-expression distributions exhibit good dynamic ranges as described in the text (*see also* **Note 57**)

data and metadata from the National Cancer Institute (*see also* Subheading 2.3).

3. The demonstration code uses judiciously sized subsets of compounds for each procedure to keep total run-time down while still providing a complete analysis. Our recent production dataset [2, 3] includes 481 small molecules, and we imagine the provided code could be easily modified to perform a global analysis of the complete dataset.
4. The National Cancer Institute (NCI) has supported multiple Cancer Target Discovery and Development (CTD^2) Centers nationwide in the USA through several rounds of funding with an evolving mission directed at improving cancer patient outcomes with basic research activities. The Cancer Therapeutics Response Portal (CTRP) is one flagship project resulting from the Broad Institute's Chemical Biology and Therapeutics Science program participating in the NCI-funded CTD^2 effort.
5. Throughout these procedures, we use a special type of MATLAB structure called a DataFrame, which was developed by Hyman Carrel in one of our laboratories (PAC) over a

decade ago, inspired by data frames in R. DataFrames are MATLAB structures with one or more fields, constrained to each contain column vectors of equal length, but which collectively may mix numeric and text data types. In more modern releases of MATLAB, the utility of DataFrames has been essentially supplanted by the new MATLAB *table* variable type.

6. On first use, we recommend simply running the subsidiary script from within *master.m*, and skipping to **step 21**. Doing so will ensure the user can get to Fig. 1 more quickly and validate that *procedure1CurveFit.m* runs to completion in their environment. Detailed exploration of the inner workings of *procedure1CurveFit.m* (**steps 3–20**) can be saved for later exploration.
7. In several steps, we make use of a special indexing function for DataFrames (*DFindex.m*) that allows for rapid conversion of tabular data to matrices and without requiring complete data or that tabular data be presorted. We recommend studying the documentation within *DFindex.m* (and other DataFrame functions) to learn how it operates in detail.
8. During our cell-sensitivity profiling studies, we envisioned profiling data acquisition as a matrix of tests representing compounds by cell lines. As described in the relevant publications [2, 3], however, the reality was less tidy. Checking the identity of cell lines by single-nucleotide-polymorphism (SNP) fingerprinting [16] revealed that sample-handling issues had resulted in a small fraction of intended cell lines being omitted, while others were inadvertently tested twice (or three times in one case). An important consequence of these practical considerations for data analysis is that the relationship of an experiment (a specific cell-line sample exposed to a compound collection) to a cell line (an abstract entity annotated with prior information about lineage, mutation, or basal gene expression) is not one-to-one. The public metadata reflects these details, and the exhibition code accounts for them.
9. In the demonstration, we select a single compound at random and fit all curves available for that compound. The code could easily be modified to select a specific compound of the user's choice, and we recommend such a modification as a first step in customizing the analysis. More aggressive modifications might include studying multiple compounds, a single cell line across all compounds, or all possible curves in the dataset. However, such modifications will also require modification of the visualization code for Fig. 1, since the current visualization code for curve-fitting expects a single compound.
10. In our earlier studies [1–3], we set limits of integration for area-under-curve (AUC) that were based on the

concentrations tested for each compound individually, making the comparison of AUC values across compounds potentially problematic. In this chapter, consistent with our current best practices, we define a single set of integration limits across all compounds in the dataset, normalized from 0 (complete killing) to 1 (equivalent to untreated controls).

11. We seed three of the four possible sigmoid curve parameters here; the concentration parameter, $\log_2(EC_{50})$, is seeded later, using the response data to improve the initial guess.
12. Within the *for* loop, data for each curve are collected and processed in a set of temporary variables that are reset with each loop iteration. Data kept for output are stored in a growing MATLAB structure before the loop ends.
13. We seed the concentration parameter, $\log_2(EC_{50})$, at this stage, using the response data to improve the initial guess by choosing either the lowest concentration at which 50% cell killing is achieved, or the median percent killing if 50% cell killing is not achieved at any concentration.
14. In practice, we observed a number of different issues with data quality in our experiments, and therefore defined in data preprocessing [2] some quality-control measures ("QC-types"). Most curves either stayed flat (no compound effect) or relatively smoothly descended from no cell killing at low concentration to maximal cell killing at high concentration (QC-type 0). We observed cases where the top one (QC-type 1) or top two (QC-type 2) concentrations returned to the "no effect" baseline after observing concentration-dependent cell killing at lower concentrations. These aberrant data points are likely due to compound precipitation in the assay plate and were therefore omitted. We also observed cases where fluctuations in the data were more complex, presumably due to liquid-handling and other plate-reader artifacts (QC-type 3). In these cases, we used standard methods to censor individual data points (see the author-included function *cooksdist.m* for details).
15. In practice, we only fit curves with at least five data points passing prefit quality control. We regard this as a permissive choice, particularly for 16-point concentration-response experiments.
16. While curve-fits whose right asymptote is between 1 (no killing) and 0 (complete killing) are relatively common and may reflect a biological distinction between cytostatic effects of a compound and true cell killing, we noted a curve-fit failure mode where the predicted lower asymptote was strongly negative, which is not meaningful. This situation occurs when the predicted $\log_2(EC_{50})$ is higher than the

highest concentration tested. In such cases, we refit the curve with the lower asymptote constrained to zero.

17. To accommodate the fact that we eventually want two outputs, one with per-curve information and one with per-point information, we use an intermediate MATLAB structure to accumulate curve-fitting results and prepare the two desired outputs in a separate step.
18. We create two output DataFrames simultaneously. The first is for per-curve data, which is created directly by looping over the intermediate MATLAB structure, and accounts both for missing curves and whether a three-parameter or two-parameter curve is reported. The second is for per-point data and is created by appending to a growing DataFrame with each turn of the loop since the number of points to be included at each turn is not known in advance.
19. During our studies [2, 3], we scanned thousands of concentration-response curves and identified multiple modes of failure. While these problematic curves represented a relatively small fraction, they fell into categories that we were able to trap computationally and exclude. Both the in-code documentation and our prior reports [2] detail the specific failure modes. These steps also illustrate the use of *DFkeeprow.m*, which applies the typical MATLAB logical or linear indexing to DataFrames.
20. To allow connection to cell-line metadata (e.g., the cell-line name), we reconcile the experiment number with the cell-line identifier at this stage (*see also* **Note 8**).
21. Each of our prior studies [1–3] uses a different subset of cell lines from the Cancer Cell Line Encyclopedia [4], and they also consider overlapping but not identical sets of compounds. To aid in reconciliation between datasets, we use global identifiers *master_cpd_id* (for compounds) and *master_ccl_id* (for cell lines) that have a shared meaning across all CTRP datasets.
22. To illustrate differential sensitivity, we choose a cell line among the top 5% of responders (sensitive) and a cell line near the median responder. These choices could easily be modified to display, for example, the most and least responsive cell lines.
23. The figure code is deliberately included in the calling script *master.m* for transparency and to allow facile modification by users without disrupting the scripts that do the calculations. Users are encouraged to further customize the appearance of figures according to their preferences.
24. In this procedure, two types of categorical variable are introduced, one describing the provenance of cancer cell lines as context to understand their sensitivity (primary site or lineage,

histology terms, and other demographic information), and the other describing their mutational status.

25. On first use, we recommend simply running the subsidiary script from within *master.m*, and skipping to **step 16**. Doing so will ensure the user can get to Fig. 2 more quickly and validate that *procedure2Enrichment.m* runs to completion in their environment. Detailed exploration of the inner workings of *procedure2Enrichment.m* (**steps 3–15**) can be saved for later exploration.
26. The original study [2] uses consecutive internal indices for compounds and cell lines in addition to the master identifiers (*see also* **Note 21**).
27. We anticipate missing data in the matrix of sensitivity scores by first seeding an appropriately sized matrix with *NaN* (not a number) values, then filling in known values in a *for* loop over the indexed DataFrame.
28. In the present procedure, we use lineage and histology information about cell lines as a context feature to pick a subset of cell lines to study. We note that one could as easily check for enrichment of a single cell lineage versus all other lineages by treating lineage as a feature analogous to the way mutations are treated in the reference code.
29. We select a lineage, histology, or demographic term with at least 16 representative cell lines, but fewer than 25% of all cell lines, for illustration purposes. The user can modify these choices to expand the set of terms available or modify the code to specify a lineage of interest.
30. In the present procedure, we use mutation feature information as the primary type of feature for enrichment analysis, but we note that one could as easily use mutations for context (as we do here with lineage and histology information) to derive new and potentially interesting groups of cell lines, for example, for enrichment or correlation analyses.
31. We choose among mutation features with at least three representative cell lines, but fewer than 50% of all cell lines, for illustration purposes. The user can modify these choices to expand the set of features available or modify the code to specify a mutation of interest.
32. We choose 12 random compounds, strictly to keep demonstration run-times low. Increasing the number of compounds, including choosing specific subsets of a user's interest, is an obvious and recommended starting point for user customization.
33. The DataFrame to hold enrichment results will be grown by appending new rows because the number of rows to be

appended will not, in general, be known in advance. Therefore, each field in the DataFrame is defined in advance and populated with an empty array.

34. The primary enrichment analysis at this stage is to perform many Fisher's exact tests both to detect the optimal AUC cutoff for each compound and to iterate over candidate mutation features. We implement this feature using the custom function *sensenfex.m*. Importantly, this function avoids redundant calculations by first building an array of unique 2×2 contingency tables and tracking their relationship to metadata indices. We recommend studying the documentation within *sensenfex.m* to learn how it operates in detail.
35. The first output variable from *sensenfex.m* allows direct appending of mutational feature labels to the growing DataFrame along with the compound and lineage under consideration.
36. Experience has shown that statistical significance of an enrichment result is necessary but not sufficient to warrant continued biological interest. Accordingly, we filter on several other parameters, such as the minimum AUC (the compound must reliably kill at least one cell line), the enrichment confidence (fraction of mutant cell lines killed by the compound), the enrichment purity (fraction of sensitive cell lines harboring the mutation), and the enrichment overlap (at least two mutant cell lines must be sensitive). To ensure that the code produces at least one output for visualization, the enrichment with the best p-value is retained, even if it fails all the other criteria.
37. This step illustrates a relatively simple use of *DFkeeprow.m* to apply an accumulated set of filters (*see also* **Note 19**). To see output corresponding to those results passing each filter, a user could call *DFkeeprow.m* using each of the separate components (*procedure2Enrichment.m*, lines 71–76) in turn.
38. To use the consecutive compound index applied to the data from the original study [2], this procedure leverages the metadata file specific to that study (*see also* **Note 26**).
39. With the extensive prefiltering of enrichment results based on confidence, purity, and overlap, the selection of data for visualization simply takes the most statistically significant result remaining after applying the filters. However, multiple parameters besides the p-value are retrieved from the results record for use in the visualization.
40. We note that human-readable context names and cell-line features are procured for use in the visualization directly from the result table, while compound names are procured from the master metadata. In general, we prefer a discipline where each human-readable string is stored exactly once, and database-like identifiers are used to represent data as far into a procedure as

possible (e.g., until needed for visualization). We deliberately employed a mixed strategy here for illustration purposes.

41. The figure code is deliberately included in the calling script *master.m* (*see also* **Note 23**).

42. In this procedure, new numeric data are imported for AUCs, as well as for basal gene-expression values corresponding to our prior study of their cross-correlations [3]. However, we reuse the compound metadata from *Procedure 1* and the cellular provenance information (lineage, histology, demographic) data from *Procedure 2*. Thus, if *Procedure 3* is run in isolation, users should still load all data and metadata files specified in *master.m*.

43. On first use, we recommend simply running the subsidiary script from within *master.m*, and skipping to **step 17**. Doing so will ensure the user can get to Fig. 3 more quickly and validate that *procedure3Correlation.m* runs to completion in their environment. Detailed exploration of the inner workings of *procedure3Correlation.m* (**steps 3–16**) can be saved for later exploration.

44. Unlike *Procedure 2*, here we use the master identifiers for compounds and cell lines (*see also* **Notes 21** and **26**).

45. Again, we seed an appropriately sized matrix with *NaN* (not a number) values to anticipate missing data (*see also* **Note 27**).

46. Indexing a DataFrame of gene-expression scores by cell lines works just as it does for AUCs by cell lines, but instead using a unique numeric identifier for gene names (*see also* **Note 7**).

47. As complete coverage of all cell lines with gene-expression data is not guaranteed, we start by seeding an appropriately sized matrix with *NaN* (not a number) values (*see also* **Notes 27** and **45**).

48. We select a lineage (primary site) with at least 16 representative cell lines for illustration purposes. The user can modify this choice to expand the set of terms available or modify the code to specify a lineage of interest.

49. Though our correlation procedure can handle missing values, we save some computation time by eliminating in advance those cell lines that have either no expression data or no sensitivity data. This step has the added benefit, as implemented, of aligning our sensitivity and gene-expression matrices so their columns correspond to the same cell-line identities as each other, in the same order.

50. An important idea in correlation analysis is that a gene whose expression correlates with small-molecule sensitivity has sufficient dynamic range to qualify as a potentially useful biomarker. Strong correlations with low effect sizes are less interesting. In

practice, we save computation time by ruling out genes with low dynamic ranges in advance of computing correlations, but after the set of cell lines under consideration is known.

51. An important idea in correlation analysis is that compounds under consideration evince differential sensitivity across a set of cell lines, related to the idea of the "therapeutic window" between efficacy and toxicity. In practice, we save computation time by ruling out compounds with low dynamic ranges (or with little killing at all) in advance of computing correlations, but after the set of cell lines under consideration is known.

52. We choose 12 random compounds, strictly to keep demonstration run-times low (*see also* **Note 32**).

53. The primary correlation analysis at this stage is to perform many pairwise correlations between compound sensitivities and gene-expression levels. While we do take advantage of MATLAB's powerful built-in pairwise similarity infrastructure, we note that accounting for missing values requires that we normalize correlation coefficients using Fisher's *z*-transformation [37] to account for different numbers of cell lines participating in different comparisons. We implement these steps using the custom function *nanpw2fishz.m* and custom distance measure *nanpwcor.m*. We recommend studying the documentation within these two functions to learn how they operate in detail.

54. Since the output of *nanpw2fishz.m* is still a (potentially large) matrix of sensitivity-expression cross-correlations, we perform initial basic statistical filtering here, separately from quality-control filtering conducted downstream (in contrast to *Procedure 2*, where we performed both together).

55. We deliberately refrain from resolving compound and gene identifiers to human-readable names at this stage (*see also* **Note 40**).

56. For use in later visualizations, we record several (nonindependent) expressions of the correlation, including the number of cell lines involved, the raw correlation coefficient, the correlation *z*-score from Fisher's *z*-transformation [37], and a p-value derived from the Fisher's *z*-transformation.

57. Experience has shown that statistical significance of a correlation result is necessary but not sufficient to warrant continued biological interest. Accordingly, we apply several additional filters on results to be output, including a minimum of eight involved cell lines in the reference code (a relatively arbitrary value that is easy to modify). Most importantly, we have previously noticed many cases where a single cell line is responsible for the dynamic range of either sensitivity or

gene-expression levels, and we are wary of investing much energy on such results even if their nominal p-values appear satisfactory. For nominally significant results, therefore, we censor the most extreme-valued cell line at each end of both the sensitivity and gene-expression distributions, then recheck whether the dynamic range of each vector satisfies our original criteria (*see also* **Notes 50** and **51**). To ensure that the code produces at least one output for visualization, the correlation with the best p-value is retained, even if it fails these additional criteria.

58. With the extensive prefiltering of correlation results based on dynamic-range considerations, the selection of data for visualization simply takes the largest raw (absolute) correlation coefficient result remaining after applying the filters (*see also* **Note 39**).
59. We note that human-readable context names are procured for use in the visualization directly from the result table in this case, while compound and gene names are procured from the master metadata (*see also* **Notes 40** and **55**).
60. The figure code is deliberately included in the calling script *master.m* (*see also* **Note 23**).

Acknowledgments

Development of the code presented in the chapter was supported by the National Cancer Institute (NCI) through the Cancer Target Discovery and Development (CTD2) Network (grant numbers U01CA176152 and U01CA217848). The authors are grateful to Shubhroz Gill, Brittany Petros, and Bridget Wagner for helpful discussions on the manuscript.

References

1. Basu A, Bodycombe NE, Cheah JH, Price EV, Liu K, Schaefer GI, Ebright RY, Stewart ML, Ito D, Wang S, Bracha AL, Liefeld T, Wawer M, Gilbert JC, Wilson AJ, Stransky N, Kryukov GV, Dancik V, Barretina J, Garraway LA, Hon CS, Munoz B, Bittker JA, Stockwell BR, Khabele D, Stern AM, Clemons PA, Shamji AF, Schreiber SL (2013) An interactive resource to identify cancer genetic and lineage dependencies targeted by small molecules. Cell 154(5):1151–1161. https://doi.org/10.1016/j.cell.2013.08.003
2. Seashore-Ludlow B, Rees MG, Cheah JH, Cokol M, Price EV, Coletti ME, Jones V, Bodycombe NE, Soule CK, Gould J, Alexander B, Li A, Montgomery P, Wawer MJ, Kuru N, Kotz JD, Hon CS, Munoz B, Liefeld T, Dancik V, Bittker JA, Palmer M, Bradner JE, Shamji AF, Clemons PA, Schreiber SL (2015) Harnessing connectivity in a large-scale small-molecule sensitivity dataset. Cancer Discov 5(11):1210–1223. https://doi.org/10.1158/2159-8290.CD-15-0235
3. Rees MG, Seashore-Ludlow B, Cheah JH, Adams DJ, Price EV, Gill S, Javaid S, Coletti ME, Jones VL, Bodycombe NE, Soule CK, Alexander B, Li A, Montgomery P, Kotz JD, Hon CS, Munoz B, Liefeld T, Dancik V, Haber DA, Clish CB, Bittker JA, Palmer M, Wagner BK, Clemons PA, Shamji AF, Schreiber SL (2016) Correlating chemical sensitivity and basal gene expression reveals mechanism of action. Nat Chem Biol 12(2):109–116. https://doi.org/10.1038/nchembio.1986

4. Barretina J, Caponigro G, Stransky N, Venkatesan K, Margolin AA, Kim S, Wilson CJ, Lehar J, Kryukov GV, Sonkin D, Reddy A, Liu M, Murray L, Berger MF, Monahan JE, Morais P, Meltzer J, Korejwa A, Jane-Valbuena J, Mapa FA, Thibault J, Bric-Furlong E, Raman P, Shipway A, Engels IH, Cheng J, Yu GK, Yu J, Aspesi P Jr, de Silva M, Jagtap K, Jones MD, Wang L, Hatton C, Palescandolo E, Gupta S, Mahan S, Sougnez C, Onofrio RC, Liefeld T, MacConaill L, Winckler W, Reich M, Li N, Mesirov JP, Gabriel SB, Getz G, Ardlie K, Chan V, Myer VE, Weber BL, Porter J, Warmuth M, Finan P, Harris JL, Meyerson M, Golub TR, Morrissey MP, Sellers WR, Schlegel R, Garraway LA (2012) The cancer cell line encyclopedia enables predictive modelling of anticancer drug sensitivity. Nature 483(7391):603–607. https://doi.org/10.1038/nature11003

5. Garnett MJ, Edelman EJ, Heidorn SJ, Greenman CD, Dastur A, Lau KW, Greninger P, Thompson IR, Luo X, Soares J, Liu Q, Iorio F, Surdez D, Chen L, Milano RJ, Bignell GR, Tam AT, Davies H, Stevenson JA, Barthorpe S, Lutz SR, Kogera F, Lawrence K, McLaren-Douglas A, Mitropoulos X, Mironenko T, Thi H, Richardson L, Zhou W, Jewitt F, Zhang T, O'Brien P, Boisvert JL, Price S, Hur W, Yang W, Deng X, Butler A, Choi HG, Chang JW, Baselga J, Stamenkovic I, Engelman JA, Sharma SV, Delattre O, Saez-Rodriguez J, Gray NS, Settleman J, Futreal PA, Haber DA, Stratton MR, Ramaswamy S, McDermott U, Benes CH (2012) Systematic identification of genomic markers of drug sensitivity in cancer cells. Nature 483 (7391):570–575. https://doi.org/10.1038/nature11005

6. Yang W, Soares J, Greninger P, Edelman EJ, Lightfoot H, Forbes S, Bindal N, Beare D, Smith JA, Thompson IR, Ramaswamy S, Futreal PA, Haber DA, Stratton MR, Benes C, McDermott U, Garnett MJ (2013) Genomics of Drug Sensitivity in Cancer (GDSC): a resource for therapeutic biomarker discovery in cancer cells. Nucleic Acids Res 41 (Database issue):D955–D961. https://doi.org/10.1093/nar/gks1111

7. Iorio F, Knijnenburg TA, Vis DJ, Bignell GR, Menden MP, Schubert M, Aben N, Goncalves E, Barthorpe S, Lightfoot H, Cokelaer T, Greninger P, van Dyk E, Chang H, de Silva H, Heyn H, Deng X, Egan RK, Liu Q, Mironenko T, Mitropoulos X, Richardson L, Wang J, Zhang T, Moran S, Sayols S, Soleimani M, Tamborero D, Lopez-Bigas N, Ross-Macdonald P, Esteller M, Gray NS, Haber DA, Stratton MR, Benes CH, Wessels LFA, Saez-Rodriguez J, McDermott U, Garnett MJ (2016) A landscape of pharmacogenomic interactions in cancer. Cell 166 (3):740–754. https://doi.org/10.1016/j.cell.2016.06.017

8. Adams DJ, Ito D, Rees MG, Seashore-Ludlow-B, Puyang X, Ramos AH, Cheah JH, Clemons PA, Warmuth M, Zhu P, Shamji AF, Schreiber SL (2014) NAMPT is the cellular target of STF-31-like small-molecule probes. ACS Chem Biol 9(10):2247–2254. https://doi.org/10.1021/cb500347p

9. Aldrich LN, Kuo SY, Castoreno AB, Goel G, Kuballa P, Rees MG, Seashore-Ludlow BA, Cheah JH, Latorre IJ, Schreiber SL, Shamji AF, Xavier RJ (2015) Discovery of a small-molecule probe for V-ATPase function. J Am Chem Soc 137(16):5563–5568. https://doi.org/10.1021/jacs.5b02150

10. Stewart ML, Tamayo P, Wilson AJ, Wang S, Chang YM, Kim JW, Khabele D, Shamji AF, Schreiber SL (2015) KRAS genomic status predicts the sensitivity of ovarian cancer cells to decitabine. Cancer Res 75(14):2897–2906. https://doi.org/10.1158/0008-5472.CAN-14-2860

11. de Waal L, Lewis TA, Rees MG, Tsherniak A, Wu X, Choi PS, Gechijian L, Hartigan C, Faloon PW, Hickey MJ, Tolliday N, Carr SA, Clemons PA, Munoz B, Wagner BK, Shamji AF, Koehler AN, Schenone M, Burgin AB, Schreiber SL, Greulich H, Meyerson M (2016) Identification of cancer-cytotoxic modulators of PDE3A by predictive chemogenomics. Nat Chem Biol 12(2):102–108. https://doi.org/10.1038/nchembio.1984

12. Han T, Goralski M, Gaskill N, Capota E, Kim J, Ting TC, Xie Y, Williams NS, Nijhawan D (2017) Anticancer sulfonamides target splicing by inducing RBM39 degradation via recruitment to DCAF15. Science 356(6336): eaal3755. https://doi.org/10.1126/science.aal3755

13. Herold N, Rudd SG, Sanjiv K, Kutzner J, Bladh J, Paulin CBJ, Helleday T, Henter JI, Schaller T (2017) SAMHD1 protects cancer cells from various nucleoside-based antimetabolites. Cell Cycle 16(11):1029–1038. https://doi.org/10.1080/15384101.2017.1314407

14. Viswanathan VS, Ryan MJ, Dhruv HD, Gill S, Eichhoff OM, Seashore-Ludlow B, Kaffenberger SD, Eaton JK, Shimada K, Aguirre AJ, Viswanathan SR, Chattopadhyay S, Tamayo P, Yang WS, Rees MG, Chen S, Boskovic ZV, Javaid S, Huang C, Wu X, Tseng YY, Roider EM, Gao D, Cleary JM, Wolpin BM, Mesirov

JP, Haber DA, Engelman JA, Boehm JS, Kotz JD, Hon CS, Chen Y, Hahn WC, Levesque MP, Doench JG, Berens ME, Shamji AF, Clemons PA, Stockwell BR, Schreiber SL (2017) Dependency of a therapy-resistant state of cancer cells on a lipid peroxidase pathway. Nature 547(7664):453–457. https://doi.org/10.1038/nature23007

15. Cheung HW, Cowley GS, Weir BA, Boehm JS, Rusin S, Scott JA, East A, Ali LD, Lizotte PH, Wong TC, Jiang G, Hsiao J, Mermel CH, Getz G, Barretina J, Gopal S, Tamayo P, Gould J, Tsherniak A, Stransky N, Luo B, Ren Y, Drapkin R, Bhatia SN, Mesirov JP, Garraway LA, Meyerson M, Lander ES, Root DE, Hahn WC (2011) Systematic investigation of genetic vulnerabilities across cancer cell lines reveals lineage-specific dependencies in ovarian cancer. Proc Natl Acad Sci U S A 108 (30):12372–12377. https://doi.org/10.1073/pnas.1109363108

16. Cowley GS, Weir BA, Vazquez F, Tamayo P, Scott JA, Rusin S, East-Seletsky A, Ali LD, Gerath WF, Pantel SE, Lizotte PH, Jiang G, Hsiao J, Tsherniak A, Dwinell E, Aoyama S, Okamoto M, Harrington W, Gelfand E, Green TM, Tomko MJ, Gopal S, Wong TC, Li H, Howell S, Stransky N, Liefeld T, Jang D, Bistline J, Hill Meyers B, Armstrong SA, Anderson KC, Stegmaier K, Reich M, Pellman D, Boehm JS, Mesirov JP, Golub TR, Root DE, Hahn WC (2014) Parallel genome-scale loss of function screens in 216 cancer cell lines for the identification of context-specific genetic dependencies. Sci Data 1:140035. https://doi.org/10.1038/sdata.2014.35

17. Tsherniak A, Vazquez F, Montgomery PG, Weir BA, Kryukov G, Cowley GS, Gill S, Harrington WF, Pantel S, Krill-Burger JM, Meyers RM, Ali L, Goodale A, Lee Y, Jiang G, Hsiao J, Gerath WFJ, Howell S, Merkel E, Ghandi M, Garraway LA, Root DE, Golub TR, Boehm JS, Hahn WC (2017) Defining a cancer dependency map. Cell 170(3):564–576 e516. https://doi.org/10.1016/j.cell.2017.06.010

18. McDonald ER 3rd, de Weck A, Schlabach MR, Billy E, Mavrakis KJ, Hoffman GR, Belur D, Castelletti D, Frias E, Gampa K, Golji J, Kao I, Li L, Megel P, Perkins TA, Ramadan N, Ruddy DA, Silver SJ, Sovath S, Stump M, Weber O, Widmer R, Yu J, Yu K, Yue Y, Abramowski D, Ackley E, Barrett R, Berger J, Bernard JL, Billig R, Brachmann SM, Buxton F, Caothien R, Caushi JX, Chung FS, Cortes-Cros M, de Beaumont RS, Delaunay C, Desplat A, Duong W, Dwoske DA, Eldridge RS, Farsidjani A, Feng F, Feng J, Flemming D, Forrester W, Galli GG, Gao Z, Gauter F, Gibaja V, Haas K, Hattenberger M, Hood T, Hurov KE, Jagani Z, Jenal M, Johnson JA, Jones MD, Kapoor A, Korn J, Liu J, Liu Q, Liu S, Liu Y, Loo AT, Macchi KJ, Martin T, McAllister G, Meyer A, Molle S, Pagliarini RA, Phadke T, Repko B, Schouwey T, Shanahan F, Shen Q, Stamm C, Stephan C, Stucke VM, Tiedt R, Varadarajan M, Venkatesan K, Vitari AC, Wallroth M, Weiler J, Zhang J, Mickanin C, Myer VE, Porter JA, Lai A, Bitter H, Lees E, Keen N, Kauffmann A, Stegmeier F, Hofmann F, Schmelzle T, Sellers WR (2017) Project DRIVE: a compendium of cancer dependencies and synthetic lethal relationships uncovered by large-scale, deep RNAi screening. Cell 170(3):577–592 e510. https://doi.org/10.1016/j.cell.2017.07.005

19. Jerby-Arnon L, Pfetzer N, Waldman YY, McGarry L, James D, Shanks E, Seashore-Ludlow B, Weinstock A, Geiger T, Clemons PA, Gottlieb E, Ruppin E (2014) Predicting cancer-specific vulnerability via data-driven detection of synthetic lethality. Cell 158 (5):1199–1209. https://doi.org/10.1016/j.cell.2014.07.027

20. Hanaford AR, Archer TC, Price A, Kahlert UD, Maciaczyk J, Nikkhah G, Kim JW, Ehrenberger T, Clemons PA, Dancik V, Seashore-Ludlow B, Viswanathan V, Stewart ML, Rees MG, Shamji A, Schreiber S, Fraenkel E, Pomeroy SL, Mesirov JP, Tamayo P, Eberhart CG, Raabe EH (2016) DiSCoVERing innovative therapies for rare tumors: combining genetically accurate disease models with in silico analysis to identify novel therapeutic targets. Clin Cancer Res 22 (15):3903–3914. https://doi.org/10.1158/1078-0432.CCR-15-3011

21. Speyer G, Mahendra D, Tran HJ, Kiefer J, Schreiber SL, Clemons PA, Dhruv H, Berens M, Kim S (2017) Differential pathway dependency discovery associated with drug response across cancer cell lines. Pac Symp Biocomput 22:497–508. https://doi.org/10.1142/9789813207813_0046

22. Kim JW, Abudayyeh OO, Yeerna H, Yeang CH, Stewart M, Jenkins RW, Kitajima S, Konieczkowski DJ, Medetgul-Ernar K, Cavazos T, Mah C, Ting S, Van Allen EM, Cohen O, McDermott J, Damato E, Aguirre AJ, Liang J, Liberzon A, Alexe G, Doench J, Ghandi M, Vazquez F, Weir BA, Tsherniak A, Subramanian A, Meneses-Cime K, Park J, Clemons P, Garraway LA, Thomas D, Boehm JS, Barbie DA, Hahn WC, Mesirov JP, Tamayo P (2017) Decomposing oncogenic transcriptional signatures to generate maps of divergent cellular states. Cell Syst 5(2):105–118 e109. https://doi.org/10.1016/j.cels.2017.08.002

23. Basu A, Mitra R, Liu H, Schreiber SL, Clemons PA (2018) RWEN: response-weighted elastic net for prediction of chemosensitivity of cancer cell lines. Bioinformatics.:bty199-bty199. https://doi.org/10.1093/bioinformatics/bty199
24. Li J, Lu Y, Akbani R, Ju Z, Roebuck PL, Liu W, Yang JY, Broom BM, Verhaak RG, Kane DW, Wakefield C, Weinstein JN, Mills GB, Liang H (2013) TCPA: a resource for cancer functional proteomics data. Nat Methods 10 (11):1046–1047. https://doi.org/10.1038/nmeth.2650
25. Li J, Akbani R, Zhao W, Lu Y, Weinstein JN, Mills GB, Liang H (2017) Explore, visualize, and analyze functional cancer proteomic data using the cancer proteome atlas. Cancer Res 77 (21):e51–e54. https://doi.org/10.1158/0008-5472.CAN-17-0369
26. Cokelaer T, Chen E, Iorio F, Menden MP, Lightfoot H, Saez-Rodriguez J, Garnett MJ (2017) GDSCTools for mining pharmacogenomic interactions in cancer. Bioinformatics 34(7):1226–1228. https://doi.org/10.1093/bioinformatics/btx744
27. Hafner M, Niepel M, Chung M, Sorger PK (2016) Growth rate inhibition metrics correct for confounders in measuring sensitivity to cancer drugs. Nat Methods 13(6):521–527. https://doi.org/10.1038/nmeth.3853
28. Harris LA, Frick PL, Garbett SP, Hardeman KN, Paudel BB, Lopez CF, Quaranta V, Tyson DR (2016) An unbiased metric of antiproliferative drug effect in vitro. Nat Methods 13(6):497–500. https://doi.org/10.1038/nmeth.3852
29. Geeleher P, Cox NJ, Huang RS (2016) Cancer biomarker discovery is improved by accounting for variability in general levels of drug sensitivity in pre-clinical models. Genome Biol 17 (1):190. https://doi.org/10.1186/s13059-016-1050-9
30. Nikolova O, Moser R, Kemp C, Gonen M, Margolin AA (2017) Modeling gene-wise dependencies improves the identification of drug response biomarkers in cancer studies. Bioinformatics 33(9):1362–1369. https://doi.org/10.1093/bioinformatics/btw836
31. Chen B, Ma L, Paik H, Sirota M, Wei W, Chua MS, So S, Butte AJ (2017) Reversal of cancer gene expression correlates with drug efficacy and reveals therapeutic targets. Nat Commun 8:16022. https://doi.org/10.1038/ncomms16022
32. Wang L, Li X, Zhang L, Gao Q (2017) Improved anticancer drug response prediction in cell lines using matrix factorization with similarity regularization. BMC Cancer 17(1):513. https://doi.org/10.1186/s12885-017-3500-5
33. Haibe-Kains B, El-Hachem N, Birkbak NJ, Jin AC, Beck AH, Aerts HJ, Quackenbush J (2013) Inconsistency in large pharmacogenomic studies. Nature 504(7480):389–393. https://doi.org/10.1038/nature12831
34. Cancer Cell Line Encyclopedia C, Genomics of Drug Sensitivity in Cancer C (2015) Pharmacogenomic agreement between two cancer cell line data sets. Nature 528(7580):84–87. https://doi.org/10.1038/nature15736
35. Safikhani Z, El-Hachem N, Quevedo R, Smirnov P, Goldenberg A, Juul Birkbak N, Mason C, Hatzis C, Shi L, Aerts HJ, Quackenbush J, Haibe-Kains B (2016) Assessment of pharmacogenomic agreement. F1000Res 5:825. https://doi.org/10.12688/f1000research.8705.1
36. Safikhani Z, Smirnov P, Freeman M, El-Hachem N, She A, Rene Q, Goldenberg A, Birkbak NJ, Hatzis C, Shi L, Beck AH, Aerts H, Quackenbush J, Haibe-Kains B (2016) Revisiting inconsistency in large pharmacogenomic studies. F1000Res 5:2333. https://doi.org/10.12688/f1000research.9611.3
37. Dancik V, Carrel H, Bodycombe NE, Seiler KP, Fomina-Yadlin D, Kubicek ST, Hartwell K, Shamji AF, Wagner BK, Clemons PA (2014) Connecting small molecules with similar assay performance profiles leads to new biological hypotheses. J Biomol Screen 19(5):771–781. https://doi.org/10.1177/1087057113520226

Chapter 15

Web-Based Tools for Polypharmacology Prediction

Mahendra Awale and Jean-Louis Reymond

Abstract

Drug promiscuity or polypharmacology is the ability of small molecules to interact with multiple protein targets simultaneously. In drug discovery, understanding the polypharmacology of potential drug molecules is crucial to improve their efficacy and safety, and to discover the new therapeutic potentials of existing drugs. Over the past decade, several computational methods have been developed to study the polypharmacology of small molecules, many of which are available as Web services. In this chapter, we review some of these Web tools focusing on ligand based approaches. We highlight in particular our recently developed polypharmacology browser (PPB) and its application for finding the side targets of a new inhibitor of the TRPV6 calcium channel.

Key words Polypharmacology, Target prediction, Drug–target interactions, Similarity searching, Molecular fingerprints

1 Introduction

Most biologically active small molecules exhibit polypharmacology, that is, interact with multiple protein targets simultaneously [1–3]. It has been reported that on average any drug molecule interacts with as many as six different target proteins [4]. These multiple interactions, particularly with unintended side targets, often lead to adverse drug reactions, which is one of the most common causes of failure of potential drug molecules in clinical trials and of postmarket withdrawal of drugs [5, 6]. To date, more than 500 drugs have been reported to be withdrawn from the market due to safety concerns [7]. On the other hand, polypharmacology is not always unwelcome, as it may reveal new opportunities for an existing drug (drug repositioning) or provide a starting point for the development of a new drug, in both cases by addressing side targets known to have therapeutic roles in different disease conditions [8, 9]. Moreover, certain disease conditions such as cancers and neuropsychiatric diseases actually demand drugs with

Slava Ziegler and Herbert Waldmann (eds.), *Systems Chemical Biology: Methods and Protocols*, Methods in Molecular Biology, vol. 1888, https://doi.org/10.1007/978-1-4939-8891-4_15,

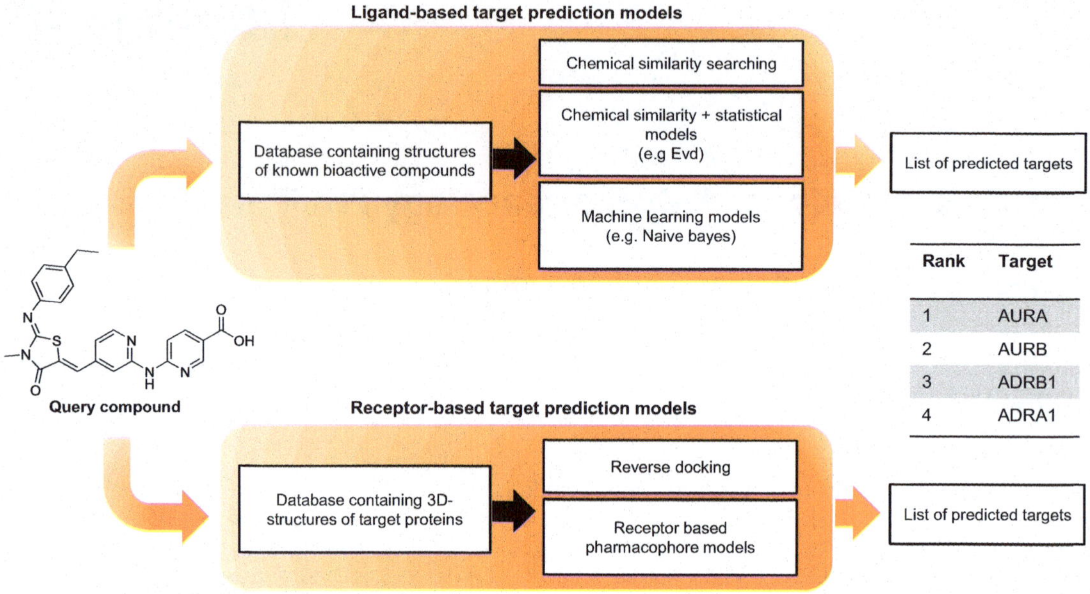

Fig. 1 Overview of target prediction methods

multitarget activities [10, 11]. Therefore tracking the polypharmacology profile of any hit compound or drug candidate is crucial in the course of drug discovery.

In principle, any compound should be tested against all possible target proteins in biological space, which would be daunting and expensive, if possible at all. To circumvent this problem a variety of computational methods have been developed to predict polypharmacology on the basis of known drug–target interactions and thereby focus experimental validation on the most probable targets [3, 12]. These methods can be broadly grouped into "ligand-based" [13–37] and "receptor-based" [38–40] methods (Fig. 1). The ligand-based methods exploit structural information of known bioactive small molecules such as those collected in various public databases like DrugBank [41], BindingDB [42], ChEMBL [43], and some of the commercially available databases like MDDR and WOMBAT [44]. On the other hand, receptor-based methods rely on the three-dimensional structural information of target proteins as provided by the Protein Data Bank (PDB) [45]. Ligand-based methods are particularly attractive due to their speed, efficiency and applicability to a wide range of target proteins, including many targets wherein the receptor based methods are not applicable due to lack of experimental 3D-structures. In light of the growing prevalence of Web applications [46], several of these ligand-based methods are now accessible by Web browsers (Table 1). In this chapter, we review selected Web-based tools that are freely available, including our recently developed

Table 1
Publicly accessible Web-based target prediction tools

Website	Method	Fingerprints	Chemogenomic Database
www.gdb.unibe.ch [13]	Similarity searching and statistical model	APfp, SMIfp, MQN, Xfp, Sfp, ECfp4, and data fusion	ChEMBL 21
http://targetnet.scbdd.com/calcnet/index/ [14]	Naive Bayes models	ECfp2, ECfp4, ECfp6, Estate, FP2, MACCS and Sfp	BindingDB
http://potentia.cbs.dtu.dk/ChemProt [15]	Similarity searching and statistical model	Sfp	ChEMBL 14, BindingDB, DrugBank, PharmGKB, PubChem bioassay, WOMBAT, IUPHAR, CTD, and STITCH
www.dddc.ac.cn/tarpred [16]	Similarity searching	ECfp4	BindingDB
http://modlab-cadd.ethz.ch/software/spider/ [17]	Self-organizing maps, similarity searchingand statistical model	CATS and MOE physiochemical descriptors	COBRA
www.swisstargetprediction.ch [18]	Similarity searching and Logistic regression	Openbabel FP2 and Electroshape descriptors	ChEMBL 16
www.cbligand.org/TargetHunter [19]	Similarity searching	ECfp6, ECfp4, and Openbabel FP2	ChEMBL 11 and PubChem bioassay
http://lilab.ecust.edu.cn/chemmapper/ [20]	Similarity searching and random walk algorithm	Openbabel FP2, MACSS, SHAFT, and USR	ChEMBL 14, BindingDB, DrugBank, KEGG, and PDB
http://mips.helmholtz-muenchen.de/proj/hitpick [21]	Combination of 1NN and Naive Bayes model	FCFP	STITCH
http://prediction.charite.de/index.php [22]	Similarity searching and statistical model	ECfp4	ChEMBL, SuperTarget, and BindingDB
http://sea.bkslab.org [23]	Similarity searching and statistical model	ECfp4	CHEMBL 21
http://www.pharmaexpert.ru/passonline/ [24]	Bayesian modeling	MNA descriptors	ChEMBL 19

Fig. 2 Result windows for Web-based target prediction tools discussed in this chapter. In all the cases the query molecule is the one shown in PPB result window (compound 1 in Fig. 3d). (**a**) PPB, (**b**) TarPred, (**c**) SwissTargetPrediction, (**d**) SPiDER, and (**e**) SEA. In the case of SPiDER, the result window is obtained by opening the text file in a Web browser by clicking on "Download" result option

polypharmacology browser (PPB). The chapter is organized as follows: first we introduce ligand-based methods for target prediction, second we discuss selected ligand-based Web tools individually, in each case highlighting website (Fig. 2), methodology, and whenever available the practical application in drug discovery projects.

2 Ligand-Based Target Prediction

Ligand-based target prediction methods are solely driven by the fact that similar molecules are most likely to exhibit similar bioactivities and interact with similar target proteins [47]. Thus, the targets of any compound can be predicted by relating the chemical structure of a query molecule to known bioactive compounds [48]. There are two primary components to any of these prediction methods, (1) a chemogenomic dataset containing known bioactive compounds annotated with their target proteins and (2) the approach used to relate the query compound to molecules in the chemogenomic database. The above mentioned compound databases annotated with compound-target interactions provide the basis for selecting a chemogenomic dataset, the richest sources being ChEMBL (~11 K targets, ~1.7 M compounds, and ~14.6 M bioactivities) and BindingDB (~7 K target, ~600 K compounds, and ~1.3 M bioactivities). Target prediction methods can be broadly classified into three groups based on approaches used for relating the query to the chemogenomic dataset, namely chemical similarity searching, chemical similarity coupled with statistical models, and machine learning models.

2.1 Chemical Similarity Searching

Chemical similarity searching [49] is relatively simple to implement, fast and performs particularly well for target prediction. Typically, it involves scoring all molecules in the chemogenomic dataset according to their similarity to the query molecule, followed by ranking of the targets using (1) the similarity score to the query molecule for the most similar molecule associated with each target, or (2) the average similarity score to the query molecule calculated for all molecules associated with each target, considering either a predefined number of most similar compounds or all of the known active molecules of a target [30]. The targets appearing at the top of the list are then listed as potential or predicted targets of the query molecule.

The target prediction performances of chemical similarity based methods strongly depend on the selected molecular representation (Table 2). The molecular representation, also called fingerprint or structural encoding, is a multidimensional feature vector describing molecular properties such as composition, connectivity, substructures, pharmacophores, or molecular shape. A molecular fingerprint can be scalar or binary and can be derived from the 2D or 3D structure of the molecule. Once the fingerprint is calculated, the similarity between any two molecules is quantified by applying a distance or similarity metric, most frequently the Tanimoto coefficient (*Tc*) or the city block distance (CBD). Binary substructure fingerprints such as extended connectivity fingerprint (ECfp) [50], Daylight type substructure fingerprints (Sfp) and MACCS keys

Table 2
Molecular fingerprints mentioned in this chapter

Name	Description
APfp [53]	21-D atom-pair fingerprint, perceives molecular shape
Xfp [53]	55-D atom category extended atom-pair fingerprint, perceives pharmacophores
MQN [63]	42-D molecular quantum numbers, scalar fingerprint counting atoms, bonds, polarity and ring features, perceives constitution, topology, and molecular shape
SMIfp [64]	34-D scalar fingerprint counting occurrence of characters in SMILES, perceives rings, aromaticity, and polarity
Sfp	1024-D binary daylight type substructure fingerprint, perceives detailed substructures
ECfp4 [50]	1024-D binary circular extended connectivity fingerprint, perceives detailed substructures and pharmacophores
Ffp1 [13]	Fusion fingerprint, Xfp + SMIfp+Sfp
Ffp2 [13]	Fusion fingerprint, Xfp + MQN + SMIfp
Ffp3 [13]	Fusion fingerprint, Xfp + SMIfp+Sfp + ECfp4
Ffp4 [13]	Fusion fingerprint, Xfp + MQN + SMIfp+Sfp + ECfp4
FCFP [50]	Binary circular extended connectivity fingerprint encoding topological pharmacophore
CATS [17, 52]	210-D atom category extended atom-pair fingerprint, perceives pharmacophores
Openbabel FP2	1024-D binary daylight type substructure fingerprint using path length of 7
MNA [65]	Set of descriptors generated from Multilevel Neighborhoods of Atoms
MACCS [51]	166-D binary fingerprint, describing specific structural patterns (keys)
USR [54]	12-D scalar fingerprint encoding shape using 3D-euclidean distance distributions
ElectroShape [55]	18-D describing shape, electrostatics, and lipophilicity

[51] perform particularly well for target prediction. Pharmacophore fingerprints such as CATS [52] and Xfp [53], fingerprints derived from 3D-structure of molecules such as USR [54] and Electroshape [55], and 3D-overlay based similarities such as ROCS [56, 57] are particularly useful to detect scaffold hopping similarities between compounds, which may reveal novel drug–target interactions. Moreover, data fusion approaches [58], which combine similarities from different fingerprints (e.g., combination of 2D and 3D fingerprints), generally improve the accuracy of target predictions.

2.2 Chemical Similarity Coupled with Statistical Models

Statistical models are often used to assess the significance or randomness of the observed similarity scores [13, 17, 23, 59] and correct for bias in chemical similarity calculations and for the uneven distribution of compounds across target proteins in any dataset. In these models a random similarity score distribution is constructed by calculating pairwise similarity scores for *n* random compound pairs in the dataset, followed by fitting a statistical distribution such as normal distribution, extreme value distribution (evd) or negative binomial distribution. The choice of the statistical distribution depends on the observed random distribution and the nature of the similarity scoring function, i.e. whether it is continuous or discrete. Using these models, the similarity scores between query and compounds are converted into significance scores such as *P*-values or *E*-values indicating the probabilities of random association between the query and the ligands of targets, which then serve to rank the targets.

2.3 Machine Learning Models

From the point of view of machine learning (ML) target prediction is a typical classification problem, wherein the fingerprints of molecules are predictor variables and the targets of molecules are class labels [60]. During the model building process, various pattern recognition algorithms are applied in order to find the mathematical function which can correctly map the fingerprint space to target labels. One of the advantages of ML models is that once the model is built, prediction of targets for any query molecule can be performed extremely fast. Classifiers built using *k*-nearest neighbors (*k*-NN), Naive bayes, Random forest and Support vector machines (SVM) perform extremely well for target predictions [29, 36, 37, 61]. *k*-NN is one of the most efficient machine learning models. *k*-NN does not involve a training phase and simply retrieves a predefined *k* number of most similar molecules (nearest neighbors) from a chemogenomic dataset and predicts targets based on a voting scheme, wherein the target most common to the nearest neighbors is assigned first rank to the query molecule.

2.4 Validation of Target Prediction Methods

Whether it relies on chemical similarity searching or on ML models, any target prediction method must be validated to assess its predictive power and reliability. For validation, the chemogenomic dataset is split into training and test sets and the performance of the model built on the training set is assessed by its ability to predict drug–target interactions in the test set. Although computationally expensive, a rigorous *k*-fold cross-validation should be performed by partitioning the chemogenomic dataset into *k* subsets and building *k* models, each using one of the *k* subsets as test set and the remaining *k*-1 subsets as training set. The leave one out method (LOO) takes the *k*-fold cross validation to its extreme, where *k* is equal to the number of compounds in the chemogenomic dataset. Often negative target interactions, [28] i.e., compounds known to be inactive against specific targets, are included in both training and test sets to evaluate model specificity.

3 Web-Based Tools

3.1 Polypharmacology Browser (PPB)

Website: PPB [13] is accessible at www.gdb.unibe.ch. PPB searches through a database containing the ligands of 4613 targets from ChEMBL, including single proteins, protein-family, protein-complexes, cell-lines, and organisms from different species. A query molecule can be drawn or input in SMILES, Mol, or SDF format. PPB performs target prediction using ten different fingerprints separately and returns the predefined number (by default 20) of the top predicted targets using each of them. The result is provided as a consolidated table of targets. In this table, each of the targets is annotated with the ChEMBL target ID, ChEMBL target name, probability in each of the ten fingerprint spaces (applicable whenever a target is found by a fingerprint), and the number of unique molecules of a target retrieved by PPB. In the PPB result window, the *P*-value of a target is mapped on to the probability bar (spanning *P*-value range of 0.01–0) and indicated by equivalent fill of the probability bar with the green color.

Methodology: The basic idea in PPB is to use multiple fingerprints and consensus data analysis for target prediction considering the fact that the chemical similarity calculation and hence predicted targets highly depend on the type of molecular fingerprint. PPB integrates similarity searching, data fusion and simple statistical analysis for target prediction (Fig. 3a). For similarity searching, PPB uses ten different fingerprints and the city block distance as the scoring function. Six of these fingerprints are APfp encoding shape, Xfp encoding pharmacophore, MQN encoding molecular composition, SMIfp encoding aromaticity, rings, and composition of the molecule, Sfp, and ECfp4 encoding substructures. The remaining four are fused fingerprints Ffp1, Ffp2, Ffp3, and Ffp4, generated by combining information from some of the individual fingerprints mentioned previously (Table 2). In case of a fused fingerprint, for instance, $\text{FP4}_{(\text{Xfp + MQN + SMIfp + Sfp + ECfp4})}$, the city block distance between a query and database compound is computed by first calculating the CBDs using each of the Xfp, MQN, SMIfp, Sfp, and ECfp4 fingerprints, followed by scaling and summation of CBDs as $(\text{Xfp}_{\text{cbd}} \times 1) + (\text{MQN}_{\text{cbd}} \times 24.9) + (\text{SMIfp}_{\text{cbd}} \times 60.7) + (\text{Sfp}_{\text{cbd}} \times 4.46) + (\text{ECfp4}_{\text{cbd}} \times 18.9)$. The numbers indicate scaling factors, which were derived after analyzing the CBD distribution for 50 M random pairs of compounds in each fingerprint space and scaling CBD values so as to match the most frequent CBD in each fingerprint space to the most frequent CBD in Xfp fingerprint space.

Given a query molecule PPB performs the following steps for each of the 10 fingerprints: (1) compute the city block distance of each of the compounds in the chemogenomic dataset to the query molecule (2) sort the dataset compounds by CBD and (3) traverse through the sorted compound list and collect targets associated

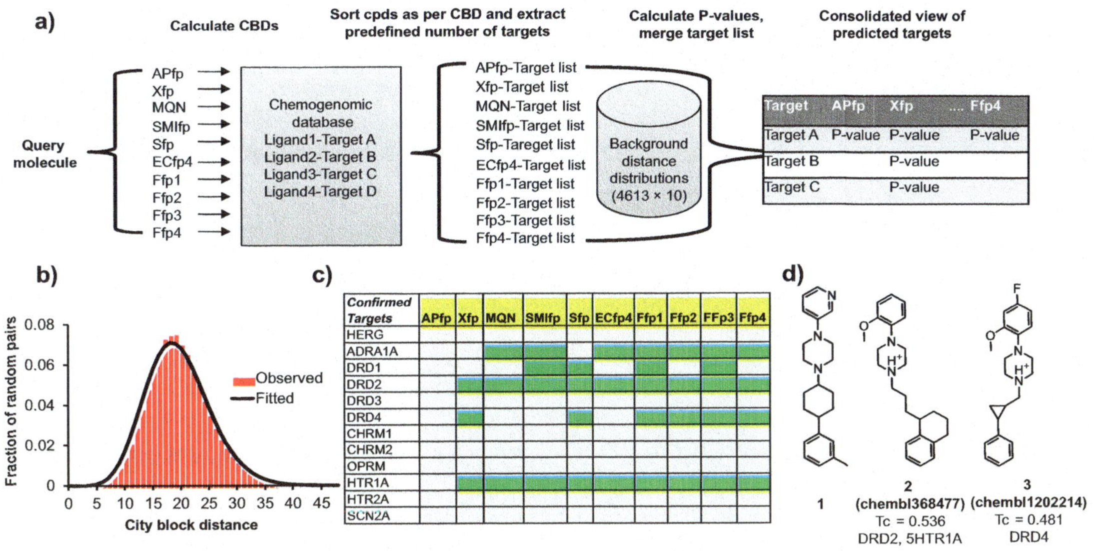

Fig. 3 (**a**) Overview of PPB target prediction workflow. (**b**) An example of observed (red) and fitted (black) distance distribution used for the calculation of a *P*-value for a target. The case is shown for CHRM1 in MQN fingerprint space. MQN city block distances were calculated for 1788 ligands of CHRM1 with respect to random compounds from ZINC database. Negative binomial distribution was used for fitting. PPB contains 46′130 distance distributions (4′613 targets × 10 fingerprints). (**c**) Confirmed side targets of compound 1. Green color indicates targets predicted by PPB. (**d**) Structure of compound 1 and ChEMBL compounds used by Xfp to link the targets DRD2, DRD4, and 5HTR1A to compound 1. *Tc*: Tanimoto coefficient calculated using Sfp. Target full names: Adrenergic α1A (ADRA1A), cholinergic muscarinic receptor 1 (CHRM1) and 2 (CHRM2), dopamine receptor subtypes D1–4 (DRD1–4), 5-hydroytryptamine receptor 1A (HTR1A) and 2A (HTR2A), voltage-gated potassium channel subfamily H member 2 (HERG), μ opioid receptor (OPRM), and voltage-gated Na + channel (SCN2A)

with each compound. PPB stops passing through the compound list when a predefined number of unique targets are collected. PPB finally merges the 10 lists of predicted targets and generates a consolidated view, wherein targets are ranked by the number of fingerprints which predict a target and the sum of *P*-values for a target across these fingerprints. PPB calculates the *P*-value for each predicted target based on the CBD of the most similar compound of a target to the query and the empirically fitted CBD distance distribution generated by computing CBDs of known ligands of a target to randomly selected compounds from the ZINC database (Fig. 3b). A *P*-value indicates the degree of randomness of the CBD and hence the probability for by chance association between a query and a predicted target.

Applications: We have used PPB to study the polypharmacology profile of compound 1 [13, 62] (Fig. 3d), a first potent inhibitor of TRPV6 transmembrane calcium channel recently reported from our group. TRPV6 is known to be overexpressed in breast and prostate cancers. We tested the activity of compound 1 against 24 out of 44 targets available in the "safety screen" panel of Cerep Pvt. Ltd. We selected these 24 targets based on the PPB target prediction result for compound 1. Five of these targets ADRA1A, DRD1, DRD2, DRD4, and HTR1A were chosen as they were predicted by multiple fingerprints, while the remaining targets were chosen considering subtypes of these five targets and the presence of other ion channels in the safety screen panel. In vitro profiling confirmed the activity of compound 1 (>50% inhibition at 10 μM) against 12 targets, 5 of which were directly proposed by PPB. Interestingly, in the case of Xfp, bioactive compounds which link the targets to a query showed a significantly lower substructure similarity to compound 1, which is particularly interesting to detect nonobvious query–target interactions.

3.2 TarPred

Website: TarPred [16] (www.dddc.ac.cn/tarpred/) performs a target prediction using data from Binding DB on 1703 targets and their associated small molecule ligands. A query molecule can be input in SMILES format or drawn in the provided molecular sketcher window. As a result, the top 30 predicted targets are shown, each of which is annotated with the 3NN (3 nearest neighbors) score. The 3NN score indicates the average similarity of the top three most similar ligands of a target to a query. The structures of these three nearest neighbors and disease conditions associated with a target can be obtained by clicking on the respective 3NN score.

Methodology: TarPred is based on chemical similarity searching and data fusion. TarPred uses the ECfp4 Tanimoto coefficient to quantify the similarities between compounds. Given a query molecule, TarPred calculates the pairwise Tanimoto similarities between a query and all of the compounds associated with each of the targets in the database. Following, TarPred identifies the top 3 most similar compounds of each target to a query and calculates the average Tanimoto similarity score (3NN score). Finally, TarPred ranks the targets by the 3NN score (high to low) and returns the top 30 targets as predicted targets of a query molecule. The choice of 3NN was based on a previous tenfold cross-validation study, in which 3NN showed superior performance to 1NN, 5NN and centroid score (average similarity considering all of the compounds of a target).

3.3 Self-Organizing Map Based Prediction of Drug Equivalence Relationship (SPiDER)

Website: SPiDER [17] is accessible at http://modlabcadd.ethz.ch/software/spider. SPiDER takes the molecule name and the SMILES of a query molecule as input. The result is provided as a plain text file containing predicted targets annotated with *P*-values. SPiDER only returns the predicted targets for which *P*-values are <0.05. SPiDER uses COBRA, a Collection of Reference Bioactive Compounds as it reference database.

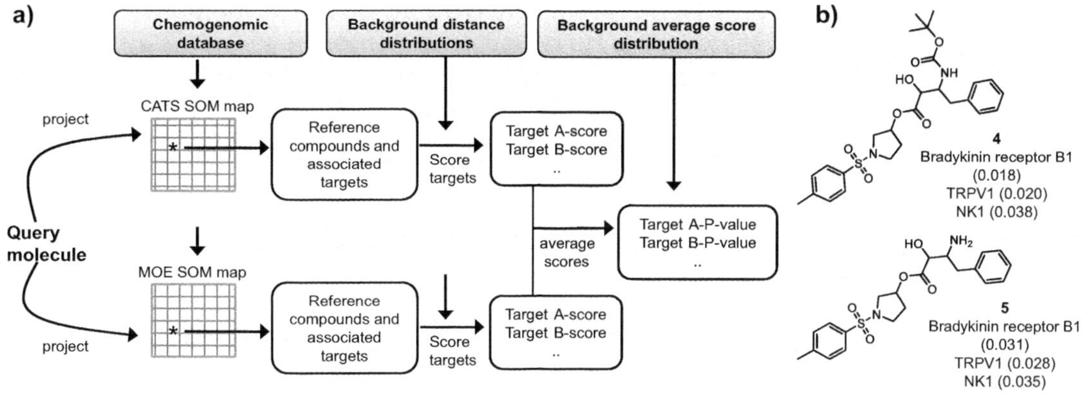

Fig. 4 (**a**) Overview of SPiDER target prediction. (**b**) Structures of de novo designed compounds and their predicted targets by SPiDER. Values in parenthesis indicate *P*-values

Methodology: SPiDER combines self-organizing maps (SOM), similarity calculation and statistical analysis for target prediction (Fig. 4a). One of the key aspects of the SPiDER model is that internally it employs two independent models for target prediction and makes the final prediction by data fusion. These two models were created using two distinct molecular fingerprints, namely the 210-dimensional CATS (fuzzy pharmacophore fingerprint) and 186 physiochemical descriptors from MOE software. Each of these internal models is composed of a 2D SOM map (13×10) trained on the respective fingerprint space of a chemogenomic database and an empirically fitted Euclidean distance distribution generated by computing pairwise Euclidean distances between ligands of different targets (inter targets similarities) in the database. This fitted distance distribution is expected to account for the false positive discovery rate in target prediction. In each of these models, a query molecule is projected onto the SOM map to define its local neighborhood (reference molecules). Next, targets defined by these reference molecules are scored as follows: (1) for each target, the pairwise Euclidean distances between a query and reference ligands of a respective target in the original fingerprint space are calculated, (2) the observed Euclidean distances are converted to *P*-values by applying an empirically fitted distribution mentioned above and, (3) the score for a target is computed as 1 − (average of *P*-values). Next, SPiDER combines the two lists of targets (one for each molecular representation) and calculates the arithmetic average score for each target. Finally, SPiDER converts the average target scores to *P*-values with the use of statistically fitted distribution of average scores obtained for known reference compounds and the likelihood of certain confidence in prediction of known drug targets in a chemogenomic database.

Application: The SPiDER model was designed to predict the targets for novel chemical entities (NCEs) or any query compound sharing low structural similarity to bioactive compounds. This is facilitated using fuzzy molecular representation (CATS), which allows to find subtle functional similarities between compounds. Reker et al. applied SPiDER to identify primary targets of de novo designed compounds 4 and 5 [17] (Fig. 4b). Initially, both compounds were designed to mimic functional similarity to Amprenavir and were thus expected to inhibit HIV protease. However, an experimental study showed that compounds 4 and 5 do not bind HIV protease. Later, with the use of SPiDER, the authors identified and experimentally confirmed binding of compounds 4 and 5 to Bradykinin B1, Vanilloid 1 (TRPV1), and Neurokinin 1 (NK1) receptors.

3.4 Swiss Target Prediction (SwissTarget-Prediction)

Website: SwissTargetPrediction [18] (http://www.swisstargetprediction.ch/) performs predictions using data from ChEMBL 16 on over 2000 target proteins from five different organisms (*Homo sapiens*, *Mus musculus*, *Rattus norvegicus*, *Bos taurus*, and *Equus caballus*). A query molecule can be input in SMILES format or drawn in the provided molecule sketcher window, and human targets are selected by default for prediction. SwissTargetPrediction returns the top 15 predicted targets, each of which is annotated with name, UniProt ID, ChEMBL target ID, prediction probability, and a number of compounds (up to predefined similarity cutoffs) of a target similar to a query molecule as per 3D and 2D similarity measures. The structures of compounds can be visualized by clicking on the respective numbers.

Methodology: SwissTargetPrediction combines 3D (ElectroShape fingerprint) and 2D (Open Babel FP2 fingerprint) similarities to improve the prediction accuracy, particularly for cases in which query molecules do not belong to well-defined series of known bioactive compounds (Fig. 5a). Given a query molecule,

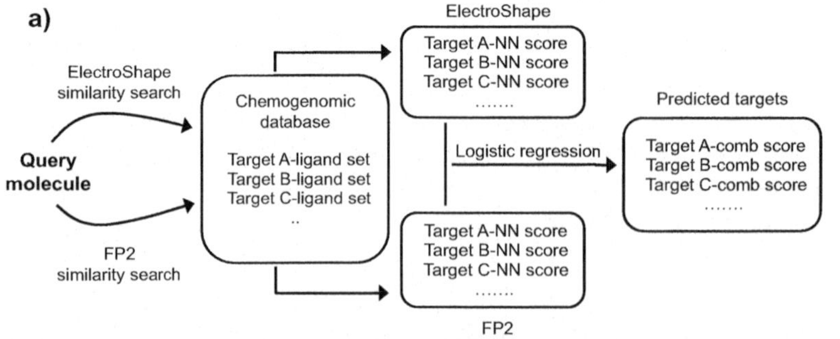

$$s1' = \frac{1}{1 + \frac{1}{18} CBD_{ij}}$$

$$s1 = \frac{(s1' - 0.65)}{1 - 0.65}$$

$$s2 = \frac{(s2' - 0.3)}{1 - 0.3}$$

$$f(s1, s2) = \frac{1}{1 + exp[-a_0 - a_1 s1 - a_2 s2]}$$

Fig. 5 (**a**) Overview of SiwssTargetPrediction method. (**b**) Formulas for computation of ElectroShape similarity score (s1'), scaled similarity scores s1 (ElectroShape) and s2 (FP2), and logistic regression model to compute the combined score. s1' is derived from ElectroShape city block distance between the pair of compounds (*i*, *j*). s2' is FP2 Tanimoto coefficient similarity between a pair of compounds. Parameters a_0, a_1, and a_2 are learned by logistic regression model on the training set. NN: Nearest neighbor

SwissTargetPrediction retrieves the most similar ligand of each of the targets in the database using the ElectroShape similarity score (Fig. 5b) and the FP2 Tanimoto coefficient, independently. This results in two similarity scores (s1', s2') for each target. Next, SwissTargetPrediction scales similarity scores for each target and combines them using a logistic regression model (Fig. 5b). SwissTargetPrediction uses 51 different logistic regression models, each of which was trained on molecules of specific heavy atom count ($\leq$10, 11, 12...59, $\geq$60) in a chemogenomic database. A ligand size specific model was created considering its influence on logistic regression coefficients. The input feature vector for the logistic regression model (selected as per the size of a query molecule) consists of two scaled similarity scores (s1, s2) and the output is the combined score. SwissTargetPrediction ranks the targets based on these combined scores, which are later converted to probabilities to give an estimate of the likelihood of correct predictions. One of the limitations of these probabilities is that they are derived from a cross-validation study indicating the average precision of prediction at given combined scores and are thus directly affected by the composition and therefore a potential bias in the training set.

3.5 Similarity Ensemble Approach (SEA)

Website: SEA [23] (http://sea.bkslab.org/) takes the SMILES of a query molecule as the only input and performs the prediction for 4161 targets proteins using data on associated small molecule ligands from ChEMBL 21 database. The output of SEA is a list of predicted targets annotated with *P*-values and Tanimoto coefficients of the most similar ligand (Max. *Tc*) of each of the predicted targets to a query molecule. The most similar ligand of each target can be accessed by clicking on the associated Max *Tc*. value. SEA only returns predicted targets for which *P*-values are significantly low.

Methodology: SEA uses the concept of raw similarity scores coupled with a statistical model to compare a single query molecule or one set of ligands (e.g., set A) to another set of ligands (e.g., set B) independent of set size, which is particularly important considering differences in sizes of known target associated ligand sets in chemogenomic databases. Given any two ligand sets (A, B) the raw similarity score can be computed by first calculating pairwise ECfp4 Tanimoto coefficient similarity for each of the compounds in set A to each of the compounds in set B and then summing up all the *Tc* values which are $\geq$0.57. To assess the significance of this raw similarity score, SEA uses a statistical model built on a random distribution of raw scores (Fig. 6a). In this statistical model, SEA creates *n* random ligand sets of varying size, representing sizes of known ligand sets one may encounter during similarity calculation. SEA calculates then raw scores for *m* ligand set pairs (set A-set B, set D-set R, etc.) and plots the means and standard deviations (S.D) of raw scores as a function of the product of ligand set sizes (e.g., size ligand set A $\times$ size of ligand set B). It has been previously observed

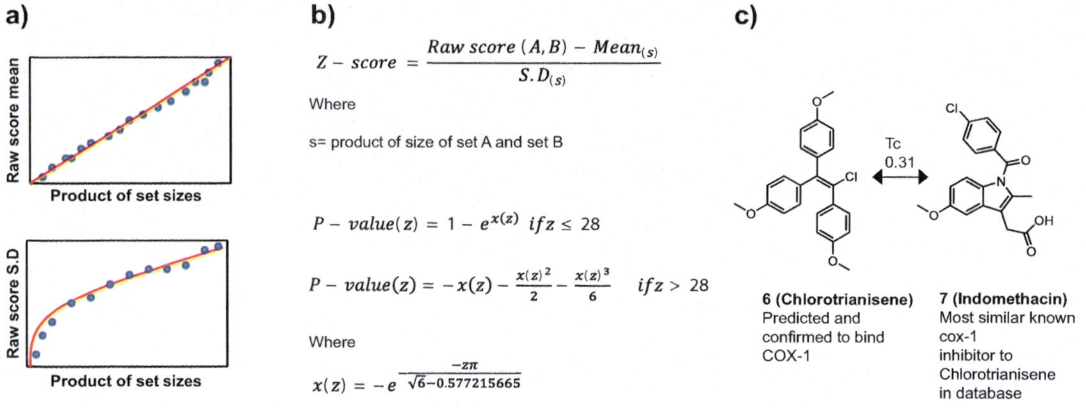

Fig. 6 Overview of SEA. (**a**) Schematic plots showing the distribution of mean and standard deviation (S.D) of raw scores as a function of the product of ligand set sizes. Blue: observed, Red: fitted (**b**) Formulas for converting raw score to *Z*-score to *P*-value. Raw score (A, B) is the observed raw score for ligand set A-ligand set B pair. Mean(s) and SD(s) can be retrieved from fitted plots (on left) given the product of ligand set sizes (size of set A $\times$ size of set B). When $z > 28$ to avoid exceeding numerical precision Taylor expansion approximation can be used. (**c**) An example of SEA prediction. For more details refer to the main text

that the mean can be fitted linearly, the S.D. can be fitted non-linearly, and that the histogram of *Z*-scores (calculated for raw scores of random sets) can be confined to an extreme value distribution (evd). This enables the computation of the probability of a *Z*-score to occur by chance alone (*P*-value or *E*-value) using the formulas given in Fig. 6b. The chosen *Tc* cutoff of 0.57 was based on the observation that at this cutoff value the evd can be best fitted to the *z*-score distribution. To predict targets of a query molecule, SEA calculates *Tc* similarities of a query to the ligand set of each of the target proteins in the database and then computes a raw similarity score for each target. Finally, for each target, the raw score is converted to a *Z*-score and finally to a *P*-value. The *P*-value for a target indicates the probability of by chance similarity between ligands of a target and a query molecule.

Application: SEA was initially used to find relationships among different target proteins by comparing their ligand sets [23]. Later, SEA was used to identify new targets for known drugs [9]. Out of 30 novel drug–target interactions suggested by SEA, 23 were confirmed including interactions of the adrenergic beta 1-receptor with Prozac, 5-hydroxytryptamine (5-HT) transporter with the drug Vadilex and the histamine H4 receptor with Rescriptor. SEA was also applied in a large scale setting to evaluate the probabilities of 695 drugs to bind 73 targets (Novartis safety panel) related to adverse drug reactions [6]. In this case, out of 1042 predictions, almost half of them were confirmed. One of the advantages of the SEA method is that the relationship between a query and target protein can often be found even if no single ligand of the target protein shows significant similarity to a query (Fig. 6c).

4 Conclusions

Web-based target prediction tools provide a fast and efficient way to study the polypharmacology of compounds. One of the advantages of Web-based tools is their ease of use with minimal or no technical know-how of the underlying methods, which is particularly appealing for end users such as medicinal chemists or biologists. Over the past decade target prediction methods have been evolving continuously, considering the advancement in technology and data mining methods and growing information on drug–target interactions. Although chemical similarity searching alone performs well for target prediction, integrating multiple approaches such as data fusion (combining information from multiple nearest neighbors or multiple fingerprints), statistical analysis (taking into account biases in chemical similarity calculations) and machine learning models improve the accuracy of predictions and the ability to reveal novel drug–target interactions. The application of PPB to study the polypharmacology of compound 1, SPiDER to find the primary targets of compounds 4 and 5, and SEA to study the adverse drug reactions and repurposing of drugs highlight some of the potentials of these methods to support drug discovery projects.

Acknowledgment

This work was supported financially by the Swiss National Science Foundation, NCCR TransCure.

References

1. Overington JP, Al-Lazikani B, Hopkins AL (2006) How many drug targets are there? Nat Rev Drug Discov 5(12):993–996
2. Anighoro A, Bajorath J, Rastelli G (2014) Polypharmacology: challenges and opportunities in drug discovery. J Med Chem 57 (19):7874–7887
3. Lavecchia A, Cerchia C (2016) In silico methods to address polypharmacology: current status, applications and future perspectives. Drug Discov Today 21(2):288–298
4. Mestres J, Gregori-Puigjane E, Valverde S, Sole RV (2009) The topology of drug-target interaction networks: implicit dependence on drug properties and target families. Mol BioSyst 5 (9):1051–1057
5. Cook D, Brown D, Alexander R, March R, Morgan P, Satterthwaite G, Pangalos MN (2014) Lessons learned from the fate of AstraZeneca's drug pipeline: a five-dimensional framework. Nat Rev Drug Discov 13 (6):419–431
6. Lounkine E, Keiser MJ, Whitebread S, Mikhailov D, Hamon J, Jenkins JL, Lavan P, Weber E, Doak AK, Cote S, Shoichet BK, Urban L (2012) Large-scale prediction and testing of drug activity on side-effect targets. Nature 486(7403):361–367
7. Siramshetty VB, Nickel J, Omieczynski C, Gohlke B-O, Drwal MN, Preissner R (2016) WITHDRAWN—a resource for withdrawn and discontinued drugs. Nucleic Acids Res 44 (D1):D1080–D1086
8. Wermuth CG (2006) Selective optimization of side activities: the SOSA approach. Drug Discov Today 11(3):160–164
9. Keiser MJ, Setola V, Irwin JJ, Laggner C, Abbas AI, Hufeisen SJ, Jensen NH, Kuijer MB, Matos RC, Tran TB, Whaley R, Glennon RA, Hert J, Thomas KLH, Edwards DD,

Shoichet BK, Roth BL (2009) Predicting new molecular targets for known drugs. Nature 462 (7270):175–181
10. Besnard J, Ruda GF, Setola V, Abecassis K, Rodriguiz RM, Huang X-P, Norval S, Sassano MF, Shin AI, Webster LA, Simeons FRC, Stojanovski L, Prat A, Seidah NG, Constam DB, Bickerton GR, Read KD, Wetsel WC, Gilbert IH, Roth BL, Hopkins AL (2012) Automated design of ligands to polypharmacological profiles. Nature 492 (7428):215–220
11. Koutsoukas A, Simms B, Kirchmair J, Bond PJ, Whitmore AV, Zimmer S, Young MP, Jenkins JL, Glick M, Glen RC, Bender A (2011) From in silico target prediction to multi-target drug design: Current databases, methods and applications. J Proteome 74(12):2554–2574
12. Cereto-Massagué A, Ojeda MJ, Valls C, Mulero M, Pujadas G, Garcia-Vallve S (2015) Tools for in silico target fishing. Methods 71:98–103
13. Awale M, Reymond J-L (2017) The polypharmacology browser: a web-based multi-fingerprint target prediction tool using ChEMBL bioactivity data. J Cheminform 9(1):11
14. Yao Z-J, Dong J, Che Y-J, Zhu M-F, Wen M, Wang N-N, Wang S, Lu A-P, Cao D-S (2016) TargetNet: a web service for predicting potential drug–target interaction profiling via multitarget SAR models. J Comput Aided Mol Des 30(5):413–424
15. Kringelum J, Kjaerulff SK, Brunak S, Lund O, Oprea TI, Taboureau O (2016) ChemProt-3.0: a global chemical biology diseases mapping. Database 2016
16. Liu X, Gao Y, Peng J, Xu Y, Wang Y, Zhou N, Xing J, Luo X, Jiang H, Zheng M (2015) TarPred: a web application for predicting therapeutic and side effect targets of chemical compounds. Bioinformatics 31(12):2049–2051
17. Reker D, Rodrigues T, Schneider P, Schneider G (2014) Identifying the macromolecular targets of de novo-designed chemical entities through self-organizing map consensus. Proc Natl Acad Sci 111(11):4067–4072
18. Gfeller D, Grosdidier A, Wirth M, Daina A, Michielin O, Zoete V (2014) SwissTargetPrediction: a web server for target prediction of bioactive small molecules. Nucleic Acids Res 42(W1):W32–W38
19. Wang L, Ma C, Wipf P, Liu H, Su W, Xie X-Q (2013) TargetHunter: an in silico target identification tool for predicting therapeutic potential of small organic molecules based on chemogenomic database. AAPS J 15 (2):395–406
20. Gong J, Cai C, Liu X, Ku X, Jiang H, Gao D, Li H (2013) ChemMapper: a versatile web server for exploring pharmacology and chemical structure association based on molecular 3D similarity method. Bioinformatics 29 (14):1827–1829
21. Liu X, Vogt I, Haque T, Campillos M (2013) HitPick: a web server for hit identification and target prediction of chemical screenings. Bioinformatics 29(15):1910–1912
22. Nickel J, Gohlke B-O, Erehman J, Banerjee P, Rong WW, Goede A, Dunkel M, Preissner R (2014) SuperPred: update on drug classification and target prediction. Nucleic Acids Res 42(W1):W26–W31
23. Keiser MJ, Roth BL, Armbruster BN, Ernsberger P, Irwin JJ, Shoichet BK (2007) Relating protein pharmacology by ligand chemistry. Nat Biotechnol 25(2):197–206
24. Lagunin A, Stepanchikova A, Filimonov D, Poroikov V (2000) PASS: prediction of activity spectra for biologically active substances. Bioinformatics 16(8):747–748
25. Wang Z, Liang L, Yin Z, Lin J (2016) Improving chemical similarity ensemble approach in target prediction. J Cheminform 8(1):1–10
26. Wang X, Pan C, Gong J, Liu X, Li H (2016) Enhancing the enrichment of pharmacophore-based target prediction for the polypharmacological profiles of drugs. J Chem Inf Model 56 (6):1175–1183
27. Cao R, Wang Y (2016) Predicting molecular targets for small-molecule drugs with a ligand-based interaction fingerprint approach. ChemMedChem 11(12):1352–1361
28. Mervin LH, Afzal AM, Drakakis G, Lewis R, Engkvist O, Bender A (2015) Target prediction utilising negative bioactivity data covering large chemical space. J Cheminform 7:51
29. Lusci A, Fooshee D, Browning M, Swamidass J, Baldi P (2015) Accurate and efficient target prediction using a potency-sensitive influence-relevance voter. J Cheminform 7(1):1–13
30. Liu X, Xu Y, Li S, Wang Y, Peng J, Luo C, Luo X, Zheng M, Chen K, Jiang H (2014) In Silico target fishing: addressing a "Big Data" problem by ligand-based similarity rankings with data fusion. J Cheminform 6(1):33
31. Alvarsson J, Eklund M, Engkvist O, Spjuth O, Carlsson L, Wikberg JES, Noeske T (2014) Ligand-based target prediction with signature fingerprints. J Chem Inf Model 54 (10):2647–2653
32. Mavridis L, Mitchell JB (2013) Predicting the protein targets for athletic performance-

enhancing substances. J Cheminform 5 (1):1–13

33. Koutsoukas A, Lowe R, KalantarMotamedi Y, Mussa HY, Klaffke W, Mitchell JBO, Glen RC, Bender A (2013) In silico target predictions: defining a benchmarking data set and comparison of performance of the multiclass naïve bayes and parzen-rosenblatt window. J Chem Inf Model 53(8):1957–1966
34. Pérez-Nueno VI, Venkatraman V, Mavridis L, Ritchie DW (2012) Detecting drug promiscuity using gaussian ensemble screening. J Chem Inf Model 52(8):1948–1961
35. AbdulHameed MDM, Chaudhury S, Singh N, Sun H, Wallqvist A, Tawa GJ (2012) Exploring polypharmacology using a ROCS-based target fishing approach. J Chem Inf Model 52 (2):492–505
36. Wale N, Karypis G (2009) Target fishing for chemical compounds using target-ligand activity data and ranking based methods. J Chem Inf Model 49(10):2190–2201
37. Nidhi GM, Davies JW, Jenkins JL (2006) Prediction of biological targets for compounds using multiple-category bayesian models trained on chemogenomics databases. J Chem Inf Model 46(3):1124–1133
38. Peragovics Á, Simon Z, Tombor L, Jelinek B, Hári P, Czobor P, Málnási-Csizmadia A (2013) Virtual affinity fingerprints for target fishing: a new application of drug profile matching. J Chem Inf Model 53(1):103–113
39. Liu X, Ouyang S, Yu B, Liu Y, Huang K, Gong J, Zheng S, Li Z, Li H, Jiang H (2010) PharmMapper server: a web server for potential drug target identification using pharmacophore mapping approach. Nucleic Acids Res 38 (suppl 2):W609–W614
40. Li H, Gao Z, Kang L, Zhang H, Yang K, Yu K, Luo X, Zhu W, Chen K, Shen J, Wang X, Jiang H (2006) TarFisDock: a web server for identifying drug targets with docking approach. Nucleic Acids Res 34(suppl 2):W219–W224
41. Law V, Knox C, Djoumbou Y, Jewison T, Guo AC, Liu Y, Maciejewski A, Arndt D, Wilson M, Neveu V, Tang A, Gabriel G, Ly C, Adamjee S, Dame ZT, Han B, Zhou Y, Wishart DS (2014) DrugBank 4.0: shedding new light on drug metabolism. Nucleic Acids Res 42(D1): D1091–D1097
42. Gilson MK, Liu T, Baitaluk M, Nicola G, Hwang L, Chong J (2016) BindingDB in 2015: A public database for medicinal chemistry, computational chemistry and systems pharmacology. Nucleic Acids Res 44(Database issue):D1045–D1053
43. Gaulton A, Bellis LJ, Bento AP, Chambers J, Davies M, Hersey A, Light Y, McGlinchey S, Michalovich D, Al-Lazikani B, Overington JP (2012) ChEMBL: a large-scale bioactivity database for drug discovery. Nucleic Acids Res 40(Database issue):D1100–D1107
44. Olah M, Rad R, Ostopovici L, Bora A, Hadaruga N, Hadaruga D, Moldovan R, Fulias A, Mractc M, Oprea TI (2008) WOMBAT and WOMBAT-PK: bioactivity databases for lead and drug discovery, chemical biology: from small molecules to systems biology and drug design. Wiley-VCH Verlag GmbH:760–786
45. Rose PW, Prlić A, Bi C, Bluhm WF, Christie CH, Dutta S, Green RK, Goodsell DS, Westbrook JD, Woo J, Young J, Zardecki C, Berman HM, Bourne PE, Burley SK (2015) The RCSB Protein Data Bank: views of structural biology for basic and applied research and education. Nucleic Acids Res 43(D1):D345–D356
46. Ertl P, Selzer P, Mühlbacher J (2004) Web-based cheminformatics tools deployed via corporate Intranets. Drug Discov Today Biosilico 2(5):201–207
47. Martin YC, Kofron JL, Traphagen LM (2002) Do structurally similar molecules have similar biological activity? J Med Chem 45 (19):4350–4358
48. Jenkins JL, Bender A, Davies JW (2006) In silico target fishing: Predicting biological targets from chemical structure. Drug Discov Today Technol 3(4):413–421
49. Hagadone TR (1992) Molecular substructure similarity searching: efficient retrieval in two-dimensional structure databases. J Chem Inf Comput Sci 32(5):515–521
50. Rogers D, Hahn M (2010) Extended-connectivity fingerprints. J Chem Inf Model 50(5):742–754
51. Durant JL, Leland BA, Henry DR, Nourse JG (2002) Reoptimization of MDL keys for use in drug discovery. J Chem Inf Comput Sci 42 (6):1273–1280
52. Schneider G, Neidhart W, Giller T, Schmid G (1999) "Scaffold-Hopping" by topological pharmacophore search: a contribution to virtual screening. Angew Chem Int Ed 38 (19):2894–2896
53. Awale M, Reymond J-L (2014) Atom Pair 2D-fingerprints perceive 3D-molecular shape and pharmacophores for very fast virtual screening of ZINC and GDB-17. J Chem Inf Model 54(7):1892–1907
54. Ballester PJ, Richards WG (2007) Ultrafast shape recognition to search compound

databases for similar molecular shapes. J Comput Chem 28(10):1711–1723

55. Armstrong MS, Morris GM, Finn PW, Sharma R, Moretti L, Cooper RI, Richards WG (2010) ElectroShape: fast molecular similarity calculations incorporating shape, chirality and electrostatics. J Comput Aided Mol Des 24 (9):789–801
56. Grant JA, Gallardo MA, Pickup BT (1996) A fast method of molecular shape comparison: A simple application of a Gaussian description of molecular shape. J Comput Chem 17 (14):1653–1666
57. Hawkins PCD, Skillman AG, Nicholls A (2007) Comparison of shape-matching and docking as virtual screening tools. J Med Chem 50(1):74–82
58. Willett P (2013) Fusing similarity rankings in ligand-based virtual screening. Comput Struct Biotechnol J 5(6):1–6
59. Baldi P, Nasr R (2010) When is chemical similarity significant? The statistical distribution of chemical similarity scores and its extreme values. J Chem Inf Model 50(7):1205–1222
60. Mitchell JBO (2014) Machine learning methods in chemoinformatics. Wiley Interdiscip Rev Comput Mol Sci 4(5):468–481
61. Nigsch F, Bender A, Jenkins JL, Mitchell JBO (2008) Ligand-target prediction using winnow and naive bayesian algorithms and the implications of overall performance statistics. J Chem Inf Model 48(12):2313–2325
62. Simonin C, Awale M, Brand M, van Deursen R, Schwartz J, Fine M, Kovacs G, Häfliger P, Gyimesi G, Sithampari A, Charles R-P, Hediger MA, Reymond J-L (2015) Optimization of TRPV6 calcium channel inhibitors using a 3D ligand-based virtual screening method. Angew Chem Int Ed 54 (49):14748–14752
63. Nguyen KT, Blum LC, van Deursen R, Reymond J-L (2009) Classification of organic molecules by molecular quantum numbers. ChemMedChem 4(11):1803–1805
64. Schwartz J, Awale M, Reymond J-L (2013) SMIfp (SMILES fingerprint) chemical space for virtual screening and visualization of large databases of organic molecules. J Chem Inf Model 53(8):1979–1989
65. Filimonov D, Poroikov V, Borodina Y, Gloriozova T (1999) Chemical similarity assessment through multilevel neighborhoods of atoms: definition and comparison with the other descriptors. J Chem Inf Comput Sci 39 (4):666–670

Chapter 16

In Silico Target Prediction for Small Molecules

Ryan Byrne and Gisbert Schneider

Abstract

Drugs modulate disease states through their actions on targets in the body. Determining these targets aids the focused development of new treatments, and helps to better characterize those already employed. One means of accomplishing this is through the deployment of in silico methodologies, harnessing computational analytical and predictive power to produce educated hypotheses for experimental verification. Here, we provide an overview of the current state of the art, describe some of the well-established methods in detail, and reflect on how they, and emerging technologies promoting the incorporation of complex and heterogeneous data-sets, can be employed to improve our understanding of (poly)pharmacology.

Key words Chemoinformatics, Computer-assisted drug design, Machine-learning, Network pharmacology, Polypharmacology

1 Introduction

The increase in the breadth and depth of understanding, and concomitant annotation, of biochemical space in the past few decades has been breathtaking. Resources focusing on biological data have appeared [1, 2] (and disappeared [3]) at a rate which has necessitated the development of various meta-databases and analytical platforms [4–7]. These resources enable prospective studies through the extraction of useful biochemical information from complex, multifaceted data. Great steps forward have been made with regard to bringing these complex data into the decision-making processes employed in academia and industry, producing useful (and valuable) information, and minimizing costs in time and money in high-risk areas [8, 9]. For many years, pharmaceutical research has focused on the identification, and bringing-to-market, of high-potency, low-promiscuity, single-target drugs, operating under the quite reasonable assumption that this should produce pharmaceutically useful substances with a mild-to-moderate side-effect profile, while also recognizing that some off-target activity was largely unavoidable [10, 11]. In recent years, however, as

Slava Ziegler and Herbert Waldmann (eds.), *Systems Chemical Biology: Methods and Protocols*, Methods in Molecular Biology, vol. 1888, https://doi.org/10.1007/978-1-4939-8891-4_16, © Springer Science+Business Media, LLC, part of Springer Nature 2019

attempts have been made to address single targets in a specific manner, the issue of what we mean when we describe a "target" that interests us, and how that definition varies as we continue along the drug-discovery pathway, has become a pressing one. Most importantly, the process of assessing the implications of our target definitions in the success or failure of our endeavors to date has made it clear that a ligand's target promiscuity is often critical in its activity against disease [12–14]. Indeed, historically, pharmacological characterization of receptors was mainly accomplished by noting the tendency of some ligands to interact with several targets [15]. These target-based ontologies were later largely reconfirmed with genetic and proteomic methods [16, 17], suggesting that biochemical promiscuity has inherent value in the tree of life [18].

To get to grips with the issue of what we mean by "a target," we might begin by comparing existing definitions in a quantitative fashion. To date, several studies have provided estimates for the size of target space: these vary considerably, from around 120 [16, 19] to 14,000 [20], although in practice most provide a value in the range of 200–400 [21]. Much of the variation is due to differences in semantics and expectations of proof; whether a "target" is truly a target if it has no known role in disease, and the extent to which it is permissible to infer between sequence-similar proteins in different species, for example.

Regardless of the difficulty of defining the targetable space, we can make certain statements about the currently drugged targets. Of the average of 18 new drugs approved by the FDA every year, four act on "novel" (previously undrugged) targets, where these are defined in genetic terms [22]. A recent study [23], has concluded that the relationship between marketed drugs and their associated targets is far from the idealized one-to-one relationship initially visualized, with only 15% obeying that relationship, and 50% of drugs having more than five targets annotated [24]. Indeed, a further study [25] found that, of approximately 650 drugs approved for human use, 10% hit just under a half (45%) of Novartis' "side-effects" target list. Consequently, we must recognise that our traditional drug–target–disease model is somewhat outdated. However, this complexity has the potential to considerably enrich our attempts to predict targets for new ligands, and to fill in the target space for established medicines, some of which have resisted characterization for an extended period of time and have been established on the basis of historical use, such as chloral hydrate and levetiracetam [26]. Having data on multiple targets for a single ligand helps us discover more meaningful chemogenomic relationships, and to form hypotheses which are amenable to statistical assessment of validity [27, 28].

The manner in which deconvoluting the relationships between a ligand and its target(s) might be accomplished, however the latter is defined, is the topic of this chapter. We will focus on how to predict as many valid targets for a ligand as possible, based on the available data, rather than a "canonical" target, and will restrict the discussion, in the main sections, to how these searches are performed with proteins as their focus, disregarding the sites-of-action of nutraceuticals and ligands which act directly on nucleotides or extracellular macromolecules, diagnostic agents, and other classes such as osmotic agents [29]. This means that we will not cover some novel resources such as the L1000CDS2 [30] suite offered by the NIH *Library of Integrated Network-Based Cellular Signatures* (LINCS) consortium, which treats transcriptional modulation as its endpoint. In addition, we will brush past the use of these and similar tools for other purposes, such a toxicity prediction, a field for which computational predictive tools are widely approved and encouraged by the relevant regulatory bodies [31]. There have been several excellent reviews and opinion pieces published covering the subject of target prediction [32–36] to date, focusing on the specifics of individual approaches in the main; this work aims to provide a comprehensive introduction to the broad range of technologies employed thus far, and discuss future steps for the field.

2 Ligand-Based In Silico Target Prediction

2.1 What Do We Mean by Ligand Similarity?

It is fair to say that the central dogma of molecular pharmacology, in an analogous fashion to that of structural biology, is this: structure causes effect. Given this, it is commonly accepted, and with good reason, that "similar" ligands will bind to the same, or similar, targets. Indeed, much of the early, pregenomic work in protein classification was on the basis of joint biochemical or phenotypic activity profiles in the presence of a molecular probe [15]. Accepting this, we approach a somewhat philosophical problem: what do we mean when we say that two chemical entities are similar?

Essentially this is based, either implicitly or explicitly, on identifying a common ligand (sub)structure or chemical framework, and certain substituents and properties which extend that framework into a different region of chemical space [37]. The subtleties of the hard or soft demarcation problems are both obvious, and controversial. Various schemes for encoding these chemical entities in a manner similar to that in which experts incorporate data semiconsciously into their decision-making processes have been proposed, and successfully utilized over the last decades. These "descriptors" are numerical quantities, or qualities, of molecular properties which vary in complexity from simple schemes for enumerating features (e.g., the molecular weight, lipophilicity, and the number of hydrogen-bond donors) to three-to-six-dimensional representations and complex target-fingerprints [38, 39]. Some of these descriptors aim to encode specific information about the pharmacophoric features of molecules. These are defined as the "*ensemble of steric and electronic features that [are] necessary to ensure the optimal supramolecular interactions with a specific biological target and to trigger (or block) its biological response*" [40]. These are then encoded in a descriptor space, as above. In practice, some combination of these features and models often proves a more powerful tool than either alone [41].

Fundamentally, the issue is how one should represent a structure which has a complex topological, physicochemical, and geometric nature in a fashion which allows us to compare them meaningfully [42–44]. Equally, our definition of activity is important: ligand–target affinity has often been viewed as a binary classification (where a molecule has affinity for a given target, or has not) rather than as a regression (molecules have a continuously valued affinity for a target) problem [45]. This distinction also applies in our concept of ligand similarity itself: while descriptors are often described as being a means of encoding a medicinal chemist's complex, theoretically underpinned, and, to some extent, intuitive view of the chemical world in such a way that a computer can understand it, it is also true that the classification problem (deciding whether two or more molecules are similar to one another, or not) is much easier for trained humans than the ordering or regression problems (e.g., the most similar molecule from a set of 100 is number 47, then 35, etc., or that something is 35% similar). This proves to be a fundamental issue in all chemoinformatics endeavors; however, given the comparatively large volume of data a computer can hold in memory at any one point in time, it is not an intractable one, and computational efforts have proven to be increasingly helpful in assisting the efforts of lab-based chemists, biochemists, and molecular biologists.

2.2 Similarity Metrics and Indices

In the decades since its inception, chemoinformatics has produced many means of comparing such descriptors once calculated, some novel forms of which we will return to in the closing matter of this chapter, which vary in their properties to a considerable degree. In the main, though, a few of these have proved especially useful to date. For the sake of clarity and utility, we will discuss a few here at length. In the first instance, it is important to be clear with the field's terminology. "Similarity," in its applications in chemoinformatics, refers to comparing molecular representations in such a manner that two identical representations would evaluate to one, i.e., 100% similarity. For these same two representations, their "distance" would be zero. Indeed, distance is the complement of similarity, being definable as the subtraction of the similarity from a maximal value of one.

Obviously, identity comparisons are relatively straightforward, with the nuance being in the noninteger ranges of comparison. For these methods to perform well, it is necessary to characterize them in a formal manner. Certain characteristics are required to be met before a similarity or distance measure can be properly classed as a metric:

1. Distance values D for two objects A and B must be greater than or equal to zero, and the distance between an object and itself must be equal to zero:

$$D_{\mathrm{A,B}} \geq 0, D_{\mathrm{A,A}} = D_{\mathrm{B,B}} = 0$$

2. Distance values must be symmetric, that is, evaluate to the same value regardless of direction:

$$D_{\mathrm{A,B}} = D_{\mathrm{B,A}}$$

3. Distance values must obey the triangular inequality:

$$D_{\mathrm{A,B}} \leq D_{\mathrm{A,C}} + D_{\mathrm{C,B}}$$

4. Nonidentical objects must be separated by a distance greater than zero, i.e., they cannot collide in the metric space:

$$\mathrm{A} \neq \mathrm{B} \Leftrightarrow D_{\mathrm{A,B}} > 0$$

If a given measure meets the first three characteristic requirements, we call it *pseudo*-metric; it gives consistent a consistent representation in space of objects, but is not capable of separating all objects in space.

Broadly speaking, one might divide the commonly employed metrics into two camps, based on their consideration, or otherwise, of "absent" features in the molecular descriptors. The former camp includes the Minkowski distance metrics (Euclidean, Manhattan, etc.), the latter measures such as the Tanimoto similarity coefficient and its complement (for binary-valued feature vectors), the Soergel distance.

In the equations to follow, the variables are defined as such: S defines a similarity metric, D a distance metric. For the continuous versions of each function, $X_{j\mathrm{A}}$ refers to the j-th element in the descriptor vector for molecule A. For binary-valued variables, a is the number of bits which are 1-valued in molecule A, b the same for molecule B, and c the common on-bits for both molecules. In all cases, formulae are paired in their forms for continuous- and binary-valued feature vectors, respectively.

$$D_{\mathrm{A,B}} = \sqrt{\sum_{j=1}^{j=n} \left| \left(X_{j\mathrm{A}} - X_{j\mathrm{B}} \right)^2 \right|} \tag{1}$$

$$D_{\mathrm{A,B}} = \sqrt{a + b - 2c} \tag{2}$$

Equations 1 and 2 describe the form of the Euclidean (also L2 norm) distance metric.

$$D_{\mathrm{A,B}} = \sum_{j=1}^{j=n} \left| X_{j\mathrm{A}} - X_{j\mathrm{B}} \right| \tag{3}$$

$$D_{\mathrm{A,B}} = a + b - 2c \tag{4}$$

Equations 3 and 4 describe the form of the Manhattan (also Hamming, city-block, L1-norm) distance metric.

$$S_{\mathrm{A,B}} = \frac{\sum_{j=1}^{j=n} X_{j\mathrm{A}} X_{j\mathrm{B}}}{\sum_{j=1}^{j=n} (X_{j\mathrm{A}})^2 + \sum_{j=1}^{j=n} (X_{j\mathrm{B}})^2 - \sum_{j=1}^{j=n} X_{j\mathrm{A}} X_{j\mathrm{B}}} \tag{5}$$

$$S_{\mathrm{A,B}} = \frac{c}{a + b - c} \tag{6}$$

Finally, Eqs. 5 and 6 describe the form of the Tanimoto similarity coefficients. These metrics vary in their behaviors; some sources [46–48] state that the Tanimoto coefficient is a better means of assessing the similarity of two molecules, with the other metrics being of more use in placing multiple molecules in context with one another. One disadvantage of the Tanimoto similarity and Soergel distance metrics is their susceptibility to molecular size, owing to a combination of their properties and those of the underlying fingerprint descriptors themselves; the latter tend to be relatively sparse, and are unscaled, so a larger molecule necessarily has a higher probability of exhibiting a given feature, along with other, irrelevant features. This property, coupled with the Tanimoto and Soergel metrics' lack of accounting for shared absent (zero-valued) features, leads to odd size-dependency behaviors. Means of minimizing these issues have been discussed in the literature, and mostly involve introducing an additional corrective step for the normalization of the metric value [49–51].

3 Web-Based Tools that Directly Incorporate Ligand Similarity in Target Prediction

Having described the means by which we encode information on the features present in a given molecule, and defined how we compare such depictions, we proceed to a discussion of how these techniques can be used to enrich the field of target prediction. A wide variety of ligand-based target prediction software is now available online, on a free-usage basis, a selection of which will be covered here in some depth, and some indications given for the strengths and weaknesses of each method (Table 1).

3.1 Similarity Ensemble Approach (SEA)

SEA (http://sea.bkslab.org) is a tool for conducting structure-based fingerprint similarity searches [83], and returns a ranked list of predictions, and comparative information on ligands known to

Table 1
Selected target prediction tools available on the Internet

Resource	Algorithmic basis	Database(s)
SEA [52]	Ligand-based: Daylight fingerprint similarity, Kruskal's minimum-spanning tree [53]	ChEMBL [54],WOMBAT [55], MDDR [56], stARLITe [57]
SwissTarget [41, 58]	Ligand-based: Shape- and fingerprint- based similarity [46, 59] combined, optionally, with a logistic regression model	Curated ChEMBL: human-only, direct activity only, proteins and protein complexes only [60]
SPiDER [61]	Ligand-based: CATS [62, 63] topological pharmacophore and MOE [64] descriptors grouped by means of self-similarity with self-organizing maps [65–67]	COBRA [68]
SuperPred [69]	Ligand-based: ECFP4 radial fingerprint similarity [70]	SuperTarget [71], ChEMBL, BindingDB [72, 73], Therapeutic Target Database
PPB (Poly-Pharmacology Browser) [74]	Ligand-based: focus on shape-based and pharmacophoric fingerprint similarity	ChEMBL
HitPick [75]	Ligand-based: ECFP fingerprint similarity, LMNB classifier [76]	STITCH [77, 78]
PASS [79–81]	Ligand-based: MNA [82] atomic neighborhood descriptors, Bayesian analysis	Various

share those targets. SEA makes use of the MDDR, ChEMBL, and WOMBAT databases for its combined ligand–target datasets. These data are encoded using the Daylight 2D-topological fingerprints, and the target ligand-sets are compared against each other, all-against-all. It has been shown that Tanimoto coefficients (*Tc*) for sets of ligands resemble a Weibull, or extreme-value, distribution [48, 84, 85]. Given this, BLAST-like *E*-values [86] can be calculated for sets of ligands, based on their *Tc*, or more accurately, on the *z*-scores associated with those *Tc*. The E-value gives an indication of how likely it is that we would achieve a *Tc* at random which is as good or better than the one observed, given the size of the sets involved. As noted by the authors, a strong (low *E*-value) prediction has no correlation with the magnitude of activity, only its presence in a pharmacologically relevant concentration range (*see* **Note 1**).

3.2 SwissTarget

SwissTarget (http://www.swisstargetprediction.ch) returns a list of targets, with links to the relevant ChEMBL and UniProt records, ranked by the *pseudo*-probability of the ligand–target interaction being significant, and also a detailed analysis of the relative contribution of 2D and 3D similarity metrics in its assignations. We say *pseudo*-probability, in this instance, as it is conditional on the calibration of the classifier, and approaches the "true" probability only as the calibration of the model improves. This means, for example, that for ten compounds each assigned a ligand–target interaction *pseudo*-probability of 0.8 by a well-calibrated model, approximately eight will truly be valid interactions. This is further complicated by the introduction of biases owing to the employment of cross-validation on heterogenous chemical data. SwissTarget's data come from a curated subset of CheMBL, filtered to remove nonprotein or protein-complex targets, large ligands, and those annotated without a "direct-binding" tag. It compares the 2D and 3D properties of the molecules, by computing FP2 fingerprints and ElectroShape vectors [59], which are then combined and fed into a logistic regression model used to predict class membership. SwissTarget also allows one to extend target assignations between interspecies orthologs, dependent on sequence homology, on an automatic basis.

3.3 SPiDER

SPiDER (http://modlabcadd.ethz.ch/software/spider) is predicated on the basis that "fuzzy" pharmacophore searching allows one to find molecules with different scaffolds that may interact with the same, or similar, set of targets [62, 63, 87]. To allow for this, SPiDER compares query ligands to self-organizing maps [65] (SOMs), a form of unsupervised neural network, trained upon the COBRA dataset [68]. These SOMs automatically cluster the descriptor vectors for its reference molecules in two descriptor spaces, one for CATS topological pharmacophore descriptors, another for MOE's (Molecular Operating Environment, The Chemical Computing Group, Montréal) physicochemical and index descriptors. Query compounds are then assigned to one cluster in each map, and the COBRA compounds which inhabit that cluster are chosen as the reference set. Next, it considers the distribution of the Euclidean distances between the query molecule's descriptor vectors and the reference compounds, in light of the background distance distribution. This latter parameter is used to assess the likelihood of the assignation being significant. The average of the two values (one for each descriptor SOM) is used as the overall *pseudo*-probability of the ligand–target assignation (*see* **Note 2**).

3.4 SuperPred

SuperPred (http://prediction.charite.de) returns both known and predicted targets for a submitted ligand, the latter of which are ranked by a version of the *E*-score as discussed below. Rather than being updated on a continual basis, its input data are curated and updated upon each major release from the sources described in Table 1. Only "successful" targets from the Therapeutic Target Database are used for validation. With regard to the algorithmic basis for target prediction, SuperPred uses simple similarity measures based on the ECFP fingerprint set [70, 88] and the distribution of each of these in its training data. A computationally inexpensive 2D fingerprint search is performed, and the set-size normalized sum of all *Tc* values >0.45 for a given target is calculated for each target set. BLAST-like *E*-values are calculated, to get an idea of the likelihood of an equally good combined *Tc* being observed at random in a training set of equivalent size.

3.5 Polypharmacology Browser (PPB)

The PPB (http://gdbtools.unibe.ch:8080/PPB) returns ranked lists of targets, based on similarity comparisons with multiple descriptors to known ligands for individual targets. Uniquely, in addition to the usual SMILES and sketcher entry modalities, it allows one to extract bound ligands from a PDB entry (these are available from http://ligand-expo.rcsb.org, as of the time of writing) for analysis. The descriptors are calculated for all known ligands from ChEMBL. For each query molecule, the smallest Manhattan distance between the compound and the members of each target-class ligand set is computed. A *p*-value is then calculated, based on a prior computation wherein random members of the ZINC dataset were placed in each descriptor space, and their distance-to-targets calculated, up to 1 million per target.

3.6 HitPick

Hitpick (http://mips.helmholtz-muenchen.de/hitpick) returns a list of targets, the length of which is specified by the user-supplied precision cut-off criterion. Its approach, based on the Laplacian-modified naïve Bayes classifier as applied to target prediction [76], is to construct these models for each of the targets in the STITCH database, and then to find the closest molecule in its ligand set (as discussed in Sect. 4), according to the *Tc* of their FCFP fingerprints (another version of the extended-connectivity fingerprints, defined in the original paper [70]), and make predictions against the annotated targets.

3.7 Prediction of Activity Spectra for Biologically Active Substances (PASS)

Very briefly, we will discuss the PASS algorithm, which has historically been of considerable interest. While it is still available online (http://www.way2drug.com/PASSonline), it has gone through many revisions since its initial publication in 1996, without adequate discussion of the changes made to enable proper characterization. Initially, the algorithm used the SSFN [89] descriptors, later replaced by the MNA [82] fingerprints developed in the same lab, which describe the local environment of atoms in an iterative process not entirely dissimilar to that of the Morgan-type fingerprints. These were then assessed via a Bayesian-learning algorithm for each target, which are a mixture of protein and phenotypic endpoints, which are returned with their associated class probabilities.

4 Bayesian Inference

Given the popularity of the Laplacian-modified naïve Bayes classifier in chemoinformatics in general [90, 91], and in target prediction in particular [90, 92], we will discuss the properties of this sort of classification algorithm here, after providing a brief introduction to the Bayesian model, and the naïve Bayes assumption itself. Statistics, in general, is divided into two schools of thought, frequentist and Bayesian. Essentially they differ in their model of how the world works. To understand the differences between them, it is perhaps best to start with a trivial example: Given that the author has lost their keys in the house before, and they have, on every previous occasion, found them in the refrigerator, where should they begin looking when their keys go astray?

A frequentist might say that losing one's keys is a random process, and therefore they will be located in a place which obeys some simple probability distribution, conditional on underlying variables. That is, a frequentist a priori concept of the world is a simple one. In this case, every equally sized room in the house would be an equally good prospect.

A Bayesian might say that, while there is certainly an underlying distribution of probability, and that it could indeed be a uniform or normal distribution, that this is only a reasonable assumption if they had never lost their keys before. The Bayesian wishes to update their fundamental probability model of the world (or of key-losing, to be specific) to make a better guess about their location, by combining the probability model of the frequentist with additional information on where the keys had previously been found. Less informally, Bayes' theorem is as such:

$$P(C|D) = \frac{P(D|C) \times P(C)}{P(D)} \tag{7}$$

Essentially, the a posteriori likelihood of our class label, given our data, is a function of how likely the data is given the a priori probability of that class occurring, and the data density distribution (Eq. 7). If you consider this written out according to the chain rule, it becomes computationally prohibitive to calculate rather rapidly. So, from there, we move to the Naïve Bayes assumption:

$$\frac{P(C|D)}{P(C)} = \frac{P(D_1|C) \times P(D_2|C), \ldots, P(D_n|C)}{P(D_1, D_2, \ldots, D_n)} \tag{8}$$

Equation (8) is a lot easier to evaluate, as we do not have to consider all of the proceeding variables at each step. Each variable is independent of all others, and contributes individually to our eventual classification *pseudo*-probability.

The Laplacian-modified Naïve Bayes learners introduce a further speed-up: primarily, their innovation is to only consider the presence of features, rather than also considering their absence. It also makes some clever guesses/heuristic assumptions about the underlying distributions. However, this is not without its problems, on a probabilistic basis: there is little evidence that the contribution of the empty vectors to the class probability should be zero, and, in addition, the manner in which it is commonly applied considers all class-prediction problems to be equally hard [76].

Given a descriptor vector encoding some information about a molecule or set of molecules, we can use the Naïve Bayes and Laplacian-modified Naïve Bayes algorithms to make predictions for a given descriptor vector corresponding to a target or set of targets. A similar method to that employed in HitPick, without employing the Laplacian modification, has been employed extensively in ligand–target prediction [93].

5 Target-Based In Silico Prediction

As we see from the ligand-based target prediction strategies, an appreciation of the underlying ligand structure–protein structure relationship allows for the generation of sensible predictions. The methods above, especially SPiDER and PPB, touch on the notion of pharmacophores, which we mentioned previously. This is a logical point of transition to our next major topic: how to move from comparing ligands to cross-annotate their target activities, to considering how their on-target activity is established in the first instance.

Dealing with target prediction from the target side requires an appreciation of the complexity of the target landscape, and a working definition of what we mean by "a target". This has been introduced, briefly, above, but now we reach a more fundamental treatment of the problem. As implied earlier in the discussion of the manner in which variability is achieved with limited biological tools, protein translation, functionalization, and localization, among other factors, complicate the business of target definition [24, 94]. Taking a hypothetical target "T" as a subject for discussion, we will discuss the most common modes of variation. From the genetic perspective, one must consider the impact of environmentally and contextually dependent transcription in fluctuating levels of T over time. As well as transcriptional variation, alterations in the transcribed gene constructs themselves can play a role in this process of diversification. Alternative splicing, the dominant form of which in eukaryotes is a skipping of exons from the pre-mRNA read [95], allows for the production of dissimilar proteins from a single gene. This is not an uncommon issue: it is estimated that almost 95% of the multiexonic human genes exhibit splice variation [96]. Both of these modalities enable functional diversification, through carefully coordinated transcription, translation, and posttranslational modifications. Of course, one must also consider the alteration in protein functions brought about more directly through the process of mutation. Efforts have been made to quantify the involvement of individual somatic mutations in the phenotypic response of cancer cell lines to known and novel chemotherapeutic agents, with considerable success [26], as well as more general, genome-wide association studies, which look at outcomes for various single-nucleotide polymorphisms and other mutations [97–99].

At the protein level, confounding factors are the presence (or otherwise) of posttranslational modifications which significantly affect their activity and localization [100–102], the production of complexity through the combination of multiple domains [103–105], the observed promiscuity of fold families [103, 106, 107], and, dominating the rest, the fundamental conformational flexibility of proteins. Here, we will briefly introduce and discuss the controversy around two competing hypotheses for our ligand–target interactions, and the state of that debate as it stands. For a more complete review, Hammes et al. [108] and Csermeley et al. [109] are recommended.

Induced-fit, a model following on from the classical lock-and-key model of Fischer [110], states that enzymes require a precise arrangement of their catalytic groups to achieve their intended purpose, and that this spatial arrangement is achieved due to the properties of the substrate, which induces a conformational shift to bring these groups together [111, 112]. Both the induced-fit and conformational selection models are distinguished from Fisher's work by an appreciation of protein dynamics, although the latter takes a different approach. With the conformational selection model, protein dynamics are viewed as an intrinsic part of their complex nature and their interactions with the solvent bulk they sit in [113]. On this basis, and referring back to our example, protein T would be present in a wide variety of conformations, sampling from the breadth of its energy surface, among which would be a relatively few which had adopted the "active" conformation, and into which a waiting ligand might bind. The stabilization of this conformer in the presence of the ligand promotes a shift in the energy landscape owing to the Gibbs' free energy shift [114–116], increasing the proportion of this active form. Syncretic approaches have also been proposed, wherein induced-fit provides the final push to a privileged conformational state [117]. Current opinion lies mainly on the side of conformational selection [108, 115, 118–120], given that it explains kinetic and NMR relaxation data obtained for these systems in a more satisfactory manner [121], although some argue that induced-fit takes over at higher ligand concentrations [122–124] either in its own right, or as a subset of the parameters which determine the conformational ensemble [117, 122]. Regardless of the dominant mode, it suffices for our purposes to identify bound-conformations, with the caveat that often a multiplicity of ligand-bound protein conformations have been identified for a single protein [125–129].

As one can see from this brief discussion, the definition of a target as being "a protein involved in regulating a biochemical pathway connected to a disease state" is somewhat insufficient. In fact, taking into account an individual's entire panel of mutations, transcriptional and posttranscriptional modifications, we are likely to end up with more targets than patients [94], necessitating some sort of boundary to our subdivisions. As per the ligand-binding section, the aim of this chapter is to introduce the basic methods employed, and the manner in which they might be applied.

5.1 Binding Site Similarity

Proteins make use of modularity to enable multifunctionalization. This is both a bane and a boon to our efforts; it complicates efforts at target prediction and pharmacology rationalization, but also provides the means to employ exogenous substances to interfere with endogenous biochemistry in the first instance [130]. We can also take advantage of this property, along with other aspects of structural similarity, to predict to which proteins, or regions of proteins, putative ligands will bind. It has been observed that although sequence- and structure-based similarity, when the latter is based on alpha-carbon backbone overlap, are useful [131], they fail to capture the whole gamut of substrate recognition [132]. In general, these binding-site similarity methods take a simplified representation of a known binding pocket, the level of abstraction depending on the algorithm, and then proceed to align and score these representations [133, 134]. Given the profusion of these resources, readers are referred to Chaudhari et al. for a complete enumeration [11], and to Pérot et al. for discussion of the algorithms not mentioned here [134].

5.1.1 CavBase

CavBase [135] (http://relibase.ccdc.cam.ac.uk) adopts an approach used in generating representations of ligands and inverts it: It maps chemical properties to the amino acid residues, and then onto the protein surface. From here, it builds up a graph model of how these properties are distributed in space. These graphs are then compared with a clique-detection algorithm [136] to find the maximum overlap between site models. Further optimization is performed based on aligning the regions of the protein surface associated with each feature.

5.1.2 SuMo

Surfing the Molecules (SuMo) [137, 138] (http://sumo-pbil.ibcp.fr) begins by calculating the local density for each atom in an identified binding site. This measure is the atomic weight of all atoms within a radius *r* of the start atom, divided by *r*. In addition, a local center of mass is computed (the center of mass of the aforementioned sphere), and assigned as a vector directed toward the protein interior. A similar procedure is carried out for different "chemical groups," which are, loosely speaking, amino acids grouped by physicochemical nature. For the matching process, both of these sets of triangles, and associated vectors, are used in a simple, rule-based geometric system.

5.1.3 IsoMIF

IsoMIF [139, 140] (http://bcb.med.usherbrooke.ca/imf) aims to compare the regions in binding sites in which the binding-critical events take place, which they term a molecular interaction field (MIF). With regard to the characteristics of this field, it is parameterized on physicochemical functionalities, some of which are modeled as exponentially decaying relationships. Essentially, the binding cavity is filled with probes, for each of which a molecular interaction-likeness is calculated. As per CavBase, it makes use of the Bron–Kerbosch clique detection algorithm to find the maximum common subgraph when in query mode, which are then scored with a Tanimoto coefficient based on node-matching between the graphs.

5.1.4 PocketMatch

PocketMatch [141] (http://proline.physics.iisc.ernet.in/pocketmatch) takes a somewhat different approach. For binding sites, it takes atoms within 4 Å of a crystallographically solved protein–ligand structure. To encode the shape of this local environment, amino acids are represented as a combination of the location of their alpha- and beta-carbons, and a measure of the average position of the side-chain atoms. The distances between all pairs of atoms on that local surface are calculated, and the atoms are grouped according to the characteristics of their amino-acid parents (user-defined groupings are also accepted), and are then grouped and binned into a matrix representing the interpoint and intergroup distances. When querying these models, a simple greedy alignment is performed to assess the similarity of the matrices, and then scored based on the average of the matching elements over all elements in the larger of the two matrices.

5.1.5 PARIS

PARIS [142] (http://cbio.ensmp.fr/paris) is based on the principle of convolution-kernel matching. Here, we define a pocket in terms of a cloud of labelled-atoms, and perform a fuzzier version of the Tanimoto alignments described above, the underlying principle being that a functionality provided by one amino acid may reasonably be performed from a different position in another protein. This is performed using a modified Gaussian convolution kernel, with tunable sensitivity to the displacement of atoms in space, and with the capability of handling rotation and translation of the binding pockets. This requires an alignment of the point clouds, a hard computational problem, which the authors make easier by colocating the geometric centres of each, and then rotating so their principle axes are aligned, before beginning the expensive alignment approach. With regard to the labels, they may be considered as an optional kernel, or series of kernels, during the process of optimal alignment and scoring. While a website is provided, it is only for software deposition purposes: using the software requires some (minimal) usage of the command-line interface.

5.2 Inverse Pharmacophore Searching

The first of these, the inverse pharmacophore search approach, is essentially the reverse of the methods mentioned above: instead of identifying potentially important patterns of features in a set of ligands, we identify theoretical binding interactions in a pocket and look for ligands whose features match, or partially match, that description. This method can be used in virtual screening as well, as discussed earlier. To avoid the computational complexity of flexible alignment of a given ligand to target pharmacophore model, many of these methods instead take a wide sampling of the conformational space of the ligand, and try these different conformers against their sets of pharmacophore models. The size of this problem is still relatively manageable compared to the searching of millions of ligands against one target, as in virtual screening.

5.2.1 PharmMapper

PharmMapper [143] (http:// lilab.ecust.edu.cn/pharmmapper) allows the user to provide a standard ligand file (SDF, mol, mol2) to their server for analysis. If necessary, a 3D structure for the molecule is generated, and a set of conformers is produced. In its most-recent incarnation [144, 145], PharmMapper extracts nonredundant protein–ligand complexes from the PDB using the BLASTClust BLAST-and-clustering algorithm [146], with a sequence identity cut-off of 90%. The structure with the highest resolution from each cluster is chosen, going forward. The pockets are then defined and scored by means of their software CAVITY [147], and a scoring model built for determining "ligandability" and "druggability." The software POCKET [148] is then employed to pharmacophorically categorize these extracted pockets. The features that PharmMapper searches for in its analysis are the presence of hydrophobic groups, positive- and negative-charges, hydrogen-bond acceptors and donors, and, in addition, spheres of exclusion which demarcate zones we should like to avoid. A considerable improvement from the previous iteration of the software is in its scoring of potential interactions, with the adoption of a system which handles the over-emphasis of high scoring results in situations where the pharmacophore model has many features (*see* **Note 3**).

5.3 Inverse Docking

In a related fashion to inverse pharmacophore searching, it is possible to encode this information on pharmacological interactions through consideration of the energetics of binding of a ligand and potential target. This is very similar to the related virtual screening method, with the inversion of the numerical relationship between ligands and targets. These methods have similar strengths and weaknesses to their virtual screening brethren. Compared to the pharmacophoric methods, they are somewhat more difficult to interpret, especially with regard to scoring [149], but give a more "realistic" interpretation of the hypothetical protein–ligand dynamics. The two other main issues are with protein structural availability, and computational complexity. As more and more data become available [150, 151], the former problem shrinks, while the latter grows. To that end, efforts have been made to parallelize docking algorithms, enabling their utilization in large cluster setups [152, 153], and to produce more efficient versions of the existing algorithms [154].

5.3.1 INVDock

INVDock [155] is the prototypical member of this field of tailored docking software tools, based on a divide-and-conquer approach, wherein target sites are subdivided into overlapping regions and the molecule docked against these. It is well suited for small ligands, but does not have a Web server.

5.3.2 TarFisDock

TarFisDock [156] (http://dddc.ac.cn/tarfisdock) uses the Potential Drug Target Database [157] (PDTD) of curated drug–target interaction sites, in combination with a flexible-ligand docking strategy, to provide a ranked list of putative targets for the user-provided ligand sketch or file, based on their DOCK 6 [158] scores.

5.3.3 IdTarget

IdTarget [159] (http://idtarget.rcas.sinica.edu.tw) allows for the submission of ligands in a variety of formats, with user-specified protonation states and charge-models (defaults available). Its novelty is that, in one of its modes, it is based on a divide-and-conquer approach, wherein target proteins are subdivided into overlapping regions and the molecule docked against these. Therefore, in contrast to the other resources, it docks (using MEDOCK [160]) against the entire target protein, with the potential for identifying allosteric binding sites, if they are exposed in the crystal structure. To enable the use of this computationally intensive procedure, the authors of idTarget adopt what they describe as a "contraction-and-expansion strategy," essentially a breadth-first traversal of a hierarchical-clustering tree of the scPDB [161], clustered at 40% sequence identity, and choosing a subset of the leaves attached to each penultimate node. For each subset, if their docking results seem promising, all of the leaves attached to that parent are docked against. Alternatively, idTarget has a "fast" mode, wherein its operation is similar to TarFisDock.

5.3.4 DRAR-CPI

DRAR-CPI [162] (http://cpi.bio-x.cn/drar) allows for the submission of molecules using a sketcher, as SMILES, or as mol2 files. Its main innovation is in its construction of a ligand-library vs. target-library docking matrix, which allows for the calculation of *z*-values, and the subsequent assessment of docking results (scored with DOCK 6) for novel molecules within a statistical framework. It provides ranked target-lists, and also adverse-reactions noted for ligands from its library which are observed to have similar on-target profiles. Of course, the statistical model remains fundamentally dependent on the quality of the underlying docking scoring functions.

5.4 Molecular Dynamics: Back to "Basics"

One issue with docking-based approaches is the necessity of having a "good" structure of the target, ideally a set, with and without ligands. Given that the majority of ligands are buried upon complexation with a protein [163], strategies which can treat the flexibility of the protein target itself are essential to having a clear view of the underlying interaction. To that end, molecular dynamics solvers based on Newtonian mechanics remain integral to the latter-stages of the virtual screening pathway. There are many good methods for carrying out molecular dynamics and free-energy perturbation, with much of the software is free to academic users [164, 165], and, increasingly, made available through portals which greatly simplify the process [166, 167]. To date, this has been a relatively limited [168–170] field of study with regard to target prediction, likely owing to the computational complexity. Considerable efforts are being made to enable molecular dynamics simulations at a larger scale [171], so perhaps it will become a larger part of the target prediction toolset in years to come.

6 Network Pharmacology

The relative dearth of drug–target interaction data means that our current models of drug polypharmacology, are, while useful, somewhat lacking. Mestres et al. [172] adopted graph-theoretical approaches under estimate the vulnerability and variability of existing networks under addition/reannotation of drug–target information. They observe that the neatly separated modules observed from constructing such a network rapidly accumulate into one large, highly connected module (see [169] Fig. 1 for visualization) with the addition of alternative sources of drug–target data, approximating the degree of interconnectedness observed when randomly generating a network with the same number of nodes and edges but no interaction data. There is also further analysis included therein of the interconnectedness of the graphs formed when considering target families, which is recommended for motivational reading.

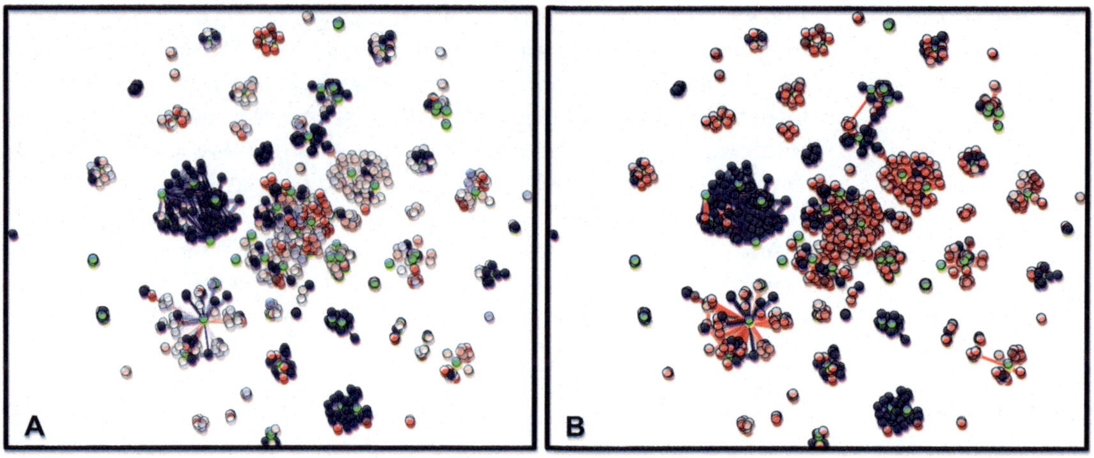

Fig. 1 Network generated from the IUPHAR/BPS Guide to Pharmacology (2017) dataset. Here, nodes in the blue–white–red spectrum indicate ligands from that set with lower-than to greater-than median Tanimoto coefficient similarity with query ligands rofecoxib (**a**) and cerivastatin (**b**). For ease of visualization, the median was adjusted for each graph. Median was 0.25 for rofecoxib, and 0.34 for cerivastatin. Green nodes represent the targets from the set. Edges between ligands and nodes represent their pIC50 values. Here the network is laid out using a Newtonian graph layout algorithm, which clusters the space by the shared affinity of different ligands. It can be seen that the more similar compounds to each ligand are unevenly distributed across the target space, which complicates target prediction approaches relying solely on substructure fingerprints

A study by the same group [24], and an excellent review by Nobeli et al. reflecting on the origins of polypharmacology in evolutionary biology [94], suggest that target and ligand promiscuity are the product of a combination of positive evolutionary pressure, a limited protein structural vocabulary (the natural amino acids, plus posttranslational modification, domain clustering [173] etc.), and the ability of regulatory systems to silence damaging mutations, "storing" the new capacities for environments in which they might be beneficial. This set of biological constraints ties in well with observed data on protein-fold level promiscuity, given that relatively small sequence variation in extended loop region can drive different binding modes [94], while maintaining similar binding sites [174]. It also, perhaps, sheds some light on why phylogenetically close proteins, for example the extended GPCR receptor family and the ion channels have such radically different network topologies, although, as previously discussed, the degree of clique formation is a function of the volume of data included, and of its quality.

From here we move to a discussion of network-based methods for target prediction. Having moved from ligands with known activities to target-based methods with approximable physical and chemical interactions, we will proceed to a representation which is, simultaneously, both more abstract, and potentially more useful.

The nascent field of network biology, and, indeed, that of network pharmacology, makes use of the astounding growth in the breadth and depth of data piling up on topics biological, chemical, and everything between. It aims to take advantage of the synergy between these largely separate data sources, and the ongoing efforts (such as SPARQL [175, 176]) to provide reliable biochemical entity resolution, to construct networks which reflect the complex interplay between some or all of the constituent systems in living organisms. Semantic Link Association Prediction [177] (SLAP) collates data on targets, such as their gene ontology (GO) [178, 179] codes, tissue data, associated disease states, and on ligands known to interact with them [20, 180–182]. These were aggregated into a semantically linked network [177, 183, 184], and the shortest paths [185] between nodes were calculated, where the number of edges in that path is less than three, for one-hundred-thousand drug pairs from DrugBank. Once this was accomplished, a statistical model was built relating the number, length, and makeup of each of these paths, allowing for weighting of the edges in a manner related to the strength of the overall commonality of the two compounds. The complexity of this information network adds rigor, in that multiple pieces of evidence are weighed to provide a probabilistic assessment of the likelihood of a target assignation, but is also likely to introduce some unusual behaviors owing to its differential weighting of path types, and the concomitant risk of overweighing noisy data. Overall, SLAP and other programs in its domain [186–188] perform well, within a relatively narrow domain of applicability as compared to some of the methods discussed previously. It performs well in clustering compounds, identifying likely side effects, and has been discussed for use in drug repurposing efforts [189].

Campillos et al. [190] showed that pairing known ligands based on observed side effects could be combined with known target data to cross-annotate target assignations [191]. The overall success rate of their predictions was around 55%. Extending this method to more generic applications is somewhat difficult, as, in its current form, it depends on the subject compound being marketed, and on having data from the postmarket surveillance. Given that structurally similar ligands have been observed to bind to different sets of the proteins within a pathway, it is perhaps unsurprising that this method had some success, although, of course, it could only predict target assignations that had previously been made to another of the ligands. A similar approach was undertaken with information taken from Japanese pharmaceutical packaging inserts, extended with genomic data on the target proteins [191]. Another work by the same group took advantage of information on adverse event reporting to extend annotations in an interesting fashion [192]. These resources are now available on the DINIES server (http://www.genome.jp/tools/dinies) [193].

To enable precisely this sort of inferential approach, resources such as PROMISCUOUS [22] (http://bioinformatics.charite.de/promiscuous) have been developed, which integrate data on known targets, side effects, KEGG pathways, and drugs to provide a useful resource. Similar approaches have been taken by other groups, extended by the integration of experimental protein–protein interaction data [23, 191, 194, 195], or focusing on a specific disease or disease state. With PROMISCUOUS, an end-user can provide target- or ligand-information, and look through the constructed graph related the aforementioned properties. In addition, and especially useful for our purposes, is the ability to search for compounds which share some of the same targets or side effect activities. However, the ligand entries are limited to known ligands, so its utility is primarily in exploring the annotation space, although one might perform a simple similarity search, using an external program, to find the most similar ligand in their network.

To demonstrate what this might look like in practice, we have prepared a visualization of a toy network (Fig. 1) constructed using data from the IUPHAR/BPS guide to pharmacology [196] database. This is a relatively small but extensively curated collection of pharmacological information on drugs and targets. Using this data, we built a graph (using the networkx, numpy, and rdkit [197–200] packages in Python2.7 [201]) from the interaction data for ligands and GPCR targets (3924 and 182, respectively). In this graph, nodes represent entities, namely ligands and associated targets, and edges represent affinities (We used the median pIC_{50} values provided.). After exporting this data, we used mechanistic network layout algorithms, provided in Gephi 0.9 [202, 203] to simulate the behavior of the graph as a Newtonian system, with the nodes as balls and the edges as springs. This allows the software to lay the graph out according to the assigned importance of each edge. The resulting image shows an uneven distribution of similarity across the network of receptor-associated ligands. We can see the clustering of receptors into tight cliques based on joint affinity with multiple ligands, and also the "orphaning" of some receptors, even in this highly curated dataset.

These sorts of efforts, to link different search methodologies, are often necessary sooner or later in chemoinformatics. However, as the meta-databases, of which PROMISCUOUS is an example, mature, it can be assumed that the resources themselves will become more user-friendly, in a similar fashion to the evolution of the UniProtKB [204] and other aggregative bioinformatics toolsets, while developing their capabilities with prospective applications. Going some way toward that, canSAR [6] (http://cansar.icr.ac.uk), a cancer-focused data-aggregation portal, provides a user-friendly interface allowing one to search their comprehensively annotated protein-interaction networks. They integrate pharmacological data to provide an assessment of the druggability of all of the "nearby" proteins in the cellular network. Importantly, it allows for the clustering (according to Murcko-scaffolds [205]) of all compounds for which it has data, extracting entries from that cluster, and then cross-referencing these with protein–protein interaction databases [206] and to COSMIC [207] and EnsEMBL [208] databases, providing feedback on known shared targets, the role of somatic mutations in modulating their activity, and also helping to identify cell-lines which might be of especial interest for phenotypic testing of proposed ligands. While canSAR is a very useful tool, it seems to be impossible to use its Murcko-scaffold functionality, for instance, to retrieve salient data for a user-provided ligand.

To extend these sorts of resources into an arena in which they might provide information on a more diverse field of targets, and for a broader range of the chemical space, approaches have been made toward integrating insights from other fields. The combination of network pharmacology and structural biology has facilitated the development of a new understanding of the complex polypharmacology of known ligands. Insights gained from this approach have been used in the construction of structural systems pharmacology approaches for the discovery of off-target effects in the activity of nelfinavir, an HIV protease inhibitor. To accomplish this, the authors searched through the structural genome [209] to find similar pockets to the known binding site of nelfinavir (PDB-ID: 1OHR) [210] using the SMAP algorithm [211], docking nelfinavir against those determined to have statistically significant similarity, filtering the results based on scoring to enable a molecular dynamics approach, and then constructing a network relating the predicted activities, which were then tested experimentally [212]. As noted by the authors, a weakness in this approach is its treatment of a biological organism as an essentially static entity. They are not alone in that regard, given the very limited dynamic behaviors considered in traditional ligand- or structure-based efforts in target fishing, but some work has been done toward integrating insights from the study of protein and cellular dynamics [213–215].

7 Systems Pharmacology

The field of systems biology aims to describe, and quantify, the manifold interactions taking place in a living organism to enable its simulation and analysis. To accomplish this, it relies on complex networks of molecular pathways, and parameterized equations which describe individual interaction kinetics, the higher-order behavior of which approximates a cell, an organ, an organism [216, 217]. These networks, constructed from the vast array of omics data now available, and validated experimentally, demonstrate complex emergent behaviors, simulating cross talk between adjacent networks, and high-level regulatory behaviors. Given the rapid increase in their quality, they have been used to simulate the metabolic effects of various mutations in man, with considerable success [218, 219]. An overview of the available Web-based systems biology toolsets is provided in Klipp et al. (2010), for interested readers [189].

These models can be subdivided into the network models, which have largely focused on identifying pharmacologically relevant targets and exploring the upstream and downstream consequences of their modulation [220–222], or "true" systems biology models, such as those stored in the EMBL BioModels [223] (https://ebi.ac.uk/biomodels-main) repository, wherein the flow of flux across the network, based on systems of stoichiometric, or, in the presence of information kinetics, ordinary differential equations (ODEs), is used to build a system for exploring the effects of individual modulations on the metabolic state of the whole cell. Flux-balance analysis (FBA), based on the stoichiometric equations, is employed in the absence of accurate kinetic information, and seeks to account for the total change in metabolite concentrations across the model network [224]. Alternatively, with kinetic information, models can be created which simulate the behavior of pathways within a model organism. This allows for estimations of the effective concentration of the compound required to induce inhibition, and has been effectively applied, in combination with FBA, in predicting the dose-response curve of an inhibitor on the growth of *Mycobacterium tuberculosis* [225].

To date, much of the work in systems pharmacology has been in the identification of novel targets which might be of interest for pharmaceutical development, rather than in predicting targets of known compounds [226]. By and large, these have been generated by combining high-throughput "omics"- data to sketch out, and then parameterize, the partial or full metabolic systems of an organism. Some novel approaches have been made to extend this genomic and functional data with higher-order considerations of how a ligand might interact with such a network. Progress in analyzing the protein fold diversity of an organism's metabolic system, namely the pharmaceutically irrelevant *Thermatoga maritima*, succeeded in demonstrating the utility of combining structural data, from resolved structures and homology-modeling, with more traditional sources for systems biology [227]. This analysis demonstrated that fold-diversity is both relatively limited, and also unevenly distributed across metabolic subsystems, lending credence to the observed effect size for greater-than-random affinity of ligands for downstream proteins in the same pathway as the known target. Interestingly, they also noted that most of the fold-level structural diversity for proteins in *T. maritima* was in the survival-critical subset, suggesting a route for the observed promiscuity of ligands for cross-family protein interactions, and also a potentially privileged subset of targets. Given the relationship between fold-similarity and pharmacological promiscuity, this concept might enable the development, and target-characterization, of intentionally promiscuous [228–230] drugs with an activity mode similar to that of nelfinavir, discussed above.

A related approach has been used to identify "safe" targets in parasitic species, which lack orthologs in humans, by means of aligning the metabolic maps of multiple species. Perhaps the most persuasive example to date lies in the constellation of work evolving around the *M. tuberculosis* pathogenic organism. Several models have been published, which have successfully emulated growth and inhibition behaviors under physiological conditions [225, 231]. These have been extended with information on known ligand–protein interactions, revealing insights into patterns of resistance development [232] and deployment of natural defensive mechanisms [233], and also in assessing the druggability of other enzymes present [234].

With regard to applications in human pharmacology-related issues, these systems have shown considerable promise [215, 235]. Interesting applications have been in the deconvolution of complex transcriptomic data, to predict which proteins in a molecular network are modulated to achieve synergistic inhibition of downstream protein–protein interactions in cancer models and promote repositioning efforts [236–239], simulating the pharmacology of Chinese traditional medicine [240], and work in the realm of pharmacokinetics using the most complete model of the human metabolism to date, Recon2 [215, 241] to assess the impact of metabolite-likeness in active transporter uptake of drugs.

Existing methods might profitably be employed in combination with existing ligand- or structure-based target prediction methods, in that, returning to our dogma, structural and functional similarity are highly correlated. This would allow extension of newly gleaned, probabilistic insights for one molecule to another, or, at the very least, a description of the theoretical targets for a class of molecules. These sets of targets could then be mapped back onto the metabolic network, or perform additional in silico and in vitro experiments to reinforce the predictions.

To the best of our knowledge, no comprehensive mapping of the systems biology methods for target predictions in known drugs has been carried out to characterize the distribution of predicted activity, nor have the predictions, for example, been mapped to the human metabolome. These might prove fruitful sources of further information, and help to predict the number of hidden targets, in combination with metabolomics efforts, by locating unexplained flux across the network. Chemogenomics efforts are predicated on the basis that the compounds targets are well annotated. If we could extend our knowledge of new and existing chemogenomic screening libraries, it could serve to increase useful data throughput significantly, by means of more formally defining what we learn from each experiment. This virtuous cycle of hypothesis generation and model enrichment stands to benefit the field of chemogenomics, by extending the field of interactions noted for each compounds, systems biology by the provision of more usefully annotated test compounds for generated hypotheses, and pharmacology, by better exploring the interactions between our compounds and the whole-organism we aim to treat or terminate.

8 Deep Learning Methods in Target Prediction

To date, most machine-learning applications in the field have been "shallow," that is, utilizing a single layer of feature transformation to achieve their goals. This class of algorithms includes many of those previously mentioned in this text, and a few we have not touched on; *k*-means clustering, support vector machines, linear discriminant analysis, logistic and kernel regression, and simple neural nets, just to mention the most prominent representatives [242, 243]. The successes of these methods in ligand-based target prediction [45, 244, 245] and lead-suggestion [245–247] are, in part, due to the work done in the field to encode useful molecular information in molecular descriptors, enabling simple architectures to make reasonable predictions for both fields [248, 249]. We accept, in the process of engineering and applying these descriptor systems, that we introduce a measure of artificiality in our depiction of the actuality of these molecules. Accordingly, novel methods based on learning directly from molecular graphs and other, physically oriented, models of the complex molecular objects have been proposed which remove some of this input-level abstraction. This, however, requires a more advanced machine-learning methodology, as the data are much less amenable to producing useful output with simple, single-layer transformation methods.

After a fashion, we may take some heart from this; after all, it reiterates that considerable progress has been made toward capturing the essential information that describes the relationship between a molecule and a potential proteinaceous target. However, the vastness of "descriptor space," and the technical and philosophical debates about the importance of different descriptor families to on-target activity, suggest that the way forward may be illuminated by shedding ourselves of these manually constructed abstractions, and instead focusing on harnessing the power of machine-learning methods to identify patterns in datasets too large for human pattern-recognition to produce useful insights. To that end, the emerging, or perhaps newly emerged, field of deep-learning [250, 251] research suggests a future for these sorts of low-abstraction-input learning systems. A full review of the theory behind deep learning and how that relates to its applications in pharmacology are beyond the scope of this chapter; interested readers are referred to a review by Gawehn et al. [252].

In brief, however, these deep-learning methods extend on the basic concepts of neural networks by adding more depth, i.e., further layers of "neurons," the interplay between which allows for the description of more complex features than are approachable with a single-layer network. For example, a single-layer network trained on the ECFP4 fingerprints discussed earlier, with activity against a given target as its classification, might produce some predictions based on the presence or absence of a given substructure element in an input molecule. Based on the same input and desired output, deeper, multilayer networks would be generating and searching for descriptions more resembling pharmacophoric models when making their predictions [253].

These methods have already had some applications in the field, focusing on the prediction of molecular physicochemical properties [254], on approximating expensive calculations [255], or in QSAR building [256–258] and target prediction [253, 259–261]. To date, these have largely relied on descriptors and fingerprinting methods, although some graph-theoretical approaches have been utilized [243, 249, 257, 262–267], with various levels of abstraction applied to the graphs themselves. To some extent the progenitor for modern interest in the field was the success of deep-learning methods for target prediction in a public competition run by Merck on the Kaggle machine-learning Web platform [268]. This involved predicting the affinity of hidden (from the training algorithms) compounds for a set of 15 targets of interest, with the training data being a relatively small set of compounds with known activity values for each target [256, 269]. The success of the winning team, composed mainly of computer scientists and statisticians, was accomplished with minimal preprocessing and data manipulation in a short time-span, with the focus having been very much on the specifics of the architectures employed [268, 270–272]. The resulting predictions were of a higher quality than Merck's own, in-house solution, and beat the industry standard accuracy benchmark by 17% overall. Their approach was to build an ensemble predictor, combining the output of three machine-learning strategies to form a sort of jury system, before judging which test compounds were likely to be active at which targets [256].

Since then, efforts at further scaling up the systems, in terms of data- and network-complexity, have been made [259]. These have analyzed the behavior of such algorithms with increasing input- and target-set size, used novel representations of molecules, and refined the architectures of the networks themselves. Some interesting work has been done toward learning directly from graph structures [262, 273, 274] treated as simple images [257, 267], utilizing state-of-the-art methods developed in the field of machine-vision similar to those employed for object recognition in Google's image search algorithms [275, 276]. It seems likely that approaches of this sort, and other low input-feature-engineering strategies, are likely to produce interesting and fruitful work in the near future. Other, related, domains of application with demonstrated feasibility are in improving structure-based methods for drug discovery by the creation of machine-learning driven empirical scoring systems [277, 278], and in analyzing the aforementioned graph-based drug–target networks [279]. Given the coupling of increasingly large, complex, and interconnected data with the radical evolution in our analytical capabilities, it seems appropriate to say that this is the era when "big data" meets "big pharma."

9 Notes

1. In one study, carried out by the team behind the SEA algorithm, a straightforward application of the program to making target predictions to be employed in a zebrafish phenotypic screen reported eleven out of twenty compounds which were active at the predicted targets, for which twenty-two of thirty-one predicted targets were valid [280]. Other works employing SEA are a large-scale prediction of activity of marketed drugs against a set of "side-effect" targets maintained by Novartis. For the 656 molecules, and 73 targets, approximately 1200 predictions were made which were not annotated in ChEMBL, of which 300 were confirmed trivially with an external database. For the remainder, all barring 200 were tested, giving a total set of approximately 1000 predictions, of which 50% were experimentally validated [25]. In addition, a guilt-by-association model was built, which enables cross-annotation of activity at these targets by ligands which share activity at another. Similar networks of activity have been built using the SEA approach for target prediction, with mixed success [281].
2. SPiDER has been experimentally validated through prospective studies wherein LOPAC (Sigma-Aldrich) compounds were mapped with predictions, a subset of which (those which were generated by compounds with ≤ 0.2 Tanimoto coefficient to the respective query molecule) were identified for further testing. Of these, the highest-confidence predictions ($p < 0.001$) included an annotation for monoamine oxygenase inhibition for fluoxetine, and sodium-channel activity for fenofibrate, both of which were experimentally confirmed. In addition, SPiDER was employed to predict the targets of a series of de novo designed compounds [282–285] based on amprenavir, correctly asserting that both analogs, and the parent compound, would have specific, different, activities at bradykinin-1 (B_1), neurokinin-1 (NK1), and vanilloid-1 (TRPV1) receptors.

3. While a useful tool, PharmMapper's main drawback is in its time complexity, with a search of all ligandable binding-sites taking up to 24 h. Its usefulness could be much improved with the addition of a less-linear search structure, such as the KDB-tree [286] implemented in the open-source Pharmer software, which, to the authors' knowledge, has not been utilized in reverse-pharmacophore searches to date. It has, however, been used to considerable effect in VS approaches, such as in ZINCPharmer [287] (http://zincpharmer.csb.pitt.edu).

Acknowledgment

This research was supported by the European Union Framework Programme for Research and Innovation (Horizon 2020, Marie Skłodowska-Curie ITN grant number 675555 'AEGIS'), and the OPO-Foundation Zurich.

References

1. Berman HM (2000) The protein data bank. Nucleic Acids Res 28:235–242
2. Piovesan D, Luigi Martelli P, Fariselli P et al (2011) BAR-PLUS: the bologna annotation resource plus for functional and structural annotation of protein sequences. Nucleic Acids Res 39:197–202
3. Villoutreix BO, Lagorce D, Labbé CM et al (2013) One hundred thousand mouse clicks down the road: Selected online resources supporting drug discovery collected over a decade. Drug Discov Today 18:1081–1089
4. Kamburov A, Wierling C, Lehrach H, Herwig R (2009) ConsensusPathDB–a database for integrating human functional interaction networks. Nucleic Acids Res 37:623–628
5. Ogata H, Goto S, Sato K et al (1999) KEGG: kyoto encyclopedia of genes and genomes. Nucleic Acids Res 27:29–34
6. Tym JE, Mitsopoulos C, Coker EA et al (2016) canSAR: An updated cancer research and drug discovery knowledgebase. Nucleic Acids Res 44:938–943
7. Chen X, Ji ZL, Chen YZ (2002) TTD: therapeutic target database. Nucleic Acids Res 30:412–415
8. Loging W, Rodriguez-Esteban R, Hill J et al (2011) Cheminformatic/bioinformatic analysis of large corporate databases: Application to drug repurposing. Drug Discov Today Ther Strateg 8:109–116
9. Bickle M (2010) The beautiful cell: High-content screening in drug discovery. Anal Bioanal Chem 398:219–226
10. Roth BL, Sheffler DJ, Kroeze WK (2004) Magic shotguns versus magic bullets: selectively non-selective drugs for mood disorders and schizophrenia. Nat Rev Drug Discov 3:353–359
11. Chaudhari R, Tan Z, Zhang S (2017) Overview of drug polypharmacology and multitargeted molecular design. In: Davis A, Edge C (eds) In silico drug discovery tools, Comprehensive medicinal chemistry III, vol 2. Elsevier, Amsterdam, pp 259–275
12. Hu Y, Bajorath J (2013) Compound promiscuity: What can we learn from current data? Drug Discov Today 18:644–650
13. Frantz S (2005) Drug discovery: playing dirty. Nature 437:942–943
14. Mencher SK, Wang LG (2005) Promiscuous drugs compared to selective drugs (promiscuity can be a virtue). BMC Clin Pharmacol 5:3
15. Keiser MJ, Irwin JJ, Shoichet BK (2010) The chemical basis of pharmacology. Biochemistry 49:10267–10276
16. Hopkins AL, Groom CR (2002) The druggable genome. Nat Rev Drug Discov 1:727–730
17. Russ AP, Lampel S (2005) The druggable genome: An update. Drug Discov Today 10:1607–1610

18. Bunnage ME (2011) Getting pharmaceutical R&D back on target. Nat Chem Biol 7:335–339
19. Overington JP, Al-Lazikani B, Hopkins AL (2006) How many drug targets are there? Nat Rev Drug Discov 5:993–996
20. Wishart DS (2006) DrugBank: a comprehensive resource for *in silico* drug discovery and exploration. Nucleic Acids Res 34:668–672
21. Santos R, Ursu O, Gaulton A et al (2016) A comprehensive map of molecular drug targets. Nat Rev Drug Discov 16:19–34
22. Von Eichborn J, Murgueitio MS, Dunkel M et al (2011) PROMISCUOUS: A database for network-based drug-repositioning. Nucleic Acids Res 39:1060–1066
23. Yildirim MA, Goh K-I, Cusick ME et al (2007) Drug-target network. Nat Biotechnol 25:1119–1126
24. Jalencas X, Mestres J (2013) On the origins of drug polypharmacology. Med Chem Commun 4:80–87
25. Lounkine E, Keiser MJ, Whitebread S et al (2012) Large-scale prediction and testing of drug activity on side-effect targets. Nature 486:361–367
26. Imming P, Sinning C, Meyer A (2006) Drugs, their targets and the nature and number of drug targets. Nat Rev Drug Discov 5:821–834
27. Bunnage ME, Gilbert AM, Jones LH, Hett EC (2015) Know your target, know your molecule. Nat Chem Biol 11:368–372
28. Jordan AM, Waddell ID, Ogilvie DJ (2015) Rethinking "academic" drug discovery: The Manchester Institute perspective. Drug Discov Today 20:525–535
29. Rask-Andersen M, Almén MS, Schiöth HB (2011) Trends in the exploitation of novel drug targets. Nat Rev Drug Discov 10:579–590
30. Duan Q, Reid SP, Clark NR et al (2016) L1000CDS2: LINCS L1000 characteristic direction signatures search engine. NPJ Syst Biol Appl 2:1–12
31. Humbeck L, Koch O (2017) What can we learn from bioactivity data? Chemoinformatics tools and applications in chemical biology research. ACS Chem Biol 12:23–35
32. Koutsoukas A, Simms B, Kirchmair J et al (2011) From *in silico* target prediction to multi-target drug design: Current databases, methods and applications. J Proteome 74:2554–2574
33. Jenkins JL, Bender A, Davies JW (2006) *In silico* target fishing: Predicting biological targets from chemical structure. Drug Discov Today Technol 3:413–421
34. Peón A, Dang CC, Ballester PJ (2016) How reliable are ligand-centric methods for target fishing? Front Chem 4:15
35. Nettles JH, Jenkins JL, Bender A et al (2006) Bridging chemical and biological space: "Target fishing" using 2D and 3D molecular descriptors. J Med Chem 49:6802–6810
36. Lavecchia A, Cerchia C (2015) *In silico* methods to address polypharmacology: Current status, applications and future perspectives. Drug Discov Today 21:288–298
37. Menziani MC, Montorsi M, De Benedetti PG, Karelson M (1999) Relevance of theoretical molecular descriptors in quantitative structure-activity relationship analysis of alpha1-adrenergic receptor antagonists. Bioorg Med Chem 7:2437–2451
38. Verma J, Khedkar V, Coutinho E (2010) 3D-QSAR in drug design–a review. Curr Top Med Chem 10:95–115
39. Brogi S, Papazafiri P, Roussis V, Tafi A (2013) 3D-QSAR using pharmacophore-based alignment and virtual screening for discovery of novel MCF-7 cell line inhibitors. Eur J Med Chem 67:344–351
40. Wermuth CG, Ganellin CR, Lindberg P, Mitscher LA (1998) IUPAC recommendations: glossary of terms used in medicinal chemistry. Pure Appl Chem 70:1129–1143
41. Gfeller D, Michielin O, Zoete V (2013) Shaping the interaction landscape of bioactive molecules. Bioinformatics 29:3073–3079
42. Schuffenhauer A, Floersheim P, Acklin P, Jacoby E (2003) Similarity metrics for ligands reflecting the similarity of the target proteins. J Chem Inf Comput Sci 43:391–405
43. Bender A, Jenkins JL, Scheiber J et al (2009) How similar are similarity searching methods? A principal component analysis of molecular descriptor space. J Chem Inf Model 49:108–119
44. Willett P (2000) Chemoinformatics–similarity and diversity in chemical libraries. Curr Opin Biotechnol 11:85–88
45. Pahikkala T, Airola A, Pietilä S et al (2015) Toward more realistic drug-target interaction predictions. Brief Bioinform 16:325–337
46. Daylight Information Systems (2008) Daylight theory: fingerprints. Accessed 8 Jul 2017
47. Bajusz D, Rácz A, Héberger K (2015) Why is Tanimoto index an appropriate choice for fingerprint-based similarity calculations? J Cheminform 7:1–13

48. Baldi P, Nasr R (2010) When is chemical similarity significant? The statistical distribution of chemical similarity scores and its extreme values. J Chem Inf Model 50:1205–1222
49. Arif SM, Holliday JD, Willett P (2009) Analysis and use of fragment-occurrence data in similarity-based virtual screening. J Comput Aided Mol Des 23:655–668
50. Arif SM, Holliday JD, Willett P (2016) The use of weighted 2D fingerprints in similarity-based virtual screening. Adv Math Chem Appl Revis Ed 1:92–112
51. Willett P, Barnard JM, Downs GM (1998) Chemical similarity searching. J Chem Inf Comput Sci 38:983–996
52. Keiser MJ, Roth BL, Armbruster BN et al (2007) Relating protein pharmacology by ligand chemistry. Nat Biotechnol 25:197–206
53. Kruskal JB (1956) On the shortest spanning subtree of a graph and the traveling salesman problem. Proc Am Math Soc 7:48–48
54. Bento AP, Gaulton A, Hersey A et al (2014) The ChEMBL bioactivity database: An update. Nucleic Acids Res 42:D1083–D1090
55. Olah M, Mracec M, Ostopovici L et al (2005) WOMBAT: world of molecular bioactivity. In: Oprea TI (ed) Chemoinformatics in drug discovery. Wiley-VCH Verlag GmbH & Co. KGaA, Weinham, pp 221–239
56. Symyx MDL Technologies, Dassault Systèmes BIOVIA (2009) MDL drug data report. Accessed 6 Jul 2017
57. Inpharmatica Ltd. (2005) StARLITe. Accessed 8 Aug 2017
58. Gfeller D, Grosdidier A, Wirth M et al (2014) SwissTargetPrediction: A web server for target prediction of bioactive small molecules. Nucleic Acids Res 42:32–38
59. Armstrong MS, Finn PW, Morris GM, Richards WG (2011) Improving the accuracy of ultrafast ligand-based screening: Incorporating lipophilicity into ElectroShape as an extra dimension. J Comput Aided Mol Des 25:785–790
60. Kramer C, Kalliokoski T, Gedeck P, Vulpetti A (2012) The experimental uncertainty of heterogeneous public K i data. J Med Chem 55:5165–5173
61. Reker D, Rodrigues T, Schneider P, Schneider G (2014) Identifying the macromolecular targets of *de novo*-designed chemical entities through self-organizing map consensus. Proc Natl Acad Sci U S A 111:4067–4072
62. Schneider N, Giller S (1999) Scaffold-hopping by topological pharmacophore search: A contribution to virtual screening. Angew Chem Int Ed Engl 38:2894–2896
63. Reutlinger M, Koch CP, Reker D et al (2013) Chemically advanced template search (CATS) for scaffold-hopping and prospective target prediction for "orphan" molecules. Mol Inf 32:133–138
64. Molecular Operating Environment (MOE), Chemical Computing Group, Montreal, Canada, 2011
65. Kohonen T (1982) Self-organized formation of topologically correct feature maps. Biol Cybern 43:59–69
66. Schneider P, Tanrikulu Y, Schneider G (2009) Self-organising maps in drug discovery: Compound library design, scaffold-hopping, repurposing. Curr Med Chem 16:258–266
67. Schneider G, Tanrikulu Y, Schneider P (2009) Self-organizing molecular fingerprints: A ligand-based view on drug-like chemical space and off-target prediction. Future Med Chem 1:213–218
68. Schneider P, Schneider G (2003) Collection of bioactive reference compounds for focused library design. QSAR Comb Sci 22:713–718
69. Nickel J, Gohlke BO, Erehman J et al (2014) SuperPred: Update on drug classification and target prediction. Nucleic Acids Res 42: W26–W31
70. Rogers D, Hahn M (2010) Extended-connectivity fingerprints. J Chem Inf Model 50:742–754
71. Günther S, Kuhn M, Dunkel M et al (2008) SuperTarget and matador: resources for exploring drug-target relationships. Nucleic Acids Res 36:919–922
72. Chen X, Liu M, Gilson MK (2001) BindingDB: a web-accessible molecular recognition database. Comb Chem High Throughput Screen 4:719–725
73. Gilson MK, Liu T, Baitaluk M et al (2016) BindingDB in 2015: A public database for medicinal chemistry, computational chemistry and systems pharmacology. Nucleic Acids Res 44:D1045–D1053
74. Awale M, Reymond JL (2017) The polypharmacology browser: a web-based multi-fingerprint target prediction tool using ChEMBL bioactivity data. J Cheminform 9:1
75. Liu X, Vogt I, Haque T, Campillos M (2013) HitPick: A web server for hit identification and target prediction of chemical screenings. Bioinformatics 29:1910–1912
76. Mussa HY, Mitchell JBO, Glen RC (2013) Full "laplacianised" posterior naive Bayesian algorithm. J Cheminform 5:37
77. Kuhn M, Szklarczyk D, Franceschini A et al (2012) STITCH 3: Zooming in on protein-

chemical interactions. Nucleic Acids Res 40: D876–D880
78. Szklarczyk D, Santos A, Von Mering C et al (2016) STITCH 5: Augmenting protein-chemical interaction networks with tissue and affinity data. Nucleic Acids Res 44: D380–D384
79. Lagunin A, Stepanchikova A, Filimonov D, Poroikov V (2000) PASS: Prediction of activity spectra for biologically active substances. Bioinformatics 16:747–748
80. Poroikov VV, Filimonov DA, Ihlenfeldt WD et al (2003) PASS biological activity spectrum predictions in the enhanced open NCI Database Browser. J Chem Inf Comput Sci 43:228–236
81. Filimonov DA, Poroikov VV (1996) PASS: Computerized prediction of biological activity spectra for chemical substances. In: Bioactive compound design: possibilities for industrial use. BIOS Scientific Publishers, Oxford, pp 47–56
82. Filimonov D, Poroikov V, Borodina Y, Gloriozova T (1999) Chemical similarity assessment through multilevel neighborhoods of atoms: Definition and comparison with the other descriptors. J Chem Inf Comput Sci 39:666–670
83. Wang Z, Liang L, Yin Z, Lin J (2016) Improving chemical similarity ensemble approach in target prediction. J Cheminform 8:20
84. Weibull W (1951) A statistical distribution function of wide applicability. J Appl Mech 18:293–297
85. Nadarajah S (2003) Extreme value theory, models and simulation. In: Shanbhag D, Rao C (eds) Stochastic processes: modelling and simulation, Handbook of Statistics, vol 21. Elsevier, Amsterdam, pp 607–691
86. Altschul SF, Gish W, Miller W et al (1990) Basic local alignment search tool. J Mol Biol 215:403–410
87. Schneider P, Stutz K, Kasper L et al (2011) Target profile prediction and practical evaluation of a biginelli-type dihydropyrimidine compound library. Pharmaceuticals 4 (9):1236–1247
88. Morgan HL (1965) The generation of a unique machine description for chemical structures–A technique developed at chemical abstracts service. J Chem Doc 5:107–113
89. Avidon VV, Pomerantsev IA, Golender VE, Rozenblit AB (1982) Structure-activity relationship oriented languages for chemical structure representation. J Chem Inf Comput Sci 22:207–214
90. Nigsch F, Bender A, Jenkins JL, Mitchell JBO (2008) Ligand-target prediction using winnow and naive bayesian algorithms and the implications of overall performance statistics. J Chem Inf Model 48:2313–2325
91. Nidhi, Glick M, Davies JW, Jenkins JL (2006) Prediction of biological targets for compounds using multiple-category bayesian models trained on chemogenomics databases. J Chem Inf Model 46:1124–1133
92. Drakakis G, Koutsoukas A, Brewerton S et al (2015) Comparing global and local likelihood score thresholds in multiclass Laplacian-modified naive Bayes protein target prediction. Comb Chem High Throughput Screen 18:323–330
93. Koutsoukas A, Lowe R, Kalantarmotamedi Y et al (2013) *In silico* target predictions: Defining a benchmarking data set and comparison of performance of the multiclass Naïve Bayes and Parzen-Rosenblatt Window. J Chem Inf Model 53:1957–1966
94. Nobeli I, Favia AD, Thornton JM (2009) Protein promiscuity and its implications for biotechnology. Nat Biotechnol 27:157–167
95. Blencowe BJ (2006) Alternative splicing: New insights from global analyses. Cell 126:37–47
96. Pan Q, Shai O, Lee LJ et al (2008) Deep surveying of alternative splicing complexity in the human transcriptome by high-throughput sequencing. Nat Genet 40:1413–1415
97. McCarthy JJ, Hilfiker R (2000) The use of single-nucleotide polymorphism maps in pharmacogenomics. Nat Biotechnol 18:505–508
98. Roden DM, George AL Jr (2002) The genetic basis of variability in drug responses. Nat Rev Drug Discov 1:37–44
99. Sauna ZE, Kimchi-Sarfaty C (2011) Understanding the contribution of synonymous mutations to human disease. Nat Rev Genet 12:683–691
100. Prabakaran S, Lippens G, Steen H, Gunawardena J (2012) Post-translational modification: Nature's escape from genetic imprisonment and the basis for dynamic information encoding. Wiley Interdiscip Rev Syst Biol Med 4:565–583
101. Bedford L, Lowe J, Dick LR et al (2011) Ubiquitin-like protein conjugation and the ubiquitin-proteasome system as drug targets. Nat Rev Drug Discov 10:29–46
102. Konstantinopoulos PA, Karamouzis MV, Papavassiliou AG (2007) Post-translational modifications and regulation of the RAS

superfamily of GTPases as anticancer targets. Nat Rev Drug Discov 6:541–555

103. Vogel C, Bashton M, Kerrison ND et al (2004) Structure, function and evolution of multidomain proteins. Curr Opin Struct Biol 14:208–216
104. Bashton M, Chothia C (2007) The generation of new protein functions by the combination of domains. Structure 15:85–99
105. Song N, Joseph JM, Davis GB, Durand D (2008) Sequence similarity network reveals common ancestry of multidomain proteins. PLoS Comput Biol 4:e1000063
106. Day R, Beck DAC, Armen RS (2003) A consensus view of fold space: Combining SCOP, CATH, and the Dali Domain Dictionary. Protein Sci 12:2150–2160
107. Cuff AL, Sillitoe I, Lewis T et al (2011) Extending CATH: increasing coverage of the protein structure universe and linking structure with function. Nucleic Acids Res 39: D420–D426
108. Hammes GG, Chang Y-C, Oas TG (2009) Conformational selection or induced fit: A flux description of reaction mechanism. Proc Natl Acad Sci U S A 106:13737–13741
109. Csermely P, Palotai R, Nussinov R (2010) Induced fit, conformational selection and independent dynamic segments: An extended view of binding events. Trends Biochem Sci 35:539–546
110. Fischer E (1894) Einfluss der Konfiguration auf die Wirkung der Enzyme. Berichte der Dtsch Chem Gesellschaft 27:2985–2993
111. Koshland DE (1960) The active site and enzyme action. Adv Enzymol Relat Areas Mol Biol 22:45–97
112. Koshland DE (1958) Application of a theory of enzyme specificity to protein synthesis. Proc Natl Acad Sci U S A 44:98–104
113. Monod J, Wyman J, Changeux JP (1965) On the nature of allosteric transitions: A plausible model. J Mol Biol 12:88–118
114. Motlagh HN, Wrabl JO, Li J, Hilser VJ (2014) The ensemble nature of allostery. Nature 508:331–339
115. Vega S, Abian O, Velazquez-Campoy A (2016) On the link between conformational changes, ligand binding and heat capacity. Biochim Biophys Acta 1860:868–878
116. Shanmugasundaram V, Zhang L, Kayastha S et al (2016) Monitoring the progression of structure-activity relationship information during lead optimization. J Med Chem 59:4235–4244
117. Grünberg R, Leckner J, Nilges M (2004) Complementarity of structure ensembles in protein-protein binding. Structure 12:2125–2136
118. Vogt AD, Di Cera E (2013) Conformational selection is a dominant mechanism of ligand binding. Biochemistry 52:5723–5729
119. Vogt AD, Di Cera E (2012) Conformational selection or induced fit? A critical appraisal of the kinetic mechanism. Biochemistry 51:5894–5902
120. Changeux JP, Edelstein S (2011) Conformational selection or induced-fit? 50 years of debate resolved. F1000 Biol Rep 3:19
121. Lange OF, Lakomek NA, Farès C et al (2008) Recognition dynamics up to microseconds revealed from an RDC-derived ubiquitin ensemble in solution. Science 320:1471–1475
122. Kar G, Keskin O, Gursoy A, Nussinov R (2010) Allostery and population shift in drug discovery. Curr Opin Pharmacol 10:715–722
123. Michel D (2016) Conformational selection or induced fit? New insights from old principles. Biochimie 128–129:48–54
124. Galburt EA, Tomko EJ (2017) Conformational selection and induced fit as a useful framework for molecular motor mechanisms. Biophys Chem 223:11–16
125. Ekroos M, Sjogren T (2006) Structural basis for ligand promiscuity in cytochrome P450 3A4. Proc Natl Acad Sci U S A 103:13682–13687
126. De Vivo M, Masetti M, Bottegoni G, Cavalli A (2016) Role of molecular dynamics and related methods in drug discovery. J Med Chem 59:4035–4061
127. Feixas F, Lindert S, Sinko W, McCammon JA (2014) Exploring the role of receptor flexibility in structure-based drug discovery. Biophys Chem 186:31–45
128. Buonfiglio R, Recanatini M, Masetti M (2015) Protein flexibility in drug discovery: From theory to computation. ChemMedChem 10:1141–1148
129. Teague SJ (2003) Implications of protein flexibility for drug discovery. Nat Rev Drug Discov 2:527–541
130. Lahti JL, Tang GW, Capriotti E et al (2012) Bioinformatics and variability in drug response: a protein structural perspective. J R Soc Interface 9:1409–1437
131. Xie L, Xie L, Bourne PE (2009) A unified statistical model to support local sequence order independent similarity searching for ligand-binding sites and its application to

genome-based drug discovery. Bioinformatics 25:305–312

132. Kahraman A, Morris RJ, Laskowski RA, Thornton JM (2007) Shape variation in protein bindign pockets and their ligands. J Mol Biol 368:283–301
133. Kellenberger E, Schalon C, Rognan D (2008) How to measure the similarity between protein-ligand binding sites? Curr Comput Aided Drug Des 4:209–220
134. Pérot S, Sperandio O, Miteva MA et al (2010) Druggable pockets and binding site centric chemical space: A paradigm shift in drug discovery. Drug Discov Today 15:656–667
135. Schmitt S, Kuhn D, Klebe G (2002) A new method to detect related function among proteins independent of sequence and fold homology. J Mol Biol 323:387–406
136. Bron C, Kerbosch J (1973) Algorithm 457: Finding all cliques of an undirected graph. Commun ACM 16:575–577
137. Jambon M, Imberty A, Deléage G, Geourjon C (2003) A new bioinformatic approach to detect common 3D sites in protein structures. Proteins 52:137–145
138. Jambon M, Andrieu O, Combet C et al (2005) The SuMo server: 3D search for protein functional sites. Bioinformatics 21:3929–3930
139. Chartier M, Najmanovich R (2015) Detection of binding site molecular interaction field similarities. J Chem Inf Model 55:1600–1615
140. Chartier M, Adriansen E, Najmanovich R (2015) IsoMIF Finder: Online detection of binding site molecular interaction field similarities. Bioinformatics 32:621–623
141. Yeturu K, Chandra N (2008) PocketMatch: a new algorithm to compare binding sites in protein structures. BMC Bioinformatics 9:543
142. Hoffmann B, Zaslavskiy M, Vert J-P, Stoven V (2010) A new protein binding pocket similarity measure based on comparison of clouds of atoms in 3D: application to ligand prediction. BMC Bioinformatics 11:99
143. Wang X, Pan C, Gong J et al (2016) Enhancing the enrichment of pharmacophore-based target prediction for the polypharmacological profiles of drugs. J Chem Inf Model 56:1175–1183
144. Liu X, Ouyang S, Yu B et al (2010) PharmMapper server: A web server for potential drug target identification using pharmacophore mapping approach. Nucleic Acids Res 38:W609–W614
145. Wang X, Shen Y, Wang S et al (2017) PharmMapper 2017 update: a web server for potential drug target identification with a comprehensive target pharmacophore database. Nucleic Acids Res 45(W1): W356–W360.
146. Altschul SF, Madden TL, Schäffer AA et al (1997) Gapped BLAST and PSI-BLAST: A new generation of protein database search programs. Nucleic Acids Res 25:3389–3402
147. Yuan Y, Pei J, Lai L (2013) Binding site detection and druggability prediction of protein targets for structure-based drug design. Curr Pharm Des 19:2326–2333
148. Chen J, Ma X, Yuan Y et al (2014) Protein-protein interface analysis and hot spots identification for chemical ligand design. Curr Pharm Des 20:1192–1200
149. Warren GL, Andrews CW, Capelli A-M et al (2006) A critical assessment of docking programs and scoring functions. J Med Chem 49:5912–5931
150. Bourne PE, Beran B, Bi C et al (2011) The evolution of the RCSB protein data bank website. Wiley Interdiscip Rev Comput Mol Sci 1:782–789
151. Zardecki C, Dutta S, Goodsell DS et al (2016) RCSB protein data bank: A resource for chemical, biochemical, and structural explorations of large and small biomolecules. J Chem Educ 93:569–575
152. Ellingson SR, Smith JC, Baudry J (2013) VinaMPI: Facilitating multiple receptor high-throughput virtual docking on high-performance computers. J Comput Chem 34:2212–2221
153. Chen HM, Liu BF, Huang HL et al (2007) SODOCK: Swarm optimization for highly flexible protein-ligand docking. J Comput Chem 28:612–623
154. Liu Y, Zhao L, Li W et al (2013) FIPSDock: A new molecular docking technique driven by fully informed swarm optimization algorithm. J Comput Chem 34:67–75
155. Chen YZ, Zhi DG (2001) Ligand-protein inverse docking and its potential use in the computer search of protein targets of a small molecule. Proteins Struct Funct Genet 43:217–226
156. Li H, Gao Z, Kang L et al (2006) TarFisDock: A web server for identifying drug targets with docking approach. Nucleic Acids Res 34: W219–W224
157. Gao Z, Li H, Zhang H et al (2008) PDTD: A web-accessible protein database for drug target identification. BMC Bioinformatics 9:104

158. Lang PT, Brozell SR, Mukherjee S et al (2009) DOCK 6: Combining techniques to model RNA-small molecule complexes. RNA 15:1219–1230
159. Wang JC, Chu PY, Chen CM, Lin JH (2012) idTarget: A web server for identifying protein targets of small chemical molecules with robust scoring functions and a divide-and-conquer docking approach. Nucleic Acids Res 40:W393–W399
160. Chang DTH, Oyang YJ, Lin JH (2005) MEDock: A web server for efficient prediction of ligand binding sites based on a novel optimization algorithm. Nucleic Acids Res 33:W233–W238
161. Kellenberger E, Muller P, Schalon C et al (2006) sc-PDB: An annotated database of druggable binding sites from the Protein Data Bank. J Chem Inf Model 46:717–727
162. Luo H, Chen J, Shi L et al (2011) DRAR-CPI: A server for identifying drug repositioning potential and adverse drug reactions via the chemical-protein interactome. Nucleic Acids Res 39:W492–W498
163. Smith RD, Hu L, Falkner JA et al (2006) Exploring protein-ligand recognition with Binding MOAD. J Mol Graph Model 24:414–425
164. Pronk S, Páll S, Schulz R et al (2013) GROMACS 4.5: A high-throughput and highly parallel open source molecular simulation toolkit. Bioinformatics 29:845–854
165. Phillips JC, Braun R, Wang W et al (2005) Scalable molecular dynamics with NAMD. J Comput Chem 26:1781–1802
166. Wassenaar TA, van Dijk M, Loureiro-Ferreira N et al (2012) WeNMR: Structural biology on the grid. J Grid Comput 10:743–767
167. Van Dijk M, Wassenaar TA, Bonvin AMJJ (2012) A flexible, grid-enabled web portal for GROMACS molecular dynamics simulations. J Chem Theory Comput 8:3463–3472
168. Hospital A, Andrio P, Fenollosa C et al (2012) MDWeb and MDMoby: An integrated web-based platform for molecular dynamics simulations. Bioinformatics 28:1278–1279
169. Chen Z, Rappert S, Sun J, Zeng AP (2011) Integrating molecular dynamics and co-evolutionary analysis for reliable target prediction and deregulation of the allosteric inhibition of aspartokinase for amino acid production. J Biotechnol 154:248–254
170. Páll S, Abraham MJ, Kutzner C et al (2015) Tackling exascale software challenges in molecular dynamics simulations with GROMACS. In: Markidis S, Laure E (eds) Lecture notes in computer science, vol 8759. Springer, Cham, pp 3–27
171. Bowers KJ, Sacerdoti FD, Salmon JK et al (2006) Molecular dynamics–Scalable algorithms for molecular dynamics simulations on commodity clusters. In: SC '06: Proceedings of the 2006 ACM/IEEE conference on supercomputing. ACM New York, Tampa, pp 84–96
172. Mestres J, Gregori-Puigjané E, Valverde S, Solé RV (2009) The topology of drug–target interaction networks: implicit dependence on drug properties and target families. Mol BioSyst 5:1051
173. Luo Q, Pagel P, Vilne B, Frishman D (2011) DIMA 3.0: Domain interaction map. Nucleic Acids Res 39:D724–D729
174. Weisel M, Kriegl JM, Schneider G (2010) Architectural repertoire of ligand-binding pockets on protein surfaces. ChemBioChem 11:556–563
175. Prud'hommeaux E, Seaborne A (2008) SPARQL query language for RDF. W3C Recomm 2009:1–106
176. Harris S, Seaborne A (2014) SPARQL 1.1 query language. W3C Recomm 2009:1–106
177. Chen B, Ding Y, Wild DJ (2012) Assessing drug target association using semantic linked data. PLoS Comput Biol 8:e1002574
178. Ashburner M, Ball CA, Blake JA et al (2000) Gene ontology: Tool for the unification of biology. Nat Genet 25:25–29
179. Blake JA, Christie KR, Dolan ME et al (2015) Gene ontology consortium: Going forward. Nucleic Acids Res 43:D1049–D1056
180. Degtyarenko K, De Matos P, Ennis M et al (2008) ChEBI: A database and ontology for chemical entities of biological interest. Nucleic Acids Res 36:D344–D350
181. de Matos P, Alcántara R, Dekker A et al (2009) Chemical entities of biological interest: An update. Nucleic Acids Res 38: D249–D254
182. Law V, Knox C, Djoumbou Y et al (2014) DrugBank 4.0: Shedding new light on drug metabolism. Nucleic Acids Res 42: D1091–D1097
183. Chen B, Dong X, Jiao D et al (2010) Chem2-Bio2RDF: A semantic framework for linking and data mining chemogenomic and systems chemical biology data. BMC Bioinformatics 11:255
184. Chen B, Ding Y, Wild DJ (2012) Improving integrative searching of systems chemical biology data using semantic annotation. J Cheminform 4:6

185. Dijkstra EW (1959) A note on two problems in connexion with graphs. Numer Math 1:269–271
186. Kong X, Cao B, Yu PS (2013) Multi-label classification by mining label and instance correlations from heterogeneous information networks categories and subject descriptors. In: Dhillon I et al (eds) Proceedings of the 19th ACM SIGKDD International conference on knowledge discovery and data mining. ACM Press, New York, pp 614–622
187. Wang W, Yang S, Zhang X, Li J (2014) Drug repositioning by integrating target information through a heterogeneous network model. Bioinformatics 30:2923–2930
188. Fakhraei S, Huang B, Raschid L, Getoor L (2014) Network-based drug-target interaction prediction with probabilistic soft-logic. In: IEEE/ACM Transactions on Computational Biology and Bioinformatics, vol 11. IEEE Computer society press, Los Alamitos, pp 775–787
189. Klipp E, Wade RC, Kummer U (2010) Biochemical network-based drug-target prediction. Curr Opin Biotechnol 21:511–516
190. Campillos M, Kuhn M, Gavin A-C et al (2008) Drug target identification using side-effect similarity. Science 321:263–266
191. Yamanishi Y, Kotera M, Kanehisa M, Goto S (2010) Drug-target interaction prediction from chemical, genomic and pharmacological data in an integrated framework. Bioinformatics 26:i246–i254
192. Takarabe M, Kotera M, Nishimura Y et al (2012) Drug target prediction using adverse event report systems: A pharmacogenomic approach. Bioinformatics 28:i611–i618
193. Yamanishi Y, Kotera M, Moriya Y et al (2014) DINIES: Drug-target interaction network inference engine based on supervised analysis. Nucleic Acids Res 42:W39–W45
194. Zhang X, Li L, Ng MK, Zhang S (2017) Drug-target interaction prediction by integrating multiview network data. Comput Biol Chem 69:185–190
195. Kanehisa M, Goto S, Furumichi M et al (2009) KEGG for representation and analysis of molecular networks involving diseases and drugs. Nucleic Acids Res 38:D355–D360
196. Pawson AJ, Sharman JL, Benson HE et al (2014) The IUPHAR/BPS Guide to PHARMACOLOGY: An expert-driven knowledgebase of drug targets and their ligands. Nucleic Acids Res 42:D1098–D1106
197. Hagberg AA, Schult DA, Swart PJ (2008) Exploring network structure, dynamics, and function using NetworkX. In: Varoquax G et al. (eds) SciPy 2008: Proceedings of the 7th python in science conference, Pasadena, 2008
198. Van Der Walt S, Colbert SC, Varoquaux G (2011) The NumPy array: A structure for efficient numerical computation. Comput Sci Eng 13:22–30
199. Tosco P, Stiefl N, Landrum G (2014) Bringing the MMFF force field to the RDKit: Implementation and validation. J Cheminform 6:37
200. Landrum G (2016) The RDKit Documentation–The RDKit 2016.09.1 documentation. Accessed 18 Aug 2017
201. Python Software Foundation (2017) The Python Language Reference–Python 2.7.13 documentation. Accessed 18 Aug 2017
202. Bastian M, Heymann S, Jacomy M et al (2009) Gephi: an open source software for exploring and manipulating networks. In: Third international AAAI conference on weblogs and social media, San Jose, 17–20 May 2009
203. Jacomy M, Venturini T, Heymann S, Bastian M (2014) ForceAtlas2, a continuous graph layout algorithm for handy network visualization designed for the Gephi software. PLoS One 9:e98679
204. Boutet E, Lieberherr D, Tognolli M et al (2007) UniProtKB/Swiss-Prot. Methods Mol Biol 406:89–112
205. Bemis GW, Murcko MA (1996) The properties of known drugs. 1. Molecular frameworks. J Med Chem 39:2887–2893
206. Franceschini A, Szklarczyk D, Frankild S et al (2013) STRING v9.1: Protein-protein interaction networks, with increased coverage and integration. Nucleic Acids Res 41:808–815
207. Forbes SA, Beare D, Gunasekaran P et al (2015) COSMIC: Exploring the world's knowledge of somatic mutations in human cancer. Nucleic Acids Res 43:D805–D811
208. Flicek P, Amode MR, Barrell D et al (2014) Ensembl 2014. Nucleic Acids Res 42: D749–D755
209. Lundstrom K (2006) The ultimate approach for rational drug design. Struct Genomics 34:205–212
210. Kaldor SW, Kalish VJ, Davies JF et al (1997) Viracept (nelfinavir mesylate, AG1343): A potent, orally bioavailable inhibitor of HIV-1 protease. J Med Chem 40:3979–3985
211. Xie L, Bourne PE (2007) A robust and efficient algorithm for the shape description of protein structures and its application in predicting ligand binding sites. BMC Bioinformatics 8:S9

212. Xie L, Evangelidis T, Xie L, Bourne PE (2011) Drug discovery using chemical systems biology: Weak inhibition of multiple kinases may contribute to the anti-cancer effect of nelfinavir. PLoS Comput Biol 7: e1002037
213. Csermely P, Korcsmáros T, Kiss HJM et al (2013) Structure and dynamics of molecular networks: A novel paradigm of drug discovery: A comprehensive review. Pharmacol Ther 138:333–408
214. Mei H, Xia T, Feng G et al (2012) Opportunities in systems biology to discover mechanisms and repurpose drugs for CNS diseases. Drug Discov Today 17:1208–1216
215. Kell DB, Goodacre R (2014) Metabolomics and systems pharmacology: Why and how to model the human metabolic network for drug discovery. Drug Discov Today 19:171–182
216. Kunkel EJ (2004) Systems biology in drug discovery. Nat Biotechnol 22:1253–1259
217. Chautard E, Thierry-Mieg N, Ricard-Blum S (2009) Interaction networks: From protein functions to drug discovery. A review. Pathol Biol 57:324–333
218. Raj A, Rifkin SA, Andersen E, van Oudenaarden A (2010) Variability in gene expression underlies incomplete penetrance. Nature 463:913–918
219. Susaki EA, Ukai H, Ueda HR (2017) Next-generation mammalian genetics toward organism-level systems biology. Syst Biol Appl 3:15
220. Jerby L, Ruppin E (2012) Predicting drug targets and biomarkers of cancer via genome-scale metabolic modeling. Clin Cancer Res 18:5572–5584
221. Folger O, Jerby L, Frezza C et al (2014) Predicting selective drug targets in cancer through metabolic networks. Mol Syst Biol 7:501
222. Azuaje FJ, Zhang L, Devaux Y, Wagner DR (2011) Drug-target network in myocardial infarction reveals multiple side effects of unrelated drugs. Sci Rep 1:52
223. Chelliah V, Juty N, Ajmera I et al (2015) BioModels: Ten-year anniversary. Nucleic Acids Res 43:D542–D548
224. Raman K, Chandra N (2009) Flux balance analysis of biological systems: Applications and challenges. Brief Bioinform 10:435–449
225. Fang X, Wallqvist A, Reifman J (2009) A systems biology framework for modeling metabolic enzyme inhibition of *Mycobacterium tuberculosis*. BMC Syst Biol 3:92
226. Martinez V, Navarro C, Cano C et al (2015) DrugNet: Network-based drug-disease prioritization by integrating heterogeneous data. Artif Intell Med 63:41–49
227. Zhang Y, Thiele I, Weekes D et al (2009) Three-dimensional structural view of the central metabolic network of *Thermotoga maritima*. Science 325:1544–1549
228. Kessler RM, Ansari MS, de Paulis T et al (1991) High affinity dopamine D2 receptor radioligands. 1. Regional rat brain distribution of iodinated benzamides. J Nucl Med 32:1593–1600
229. Hopkins AL, Mason JS, Overington JP (2006) Can we rationally design promiscuous drugs? Curr Opin Struct Biol 16:127–136
230. Denessiouk KA, Johnson MS (2000) When fold is not important: A common structural framework for adenine and AMP binding in 12 unrelated protein families. Proteins Struct Funct Genet 38:310–326
231. Lalande L, Bourguignon L, Maire P, Goutelle S (2016) Mathematical modeling and systems pharmacology of tuberculosis: Isoniazid as a case study. J Theor Biol 399:43–52
232. Raman K, Chandra N (2008) *Mycobacterium tuberculosis* interactome analysis unravels potential pathways to drug resistance. BMC Microbiol 8:234.
233. Rokem JS, Lantz AE, Nielsen J (2007) Systems biology of antibiotic production by microorganisms. Nat Prod Rep 24:1262–1287
234. Kinnings SL, Xie L, Fung KH et al (2010) The *Mycobacterium tuberculosis* drugome and its polypharmacological implications. PLoS Comput Biol 6:e1000976
235. Wist AD, Berger SI, Iyengar R (2009) Systems pharmacology and genome medicine: a future perspective. Genome Med 1:11
236. Jin G, Zhao H, Zhou X, Wong STC (2011) An enhanced Petri-Net model to predict synergistic effects of pairwise drug combinations from gene microarray data. Bioinformatics 27:310–316
237. Winter GE, Rix U, Carlson SM et al (2012) Systems-pharmacology dissection of a drug synergy in imatinib-resistant CML. Nat Chem Biol 8:905–912
238. Jin G, Fu C, Zhao H et al (2012) A novel method of transcriptional response analysis to facilitate drug repositioning for cancer therapy. Cancer Res 72:33–44
239. Ryall KA, Tan AC (2015) Systems biology approaches for advancing the discovery of effective drug combinations. J Cheminform 7:7
240. Li P, Chen J, Wang J et al (2014) Systems pharmacology strategies for drug discovery

and combination with applications to cardiovascular diseases. J Ethnopharmacol 151:93–107

241. Kell DB (2015) The transporter-mediated cellular uptake of pharmaceutical drugs is based on their metabolite-likeness and not on their bulk biophysical properties: Towards a systems pharmacology. Perspect Sci 6:66–83
242. Mitchell JBO (2014) Machine learning methods in chemoinformatics. Wiley Interdiscip Rev Comput Mol Sci 4:468–481
243. Azencott CA, Ksikes A, Swamidass SJ et al (2007) One- to four-dimensional kernels for virtual screening and the prediction of physical, chemical, and biological properties. J Chem Inf Model 47:965–974
244. Nascimento ACA, Prudêncio RBC, Costa IG (2016) A multiple kernel learning algorithm for drug-target interaction prediction. BMC Bioinformatics 17:46
245. Lima AN, Philot EA, Trossini GHG et al (2016) Use of machine learning approaches for novel drug discovery. Expert Opin Drug Discov 11:225–239
246. Ekins S, de Siqueira-Neto JL, McCall L-I et al (2015) Machine learning models and pathway genome data base for Trypanosoma cruzi drug discovery. PLoS Negl Trop Dis 9: e0003878
247. Williams K, Bilsland E, Sparkes A et al (2015) Cheaper faster drug development validated by the repositioning of drugs against neglected tropical diseases. J R Soc Interface 12:1289
248. Gasteiger J (2003) Physicochemical effects in the representation of molecular structures for drug designing. Mini Rev Med Chem 3:789–796
249. Sawada R, Kotera M, Yamanishi Y (2014) Benchmarking a wide range of chemical descriptors for drug-target interaction prediction using a chemogenomic approach. Mol Inf 33:719–731
250. Hinton GE, Salakhutdinov RR (2006) Reducing the dimensionality of data with neural networks. Science 313:504–507
251. LeCun Y, Yoshua B, Geoffrey H (2015) Deep learning. Nature 521:436–444
252. Gawehn E, Hiss JA, Schneider G (2016) Deep learning in drug discovery. Mol Inf 35:3–14
253. Unterthiner T, Ceulemans H, Steijaert M (2014) Multi-task deep networks for drug target prediction. In: Ghahramani Z et al (eds) NIPS 2014: Proceedings of the neural information processing systems foundation meeting, Montréal, 2014
254. Lusci A, Pollastri G, Baldi P (2013) Deep architectures and deep learning in chemoinformatics: The prediction of aqueous solubility for drug-like molecules. J Chem Inf Model 53:1563–1575
255. Faber FA, Hutchison L, Huang B, et al (2017) Fast machine learning models of electronic and energetic properties consistently reach approximation errors better than DFT accuracy. arXiv:1702.05532
256. Ma J, Sheridan RP, Liaw A et al (2015) Deep neural nets as a method for quantitative structure-activity relationships. J Chem Inf Model 55:263–274
257. Goh GB, Siegel C, Vishnu A et al (2017) Chemception: A deep neural network with minimal chemistry knowledge matches the performance of expert-developed QSAR/QSPR models. arXiv:1706.06689
258. Mayr A, Klambauer G, Unterthiner T, Hochreiter S (2016) DeepTox: Toxicity prediction using deep learning. Front Environ Sci 3:80
259. Ramsundar B, Liu B, Zhenqin W et al (2017) Is multitask deep learning practical for pharma? J Chem Inf Model.
260. Wang Q, Feng Y, Huang J et al (2017) A novel framework for the identification of drug target proteins: Combining stacked auto-encoders with a biased support vector machine. PLoS One 12:e0176486
261. Unterthiner T, Mayr A, Klambauer G et al (2014) In: Ghahramani Z et al (eds) NIPS 2014: Proceedings of the neural information processing systems foundation meeting, Montréal, 2014
262. Ralaivola L, Swamidass SJ, Saigo H, Baldi P (2005) Graph kernels for chemical informatics. Neural Netw 18:1093–1110
263. Azencott CA, Borgwardt K (2014) Data mining in bioinformatics–Graph mining in chemoinformatics. Accessed 6 Jul 2017
264. Gilmer J, Schoenholz SS, Riley PF et al (2017) Neural message passing for quantum chemistry. arXiv:1704.01212
265. Defferrard M, Bresson X, Vandergheynst P (2016) Convolutional neural networks on graphs with fast localized spectral filtering. Adv Neural Inform Process Syst 29:3844–3852
266. Niepert M, Ahmed M, Kutzkov K (2016) Learning convolutional neural networks for graphs. J Mach Learn Res 48:2014–2023
267. Altae-Tran H, Ramsundar B, Pappu AS, Pande V (2017) Low data drug discovery with one-shot learning. ACS Cent Sci 3:283–293

268. Merck & Co., Inc (2012) Merck molecular activity challenge, Accessed 4 Aug 2017
269. Chen B, Sheridan RP, Hornak V, Voigt JH (2012) Comparison of random forest and Pipeline Pilot naïve Bayes in prospective QSAR predictions. J Chem Inf Model 52:792–803
270. Dahl GE, Jaitly N, Salakhutdinov R (2015) Deep neural nets as a method for quantitative structure-activity relationships. J Chem Inf Model 55:263–274
271. Dahl GE, Jaitly N (2012) Post discussing authors' methodology and success in Merck Kaggle competition. Accessed 4 Aug 2017
272. Markoff J (2012) New York Times coverage of the Merck competition. Accessed 4 Aug 2017
273. Kearnes S, McCloskey K, Berndl M et al (2016) Molecular graph convolutions: moving beyond fingerprints. J Comput Aided Mol Des 30:595–608
274. Scarselli F, Gori M, Tsoi AC et al (2009) The graph neural network model. IEEE Trans Neural Netw 20:61–80
275. Szegedy C, Liu W, Jia Y et al (2015) Going deeper with convolutions. In: Conference on computer vision and pattern recognition (CVPR), Boston, MA, 7–12 Jun 2015
276. Karpathy A, Li FF (2015) Deep visual-semantic alignments for generating image descriptions. IEEE Trans Pattern Anal Mach Intell 39:664–676
277. Silla CN, Freitas AA (2011) A survey of hierarchical classification across different application domains. Data Min Knowl Discov 22:31–72
278. Ragoza M, Hochuli J, Idrobo E et al (2017) Protein-ligand scoring with convolutional neural networks. J Chem Inf Model 57:942–957
279. Wang Y, Zeng J (2013) Predicting drug-target interactions using restricted Boltzmann machines. Bioinformatics 29:i126–i134
280. Laggner C, Kokel D, Setola V et al (2012) Chemical informatics and target identification in a zebrafish phenotypic screen. Nat Chem Biol 8:144–146
281. Liggi S, Drakakis G, Koutsoukas A et al (2014) Extending *in silico* mechanism-of-action analysis by annotating targets with pathways: application to cellular cytotoxicity readouts. Future Med Chem 6:2029–2056
282. Schneider G, Fechner U (2005) Computer-based *de novo* design of drug-like molecules. Nat Rev Drug Discov 4:649–663
283. Klenner A, Hartenfeller M, Schneider P, Schneider G (2010) "Fuzziness" in pharmacophore-based virtual screening and *de novo* design. Drug Discov Today Technol 7:e237–e244
284. Hartenfeller M, Schneider G (2011) Enabling future drug discovery by *de novo* design. Wiley Interdiscip Rev Comput Mol Sci 1:742–759
285. Schneider P, Schneider G (2016) *De novo* design at the edge of chaos. J Med Chem 59:4077–4086
286. Koes DR, Camacho CJ (2011) Pharmer: Efficient and exact pharmacophore search. J Chem Inf Model 51:1307–1314
287. Koes DR, Camacho CJ (2012) ZINCPharmer: Pharmacophore search of the ZINC database. Nucleic Acids Res 40:W409–W414

Index

Slava Ziegler and Herbert Waldmann (eds.), *Systems Chemical Biology: Methods and Protocols*, Methods in Molecular Biology, vol. 1888, https://doi.org/10.1007/978-1-4939-8891-4, © Springer Science+Business Media, LLC, part of Springer Nature 2019

I

J

K

L

M

N

O

P

Q

R

S

Printed by Printforce, the Netherlands